TECHNOLOGY TRANSFER AND THE UNIVERSITY

TECHNOLOGY
TRANSFER
AND
THE UNIVERSITY

Gary W. Matkin

Foreword by Clark Kerr

National University Continuing Education Association

AMERICAN COUNCIL MACMILLAN
ON EDUCATION PUBLISHING COMPANY

NEW YORK

Collier Macmillan Canada

TORONTO

Maxwell Macmillan International

NEW YORK OXFORD SINGAPORE SYDNEY

Copyright © 1990 by American Council on Education and
Macmillan Publishing Company,
A Division of Macmillan, Inc.

Macmillan Publishing Company
866 Third Avenue, New York, N.Y. 10022

Collier Macmillan Canada, Inc.
1200 Eglinton Avenue East, Suite 200
Don Mills, Ontario M3C 3N1

Library of Congress Catalog Card Number: 90-32084

Printed in the United States of America

printing number
1 2 3 4 5 6 7 8 9 10

Library of Congress Cataloging-in-Publication Data

Matkin, Gary W.
 Technology transfer and the university / Gary W. Matkin.
  p.   cm.
 Includes bibliographical references.
 ISBN 0-02-897263-5
 1. Technology transfer—United States.   2. Research institutes—United States.   3. Patents—United States.   I. Title.
T174.3.M38 1990
338.97306—dc20   90-32084
                                                          CIP

*To my wife Maya, my son Ethan,
and my family and friends who were so supportive
during production of this book.*

Contents

List of Tables, Exhibits, and Figures

Foreword

A century ago, in the 1870s, the land-grant universities began making a heavy commitment to technology transfer into agriculture. Enormous transformations resulted. In 1870, 50 percent of the labor force was in agriculture; in 1970, 5 percent. Also in 1870, we had a nation of family farms; and in 1970, "factories in the fields."

Now a century later, in the 1970s, American research universities began a heavy commitment to technology transfer into industry. In doing so, will they help to save the economy or will they subvert the academy, or, more likely, some of both?

Gary Matkin's study is the best I have seen of experience to date at the operational level in the new era of technology transfer by the American research university. He surveys developments at MIT, Penn State, Stanford, and Berkeley.

Matkin starts out by quoting Eric Ashby to the effect that "universities have become absolutely essential to the economy." But Ashby also notes that they must "now strike a balance between an adaptation which is too flexible and an adherence to tradition which is too inflexible." Matkin in this study is rather more interested in how to be "essential," but he does also raise questions at the end about the "balance." My estimate is that, as time goes on, the issue of "balance" will become more important but, in the meantime, higher education will make great strides in becoming more "essential" along the lines that the four universities in this study have importantly helped to pioneer.

We need to keep learning from experience, and Matkin's explorations constitute a promising effort as this new and very significant episode unfolds.

Clark Kerr

Preface

American institutions of higher education, particularly research universities, face a challenge unlike any they have ever faced before. That challenge comes from the increased involvement of these institutions in technology transfer—the translation of university research into commercial products, services, or processes. Sir Eric Ashby, master of Clare College, Cambridge University, and chancellor of Queen's University, Belfast, has summed up that challenge and the response to it that universities must make:

> Now universities have become absolutely essential to the economy and to the very survival of nations. . . . Forces from outside the university, which formerly had only a marginal effect upon the evolution of the university, are now likely to exert a powerful influence on this evolution.
>
> Universities therefore have to strike a balance between an adaptation which is too pliable and an adherence to tradition which is too inflexible. To achieve this balance universities need to initiate and control their adaptation to society, not to allow it to be imposed on them from outside.[1]

Nature and Purpose of the Book

This book is intended to help university administrators understand and meet the challenge of technology transfer. It examines and compares a variety of technology transfer activities at four top-ranked American research universities: the Massachusetts Institute of Technology, Stanford University, the University of California at Berkeley, and Pennsylvania State University. It considers the historical development of these activities, the settings and academic cultures in which they exist today, the changes they have brought to the universities, and the responses made by the universities to those changes. Activities covered include patent policy and administration, university ownership of equity in research-based companies, industrial liaison programs, continuing education, and university contribution to community and state economic development (technical assistance programs, business incubators, and research parks). The book also surveys and describes faculty responses to each of these activities and to changes brought about by technology transfer. It concludes by making recommendations for designing broad-based, controlled institutional responses to technology transfer.

The examples, models, and lessons provided by the histories of the four universities described here can be applied in many other higher education settings. They are intended to form a schema for understanding technology transfer and for taking effective steps to plan technology transfer activities and responses in institutions of higher education.

This book differs from others on the subject in the comprehensiveness of its approach. It is based on collection of data and close observation of the experiences of the four selected universities, including extensive interviews with university faculty members. It attempts to show the relationships among many mechanisms and aspects of technology transfer that heretofore have been studied only separately. It not only describes past and present technology transfer activites but also considers the important issues they raise and their implications for the future of American universities.

Organization of the Book

Chapter 1 defines terms related to technology transfer and explains why four particular American research universities were chosen for study. It also describes the technology transfer mechanisms that the book examines and the aspects of them that are analyzed.

Chapters 2 and 3 provide the background necessary for a comprehensive understanding of college and university technology transfer. Chapter 2 offers a general discussion of the history of technology transfer in institutions of higher education and highlights events in the history of each of the selected universities that are related to technology transfer. A grasp of the overall historical background of each university is needed to understand later discussions of the history of technology transfer mechanisms. Chapter 3 describes the present institutional setting and current policies related to technology transfer at each of the four universities in this study.

Chapters 4 through 7 deal with the most important and symbolic of all technology transfer mechanisms: patent administration. These chapters and chapter 8 concentrate on the intellectual property aspects of technology transfer.

Chapter 4 is concerned entirely with the history of university patent administration. It traces the development of patent administration from early days, when universities felt that patent ownership was inconsistent with the ideals of the university, to the present, when universities are aggressively marketing patent licenses. One of the important points made in this chapter is that patent administration as we know it today is a relatively recent development even in the largest universities.

Chapter 5 discusses university patent policies. It describes and gives examples of policies related to patent administration and then compares these policies in the four subject universities. From this comparison conclusions are drawn about the role of patent policies in the academic culture of each of the institutions. Chapter 6 describes the many forms and functions of patent administration and considers how they may be evaluated. Chapter 7 completes the discussion of patent administration by describing the current situation in this area in each university, with an emphasis on the changes that have occurred recently or are in process and what they mean.

Chapter 8 extends the discussion of the university's management of and attempts to profit from its intellectual property to the next logical step: ownership of start-up companies. This is an important topic at present because such equity ownership, while attractive because of its high potential financial rewards, presents universities with ethical as well as practical problems. It is a symbolic issue now in the process of resolution and is appropriate for an examination of what universities should and should not do in regard to technology transfer.

Chapter 9 examines industrial liaison or affiliate programs and their recent growth, emphasizing the organizational dimension. The implications of the ways in which affiliate programs are organized, and particularly the differences in organization among the four campuses, are important for understanding the organization of other activities related to technology transfer.

The organizational dimension is also important in the examination of continuing education, the subject of chapter 10. The connection between technology transfer and continuing education often is not perceived, yet continuing education is one of the most important tools that colleges and universities have for relating to industry. The history of a college or university's continuing education organization says a great deal about the institution's attitude toward external constituencies. Furthermore, continuing education's traditonal market orientation makes it the natural vehicle for certain technology transfer activities. Many newer forms of technology transfer, such as liaison programs, contain components of continuing education.

Chapter 11 is concerned with the technology transfer mechanisms that are most closely association with economic development. It begins with a discussion of the roles that colleges and universities are playing in regional economic development efforts today and continues with examinations of university-sponsored technical assistance programs, business incubators, and research parks.

Chapter 12 deals with faculty attitudes toward technology transfer. Its description of such attitudes is based on interviews conducted with 104 faculty members from the four universities in this study. It presents a summary and examples of faculty views on a wide variety of topics related to technology transfer.

Chapter 13 concludes the book by summarizing the lessons that can be drawn from this study of technology transfer mechanisms at four research universities. It also establishes an analytic framework for planning rational institutional responses to technology transfer issues.

Taken together, these 13 chapters describe a significant trend in higher education today. College and university involvement in technology transfer is not a fad. It is producing changes in many aspects of institutions of higher education, including organizational forms and resource allocations. Most important, the increased emphasis on university technology transfer reflects a significant change in the public's view of the role of institutions of higher education and the role that such institutions accept for themselves.

Note

1. Eric Ashby, *Adapting Universities to a Technological Society* (San Francisco: Jossey-Bass, 1974), 1, 7.

Acknowledgments

This book is an adaptation of a dissertation that was almost eight years in preparation. During those eight years many people offered help, support, encouragement, patience, and tolerance. I cannot thank them all here, but a few deserve special mention.

My thesis advisers, Earl Cheit (Chair), Martin Trow, and Charles Hitch require special thanks not only for their encouragement but also for their patience. They recognized the personal circumstances that prolonged my study and did not lose faith in my ability to complete this project.

Milton Stern, dean of University Extension at the University of California, Berkeley, although not an official dissertation adviser, acted often to advise, encourage, and cajole me toward completion of my degree and this book. He was and is a constant source of intellectual stimulation to me and offers a model of the kind of intellectual and scholarly activity and inquisitiveness that will carry me beyond this phase of my lifelong education. I thank him for this and for his forbearance.

I owe special gratitude to many colleagues at University Extension in the University of California, Berkeley, who suffered my preoccupation with this project with unfailing good humor. They showed commendable willingness to shoulder some of my responsibilities and tolerate my lack of attention as this work progressed. These colleagues include Don McDaniel, Vivian Sutcher, Gerald Becker, Doyle Barfield, Stacey Kita, William Newton, Harriet Landon, and Jan Milstead. Even more directly involved were Irene Kim, Karen Engel, and particularly Carmen Edstrom, who helped me in many ways for many years.

This study could not have been completed without the help of the five skilled and tenacious interviewers who conducted many of the faculty interviews that form the basis of much of this study. Sharon Graves Floyd, Carol

Fuller, Bonna Elfant, Elizabeth Collins, and Liza Loop all interrupted their busy lives to help me with this task and I am very grateful. Lisa Yount also applied her considerable editorial skill to this manuscript in all of its stages, markedly improving its clarity and expression.

I am also grateful to those 104 faculty members who each gave up an hour of time to be interviewed for this study, as well as the many others from U.C. Berkeley, M.I.T., Penn State, and Stanford who were willing to express their views on technology transfer. I appreciate their willingness to help a researcher from another department or another university.

Finally, I am deeply honored by the help of my family and friends, who endured the writing process along with me without being able to receive the compensating personal satisfaction. The daily adjustment of plans and special arrangements that had to be made by everyone close to me so that I could complete this project add up to a debt that I can never repay.

Thank you all.

1

Introduction: Technology Transfer and Higher Education

Walking back from the graduate division office of the University of California, Berkeley, on the day I filed the Ph.D. dissertation that was to become the basis of this book, I encountered a professor of civil engineering who had been interviewed as a part of my study. He inquired about my progress toward the degree and then asked to be reminded of its subject. When I told him it was about technology transfer and universities, he said, "That's interesting. You know what bothers me? It takes at least ten years for a discovery in our labs to get into industry."

This was just one of many instances, both observed personally and recounted in the press, during the six months between the filing of my dissertation in April and the submission of the revised manuscript to the publisher in September 1989, that reassured me that the issues and questions examined in my dissertation were of continuing importance to universities. For example, on March 23, 1989, the University of Utah held a news conference to announce the discovery of "cold fusion," a process that promised a clean, unlimited source of energy. The discoverers and the university, however, fearing loss of commercial rights to the process, withheld crucial details of the experiment. This made scientific verification of the discovery difficult and greatly angered many other researchers, who viewed such withholding as a violation of scientific ethics and protocol—in a word, "unacademic" behavior on the part of the university.[1] By July, after many attempts at

1

reproducing the discovery in other laboratories (based on an almost illegible bootlegged copy of an article submitted to a scientific journal by the discoverers) had failed, most scientists came to the conclusion that the results of the cold fusion experiment were nothing more than experimental error.[2] If this opinion is sustained, the reputations of both the discovering scientists and their university will have suffered.

In another incident, which occurred in mid-June, 1989, Representative Ted Weiss (D-NY), chair of the House Governmental Affairs Subcommittee on Human Resources, attacked M.I.T.'s Industrial Liaison Program for selling Japanese corporations special access to research funded by U.S. taxpayers.[3] Rep. Weiss's attack was countered vigorously by M.I.T. and further in late June by Carnegie-Mellon University, which announced that it was abandoning its policy of associating only with American companies and would actively seek research funding from and cooperation with foreign companies. As a reason for this change, Carnegie-Mellon cited frustration with the "very shortsighted" view of American firms, which seemingly operate in "a financial environment in which there is a heavy emphasis on a business's stock-market price."[4] Similarly, on August 3 Harvard announced that its affiliated Massachusetts General Hospital had accepted $85 million from Shiseido, a Japanese cosmetics company, for a new skin research center.[5] This agreement echoed an earlier agreement between Massachusetts General and the Hoechst Corporation of Germany (signed in May 1981), which provided the university with almost $70 million over ten years to establish a Department of Molecular Biology under Professor Howard Goodman. Both agreements gave the foreign sponsors priority rights to intellectual property resulting from sponsored research. The Hoechst agreement prompted a two-year congressional investigation, which forced a public disclosure of the terms of the agreement.

These were only a few of the events, in the news and out of it, related to the subject of this book that occurred in those few months in early 1989. To some, such a relationship might seem slight, perhaps even nonexistent. Perhaps this book can dispel that notion. University technology transfer and the issues surrounding it present a most confusing picture when viewed narrowly, and such narrow views prompt ad hoc responses and generate confusion, both inside and outside the academy. I hope that this book will provide both the evidence and the impetus for university policymakers to consider the connections among and the broad implications of events such as those just described.

Reactions to Changes Brought by Technology Transfer

This book examines the changes taking place in American institutions of higher education, particularly in research universities, in response to demands that they become more directly involved in activities designed to promote economic growth. These demands have come from both inside and outside the university. They first became noticeable in the late 1970s, when the economic value of university research was demonstrated by advances in electronics and when advances in biotechnology offered the prospect of a wide range of valuable new products and processes. By 1980 the weakening position of the United States as a leader in basic research, combined with its apparent failure to realize the full benefit of commercial application of that research, focused attention on the value of research done in American universities and the process by which that research was developed into marketable products.

This awareness led to initiatives on the part of government, industry, private interest groups, and universities themselves to increase support for university research and to promote and quicken the pace of technology transfer. In turn, these initiatives have influenced policies, forms of organization, faculty attitudes, relations with industrial firms, and many other elements of the university.

The changes are encouraging signs to some observers. In 1983, Thomas Moss, an observer of the economic prospects of the United States, wrote in the *Chronicle of Higher Education:*

> One key component of a uniquely American strategy for innovation is beginning to emerge. It is the enormous burst of industry-university cooperation in research and development leading to new products and technology . . . [that] may be one of the most significant and powerful aspects of national technological-innovation strategy. . . . The traditional, relatively open relationship between our universities and the larger community in which they are located is a tremendous and uniquely American strength in mobilizing resources to stimulate industrial innovation.[6]

However, Derek C. Bok, president of Harvard, was not as enthusiastic about the changes taking place:

> With this bright promise, why does the prospect of technology transfer arouse anxiety on the campus of almost every distinguished research university? . . . [The causes for concern] stem from an uneasy sense that programs to exploit technological development are likely to confuse the university's central commitment to the pursuit of knowledge and learning by introducing into the very heart of the academic enterprise a new and powerful motive—the search for commercial utility and financial gain.[7]

Other observers go farther than Bok: they regard the new emphasis on technology transfer as a threat to the integrity and autonomy of the university, and view it as antithetical to the principles upon which universities are founded. One of these observers is Martin Kenney, professor of agricultural economics and rural sociology at Ohio State University and author of *Biotechnology: The University-Industrial Complex.* Kenney has written:

> The university, a peculiar and fragile social institution that can trace its history back to early feudalism, is being subsumed by industry, one of the very institutions with which it should, to some degree, be in conflict. When university

and industry become partners, the entire society is endangered. . . . Perhaps the cruelest irony will be experienced by U.S. industry itself. As the university is bought and parcelled out, basic science in the university will increasingly suffer. . . . Industry will then discover that by being congenitally unable to control itself and having no restraints placed on it by the public sector, it has polluted its own reservoir.[8]

These quotations demonstrate the range of views that establish the context for this study. The opinions of critics are helpful, however, in understanding the changes brought to higher education by technology transfer. Some critics claim that universities are not performing as well as they should in getting valuable ideas to those who can use them. They see the university as a place where hidden treasures are buried. Others complain that university researchers are not selecting sufficiently useful problems to work on. They see the university as an "ivory tower" whose protected inhabitants ignore what its patrons and public want. A third form of criticism simply expresses the general view that institutions of higher education should do more than create knowledge: they, themselves, should be active in the process of technology transfer.

The course of action proposed for colleges and universities depends on the view held by the critic. The hidden-treasure critics propose methods for increasing communication between the university and those who can develop products and, perhaps, awareness campaigns to educate faculty to the commercial potential of their ideas. The Congress of the United States is the home of many hidden-treasure critics—politicians who seek simple answers to the problem of international competition and economic decline. As a result, a large number of programs, most notably those sponsored by the National Science Foundation, have been directed at drawing out university inventions.

The ivory-tower critics press for increased standards of accountability between the university and research sponsors and an increased willingness on the part of universities to undertake applied, as opposed to basic, research projects. Many of these critics are inside the university—faculty members who rebel against the traditions calling for "basic" research and individual scientific effort as opposed to team problem solving.

Those who think colleges and universities should be more generally active in technology transfer urge higher education institutions to develop practices and organizational units that can help to turn the results of research into commercial products. Community leaders and, again, politicians are the largest groups in this category. However, industrial leaders also occasionally express this view.

It is useful for college and university administrators to understand, identify, classify, and assess the strength of the criticisms leveled at the institution, and the above categorization can help in the process. Within this schema, administrators can formulate appropriate responses and take steps to head off external criticisms that otherwise might build to considerable external pressure for change. Understanding the sources of criticism also contributes to the definition of technology transfer as it applies to higher education.

Definition of Terms

The terms in the title of this book have been used in many ways by different people. At times they have been confused or used interchangeably with related terms. It is important, therefore, to define these terms as they will be used in this context.

TECHNOLOGY TRANSFER

The term *technology transfer* is used because, in the public's understanding of its meaning, it comes closest to indicating the overall subject of this study. However, the term lacks precision because it is used in so many ways. This lack of precision often obscures issues and makes discussion difficult.

When it is associated with research universities, technology transfer means the transfer of the results of basic and applied research to the design, development, production, and commercialization of new or improved products, services, or processes. That which is transferred often is not really technology but rather a particular kind of knowledge that is a precursor of technology. The transfer is from the originator of the knowledge to the developer of that knowledge.

This definition of technology transfer differs from other common meanings of the term, such as the transfer of existing technology across national boundaries (for example, between the United States and the Soviet Union) or between institutions (for example, between the military and industry or between one industry and another). In university-related technology transfer, the boundaries crossed by the transfer process are those between the university and business concerns.

The technology transfer process can be simple or extraordinarily complex. In some cases the transferred product of research is fully formed and ready for market, needing only a manufacturing facility and a marketing apparatus. A nutrient for mushrooms or a hybridized plant from which tissues can be taken to produce more plants is an example of this type. In other cases—for example, most drugs—a great deal of further development and testing is needed before the fruits of research can be brought to a marketable form.

In addition to this specific definition of technology transfer there is another, more general meaning of the term that is relevant to universities. Technology transfer in this sense refers to a wide array of activities associated with commercializing research or with improving relations between the university and industry. These activities may include continuing education, industrial liaison or affiliate programs, industry-sponsored research, and many others. For instance, U.C. Berkeley recently established a technology transfer office charged with, among other tasks, "organizing and planning of industrial participation in development of outlying properties; . . . establishment and promotion of contacts with large and small industries and investors; . . . assistance to individual faculty who wish to start or participate in starting new businesses; and . . . acquisition and oversight of contractual/consultative arrangements with appropriate outside firms to assist as necessary in technology transfer."[9] The joining together of the activities subsumed under this broader meaning of technology transfer is a trend that is only now becoming visible. Prior to 1980, university patent offices were not seen as being associated with the university's efforts to relate to industry. Now they are viewed as an important part of that relationship, and this change is reflected in new organizational structures that place patent offices in the same reporting line as contract administration, development efforts, and offices charged with fostering relations with industry.

Technology transfer often will be used here in this broader sense. Indeed, an aim of the book is to help the reader view technology transfer as a broad-based and important set of changes that requires major readjustment in the academic community. Understanding the relationships among these changes helps make possible a more coherent response to them.

TERMS AND CONCEPTS RELATED TO TECHNOLOGY TRANSFER

An understanding of the meaning of technology transfer and of the trend it signifies is sharpened by an examination of a number of terms related to and sometimes confused with it. These include *knowledge transfer, economic development, basic* and *applied research,* and several others.

KNOWLEDGE TRANSFER. Institutions of higher education have been engaged in technology transfer, in its broadest sense, for a long time. Their involvement has come through the education and graduation of students who carry what they have learned into the commercial world, the publication of the results of research for use by the scientific and industrial community, and the consultation of faculty members by industry. These activities, all primary and traditional functions of the university, might best be referred to collectively as *knowledge transfer.*

Technology transfer is both a special case of knowledge transfer and a separate category. It certainly always involves some form of knowledge transfer, and in some cases it may be indistinguishable from traditional knowledge transfer activities. For instance, faculty consulting is often listed by universities as part of their technology transfer activities. Technology transfer, however, can include activities that go well beyond these traditional functions. As examples, universities may pay for product prototypes or invest in start-up companies formed to exploit university-owned intellectual property.

The distinction between knowledge transfer and technology transfer has an historical aspect. The term *technology transfer* gained currency in the late 1970s, about the same time that universities began to become more active in commercializing research and promoting relations with industry through patent and technology licensing offices, investment in start-up companies, research partnerships with industry, liaison programs, business incubators, technical assistance programs, and research parks. The term, therefore, is strongly associated with these new activities. This book is primarily concerned with them.

Confusion between *knowledge transfer* and *technology transfer* can lead to conceptual difficulties. According to the definitions set forth here, the kind of consulting often performed by business school faculty usually would not be considered technology transfer, even though it may benefit a commercial firm. In helping businesses make plans or solve accounting or system problems, business faculty are transferring knowledge but not technology. The process of knowledge transfer is considerably different from the process of technology transfer. Unlike knowledge transfer, in which knowledge is passed from the university to the receiver easily and usually with almost no follow-up, technology transfer requires considerable effort at all stages of the process. The analogy of a relay race, in which the baton is passed cleanly and quickly from one runner to the other, fits most knowledge transfer situations but does not apply to technology transfer. Technology transfer might be more appropriately compared to a basketball game, in which the university is only one player. This player may bring the ball over the half-court line, but must then quickly enlist the aid of teammates in order to score. In this game the "ball" is passed back and forth constantly among players who may include businesspeople, venture capitalists, patent attorneys, production engineers, and many others in addition to university faculty.[10]

While at times the distinction between knowledge transfer and technology transfer might be artificial, there are significant differences that, once understood, help to clarify the definition of both concepts. Although both are involved in this study, the focus is on technology transfer rather than knowledge transfer.

ECONOMIC DEVELOPMENT. The term *economic development,* as it relates to colleges and universities, generally means institutional activities that are designed to encourage or promote the economic development of a region, state, or country. Colleges and universities often are called on to aid state and local governments in programs to start new businesses, attract out-of-state firms, create jobs, provide technical assistance to local industry, or help to train or retrain people for employment. Economic development can also mean the creation of an environment that is attractive to business. The opportunity to send children to a well-known institution of higher education and the cultural and recreational benefits of living near such institutions are important features of an environment that attracts business, especially high-technology business.

It is clear that economic development involves much more than technology transfer. However, college and university technology transfer activities such as technical assistance programs, research parks, and business incubators can play important roles in an area's economic development.

BASIC AND APPLIED RESEARCH. The distinction between basic and applied research is useful in discussions about technology transfer, but at times this distinction may be more apparent than real. Traditionally, basic research referred to an investigation into first principles, the only object of which was the creation of new knowledge. The production of a tangible or useful result, by contrast, was the domain of applied research. In the early years of the twentieth century universities became the main performers of basic research, usually leaving applied research to industry. On the one hand, most universities could not find the money to keep up to date with the equipment needed to do applied research; on the other, thanks to foundation support, universities had the luxury of being able to do research that did not have an immediate payoff.

In some fields, during parts of their history, the distinction between basic and applied research was clear, but instances increasingly occurred in which the distinction could not be made. This was particularly true in fields where scientific breakthroughs could be applied rapidly to practical concerns, such as electronics or biotechnology. In fact, the distinction between basic and applied research now tends to be economic rather than intellectual: research that requires a great deal of effort and expense to put to commercial use is much more likely to be called *basic* than is research, however fundamental, that can be put to use immediately. Because technology transfer refers to the commercialization process, the distinction between basic and applied research is important in discussions of this activity. The blurring of distinctions between basic and applied research and the realization that applied research has a legitimate place in universities have helped to legitimize current university technology transfer efforts.

Another concept allied with the basic versus applied distinction is the idea of the "push" and "pull" of technology. The traditional notion is that basic research—that is, fundamental scientific breakthroughs—leads, after applied research, to useful technology. That certainly has been a common pattern. However, it is also true that applied research can lead to fundamental scientific breakthroughs. For instance, the commercial demand that electronic circuits be miniaturized set the basic research agenda for electronics engineers and materials scientists for many years.

OTHER TERMS. Certain other terms are commonly used in relation to or as substitutes for technology transfer. For example, *public service* sometimes is used to describe college and university activities related to technology transfer. When technology transfer cannot be related to either the teaching or the research mission of an institution, technology transfer activities can be subsumed under its public service mission.

Intellectual property is an important term that is related to technology transfer because the transfer process emphasizes the value and protection of the intellectual product of its researchers. Many of the issues and activities surrounding technology transfer are related to the notion of intellectual property. *Conflict of interest* has a similar relationship to technology transfer. As colleges and universities move closer to the commercial world and as the commercial value of the intellectual property developed by faculty, researchers, and students becomes more evident, members of an institution may encounter conflicts of interest between personal benefits and benefits to the institution. There is also a greater risk of conflict between the institution's own financial well-being and the public good to which it is supposed to be dedicated. These conflicts often arise in conjunction with technology transfer activities.

THE AMERICAN RESEARCH UNIVERSITY

Having examined one element of the title of this book, *technology transfer,* we now turn to the other element, *university.* More specifically, what we are discussing is the *American research university.* The extent, nature, and sponsors of university research are important to the story of technology transfer, as are current trends associated with university research.

American universities are important providers of research, particularly basic research. In 1988 about $132 billion was spent on research and development in the United States. Of that total, about $15 billion went for basic research, $90 billion for development, and $27 billion for applied research. It is estimated that over half of the $15 billion that funds basic research is spent in universities and that basic research accounts for over two-thirds of all research done in universities.[11]

This research is concentrated in a relatively small number of universities. There are about 200 research and doctorate-granting institutions in the United States. Of the research and development done in universities in 1988, valued at $13.42 billion, about 83 percent ($11.01 billion) occurred in the top 100 universities, 34 percent ($4.59 billion) in the top 20 universities, and 21 percent ($2.79 billion) in the top 10 universities.[12] In this book the term *research university* refers to the 100 universities having the largest volume of research funding.

By far the largest source of universities' research funding is the federal government. In 1988 it supplied about 61 percent of total university research support.[13] This was down slightly from the 63 percent provided by the government in 1985 and off considerably from the high point in 1967, when federal support made up 72 percent of the total.[14]

By contrast, industrial support of university research, which is important to this study because it is one indication of the relationship between industry and universities, is small but increasing. In 1980 industry spent $277 million to fund research in universities, and by 1985 that total had risen to $482 million (in constant 1982 dollars).[15] Industry supported 3.8 percent of total university research in 1981, 5.7 percent in 1985,[16] about 7.5 percent in 1986,[17] and an estimated 10 percent in 1988 (over $750 million in current dollars).[18] Most of this support comes from a few large corporations, goes to a few large universities, and is concentrated in a few scientific fields. In 1985 only 11 of the top 91 research universities reported receiving more than 10 percent of their research funding from industry.[19]

The Four Universities in This Study

Four universities have been selected for close examination in this study: the University of California, Berkeley (U.C. Berkeley), the Massachusetts Institute of Technology (M.I.T.), Pennsylvania State University (Penn State), and Stanford University. All are ranked among the top 20 research universities in terms of research funding. In 1988 Stanford ranked second in total research funding with $277 million, M.I.T. ranked fifth with $271 million, Penn State marked fifteenth with $188 million, and U.C. Berkeley ranked sixteenth with $186 million.[20] It should be noted that because U.C. Berkeley is part of a nine-campus system with at least part of its research-related functions centralized in a systemwide administration, certain aspects of this university cannot be compared with the others.

Two of these four universities were among the few that received over 10 percent of their research funding from industry. In 1986 M.I.T. derived 14.2 percent of its funding from industry, and Penn State received 11.8 percent from this source. U.C. Berkeley and Stanford, by comparison, received 3.3 and 3.8 percent, respectively.[21]

Each of these universities was selected because of its special association with technology transfer and its relationship with industry. As its name suggests, M.I.T. has a long history of close relations with industry and involvement with the development of technology, and many of the issues associated with university ownership of intellectual property have already been worked out on the M.I.T. campus. It therefore provides a model for other universities that have only lately come to understand that something should be done about technology transfer. Stanford, too, has a well-known relationship to industry, particularly the electronics industry, and played a prominent role in the development of "Silicon Valley."

To balance the experiences of these two private universities, two public universities also were selected for study. Penn State was chosen because it is deeply involved in Pennsylvania's concerted efforts to improve that state's economy. As a university that is trying to become one of the top-ranked research institutions in the country and at the same time to serve its public constituencies with economic development programs and directed technology transfer efforts, Penn State has had experiences that illustrate many of the trends and issues addressed here. U.C. Berkeley was chosen because of its unique position: it is both an elite university—in this respect much like private universities such as M.I.T. and Stanford—and a large public university.

Description of the Study

Comparison of these four universities, the identification and examination of "new" technology transfer mechanisms, and the consideration of those mechanisms from the point of view of seven analytic dimensions form the foundations for this work and the conclusions it reaches. This approach views the seemingly disparate aspects of technology transfer as an interrelated whole.

The book identifies the elements and phenomena associated with university technology transfer, describes how they relate to one another, analyzes the significance of their impact on traditional academic values, and makes a number of predictions about their future. It views technology transfer as a broad-based, multifaceted trend in higher education that is only now becoming visible.

TECHNOLOGY TRANSFER MECHANISMS

The technology transfer mechanisms discussed in this book include patent and technology licensing offices, university ownership of equity in start-up companies based on university research, liaison or affiliate programs, continuing education, technical assistance programs, business incubators, and research parks. Most of these mechanisms have grown markedly since the late 1970s, and their growth is both an index and a result of the importance of technology transfer in higher education. Each of these mechanisms is examined in the four universities selected for study. The only major "new" technology transfer mechanism omitted from this list and the book—research partnerships between industry and universities—does not appear because it has been covered extensively in other studies and because most of the technology transfer issues related to such partnerships, such as preinvention patent rights, delay in publication provisions, and potential intrusions on university autonomy, are covered under other headings.

ANALYTIC DIMENSIONS

This study is primarily a comparison of the ways each of the technology transfer mechanisms is addressed at the four institutions chosen for study. The comparisons are made along several analytic dimensions. The historical dimension is important both because it helps to explain the milieu in which change presently is taking place and because it provides some interesting parallels to and lessons for current situations. The institutional setting dimension involves the external forces currently affecting each university and the opportunities or limitations of the institution's present setting, including its geographical location, leadership, economic condition, and goals and objectives.

The organizational dimension aids this analysis by allowing examination of the ways that technology transfer activities are organized in the selected universities—what organizational forms they take and how these organizations are changing. The faculty dimension views technology transfer activities and issues from the perspective of faculty members, in part to determine how much of the overall faculty role technology transfer activities occupy now or are likely to occupy in the future. The importance of technology transfer activities and the significance of the changes that technology transfer is bringing to universities is measured by the extent of faculty involvement in them.

The resource dimension covers the flow into and out of the university of resources related to technology transfer activities. The quest for greater resources, particularly to fund research, is a major reason that universities have adopted some technology transfer activities. The amount of money these activities generate and the way that money is spent have a significant influence on the importance and continuation of technology transfer in universities.

The policy dimension focuses on an examination and comparison of institutional policies associated with technology transfer. Policies both reflect and determine university behavior in and attitudes toward technology transfer. Policy changes or additions often indicate that fundamental changes have occurred within a university.

Finally, the dimension of academic culture provides a tool for analysis of the relationship of university technology transfer to the prevailing academic culture in each university. "Academic culture" is a catchall concept that to an extent includes all the other dimensions previously mentioned as they balance each other and mingle with the other branches of the arts, humanities, and social sciences. It has been defined as "a shared way of thinking and a collective way of behaving."[22] Each of the universities selected for study has a distinctive academic culture within which technology transfer activities must fit to be effective.

Conclusion

In addition to describing the mechanisms of technology transfer at work in research universities today and the changes they are bringing to higher education, this book attempts to assess the impact and significance of those changes and to predict further changes that are likely to occur. It also attempts to establish a conceptual framework for analyzing and responding to the interconnected phenomena and events that can be grouped together under the heading of technology transfer. It treats both the operational concerns and the philosophical aspects of technology transfer, including the possible effect of this phenomenon on the autonomy of institutions of higher education. I hope that this work contributes to the understanding of university technology transfer and of the adaptations to technology transfer that universities are making and will continue to make.

Notes

1. Gilbert Fuchsberg, "Prospect of Commercial Gain from Unconfirmed Discovery Prompted Utah U. Officials to Skirt Usual Scientific Protocol," *Chronicle of Higher Education*, 17 May 1989, A5.
2. William J. Broad, "On Cold Fusion, It Looks Like Utah Stands Alone," *New York Times*, 16 July 1989, E5.
3. David L. Wheeler, "Lawmakers Hit Universities for Selling the Results of US-Financed Research to Foreign Companies," *Chronicle of Higher Education*, 21 June 1989, A16.
4. Scott Jaschik, "Frustrated by Tepid Response of U.S. Business, Carnegie-Mellon Says It Will Encourage Japanese Links to Its Federally Financed Research," *Chronicle of Higher Education*, 5 July 1989, A22.
5. Steven R. Weisman, "Harvard and Japanese in Skin Research Deal," *New York Times*, 4 August 1989, A8.
6. Thomas H. Moss, "New Partnerships a Bargain for Industry, a Boon for Colleges and Universities," *Chronicle of Higher Education*, 6 April 1983, 72.
7. Derek C. Bok, "Business and the Academy," *Harvard Magazine* 83 (1981): 26.
8. Martin Kenney, "University-Industry Partnerships Are a Danger to Society," *Chronicle of Higher Education*, 7 January 1987, 43.
9. Leonard V. Kuhi, memorandum to C. Judson King, 13 June 1988.
10. I owe this analogy to Richard Dorf of U.C. Davis. It was related to me by Mary Walshok of U.C. San Diego.
11. National Science Foundation, "Defense Major Factor Behind 7% Real Increase in 1986 National R&D Expenditures," *Science Resource Studies Highlights*, 27 December 1985.
12. "Fact File: The Top 100 Institutions in Total Research-and-Development Spending for Fiscal 1988," *Chronicle of Higher Education*, 29 November 1989, A25.
13. Ibid.
14. "Sources of University Research Funds," *Chronicle of Higher Education*, 10 December 1988, 8.
15. Robert M. Rosenzweig, "Academia: Are We Selling Out to Industry?" *SUNY Research '88*, New York: Research Foundation of the State University of New York, January-February 1988, 1–3.
16. "Sources of University Research Funds."
17. United States General Accounting Office, "R & D Funding, Foreign Sponsorship of University Research" (Washington, D.C.: United States General Accounting Office, GAO/RCED-88-89BR), 44.
18. Gilbert Fuchsberg, "Universities Said to Go Too Fast in Quest of Profit From Research," *Chronicle of Higher Education*, 12 April 1989, A28–30.

19. "Sources of University Research Funds."
20. Ibid.
21. "Fact File."
22. T. Becher, "The Cultural View," in *Perspectives in Higher Education*, ed. B. Clark (Berkeley: University of California Press, 1984), 166.

Bibliography

ASSOCIATION OF AMERICAN UNIVERSITIES. *Trends in Technology Transfer at Universities*. Washington, D.C.: Association of American Universities, 1986.

———. *University Policies on Conflict of Interest and Delay of Publication*. Washington, D.C.: Association of American Universities, 1985.

BECHER, T. "The Cultural View." In *Perspectives in Higher Education*, edited by B. Clark. Berkeley: University of California Press, 1984.

BLUMENTHAL, DAVID, MICHAEL GLUCK, KAREN SEASHORE LOUIS, and DAVID WISE. "University-Industry Research Relationships in Biotechnology: Implications for the University." *Science* 232 (1986): 1361–66.

BOK, DEREK C. *Beyond the Ivory Tower*. Cambridge, Mass.: Harvard University Press, 1982.

———. "Business and the Academy." *Harvard Magazine* 83 (1981): 23–35.

BROAD, WILLIAM J. "On Cold Fusion, It Looks Like Utah Stands Alone." *New York Times*, 16 July 1989, E5.

CORDES, COLLEEN. "Spending on R & D Seen Rising 3 Pct. This Year." *Chronicle of Higher Education*, 10 February 1988, A1.

DICKSON, DAVID. *The New Politics of Science*. New York: Pantheon Books, 1984.

DIMANCESCU, DAN, and JAMES BOTKIN. *The New Alliance: America's R&D Consortia*. Cambridge, Mass.: Ballinger, 1986.

"Fact File: The Top 100 Institutions in Total Research-and-Development Spending for Fiscal 1988." *Chronicle of Higher Education*, 29 November 1989, A25.

FUCHSBERG, GILBERT. "Prospect of Commercial Gain from Unconfirmed Discovery Prompted Utah U. Officials to Skirt Usual Scientific Protocol." *Chronicle of Higher Education*, 17 May 1989, A5.

———. "Universities Said to Go Too Fast in Quest of Profit from Research." *Chronicle of Higher Education*, 12 April 1989, A28–30.

JASCHIK, SCOTT. "Frustrated by Tepid Response of U.S. Business, Carnegie-Mellon Says It Will Encourage Japanese Links to Its Federally Financed Research." *Chronicle of Higher Education*, 5 July 1989, A22.

JOHNSON, LYNN G. *The High-Technology Connection: Academic/Industrial Cooperation for Economic Growth*. Washington, D.C.: ASHE-ERIC, 1984.

KENNEY, MARTIN. *Biotechnology: The University-Industrial Complex*. New Haven, Conn.: Yale University Press, 1986.

———. "University-Industry Partnerships Are a Danger to Society." *Chronicle of Higher Education*, 7 January 1987, 43.

LYNTON, ERNEST A., and SANDRA ELMAN. *New Priorities for the University*. San Francisco: Jossey-Bass, 1987.

Moss, Thomas H. "New Partnerships a Bargain for Industry, a Boon for Colleges and Universities." *Chronicle of Higher Education,* 6 April 1983, 72.

National Science Foundation. "Defense Major Factor Behind 7% Real Increase in 1986 National R&D Expenditures." *Science Resource Studies Highlights.* Washington, D.C.: National Science Foundation, 1985.

Nelkin, Dorothy. *Science as Intellectual Property.* New York: Macmillan, 1984.

Noble, David F. *America by Design: Science, Technology, and the Rise of Corporate Capitalism.* Oxford, England: Oxford University Press, 1977.

Peters, Lois S., and Herbert I. Fusfeld. "Current U.S. University/Industry Research Connections." In *University-Industry Research Relationships: Selected Studies.* Washington D.C.: National Science Foundation, 1982.

Rosenzweig, Robert M. "Academia: Are We Selling Out to Industry?" *SUNY Research '88.* New York: Research Foundation of the State University of New York, January-February 1988, 1–3.

Smith, Bruce L., and Joseph Karlesky. *The State of Academic Science.* New York: Change Magazine Press, 1977.

"Sources of University Research Funds." *Chronicle of Higher Education,* 10 December 1988, 8.

Stankiewicz, Rikard. *Academics and Entrepreneurs.* New York: St. Martin's Press, 1986.

United States General Accounting Office. "R & D Funding, Foreign Sponsorship of U.S. University Research." Washington, D.C.: United States General Accounting Office, GAO/RCED-88-89BR.

Wade, Nicholas. *The Science Business.* New York: Priority Press, 1984.

Wheeler, David L. "Lawmakers Hit Universities for Selling the Results of US-Financed Research to Foreign Companies." *Chronicle of Higher Education,* 21 June 1989, A16.

Weisman, Steven R. "Harvard and Japanese in Skin Research Deal." *New York Times,* 4 August 1989, A8.

2

The History of Technology Transfer in Research Universities

This chapter and the next describe three elements essential for understanding the changes occurring in institutions of higher education as they become more actively involved in commercial activities. These are the history of college and university technology transfer efforts, the present circumstances and culture surrounding college and university technology transfer, and institutional policies related to technology transfer. The description of these elements in the four selected universities provides a composite picture of the background surrounding technology transfer in each institution. The composite provides the setting in which to place the more specific and detailed descriptions and histories of technology transfer mechanisms offered in succeeding chapters.

The next section describes briefly the general history of technology transfer in research universities. It also provides an account of the history of technology transfer in the four universities of this study: M.I.T., U.C. Berkeley, Penn State, and Stanford.

The Beginnings of Research Universities and Their Ties to Business

The history of technology transfer in research universities is connected closely with the development of research as a distinguishable feature of university activities and with the history of the interaction between universities and commercial interests. The history of the American research university reveals some general trends that have a bearing on the history of technology transfer in the particular universities in this study.

BEFORE 1910

Between the Civil War and 1890 a series of events and trends combined to help a few colleges and universities become different from the typical college of the day, which was small and characterized by its limited and fixed curriculum.[1] Growing criticism of higher education after the Civil War focused on three shortcomings: the failure to include more natural science subjects in the curriculum, to teach practical subjects that would prepare students directly for careers, and to provide advanced training of the kind that could be obtained in European universities. By the 1850s the natural sciences had established a clear foothold in the American college curriculum, aided by private endowments such as those establishing the Lawrence Scientific School at Harvard (1847) and the Yale/Sheffield School (1861). By the 1860s a utilitarian cast was discernible in many college curricula, a trend encouraged by the Morrill Act of 1862. This act provided grants of land to institutions on the condition that a state maintain "at least one college where the leading object shall be, without excluding other scientific and classical studies, . . . to teach such branches of learning as are related to agriculture and the mechanic arts."

Research-based graduate training had become a permanent feature of a few universities by the 1890s as the Johns Hopkins University (founded 1876) adaptation of the German model had been introduced in a number of other institutions. The mission of the university began to change, shifting from an emphasis on cultural values to the general principle of cognitive rationality—knowing through the exercise of reason—on which universities are based today. These shifts, which had been started by the land-grant movement, combined with economic prosperity to lead to the academic boom of the 1890s. This period saw the rapid rise of state colleges as state legislatures recognized the value of the utilitarian knowledge taught in universities and as individual donors, primarily businessmen, began to favor academic institutions with gifts and bequests.

1910 TO 1920

By 1910 a new type of university, the research university, was clearly established. These institutions, of which there were perhaps 15, could be distinguished from others primarily by the greater amount of research performed by their faculties, an activity partly reflected in a lighter teaching load. A survey conducted in 1908 found that research university faculty members taught an average of 8 to 10 hours per week, while their counterparts in other universities taught 15 to 18 hours per week.[2] Research universities were also characterized by competition with one another for the same students and faculty.

Despite the rise in importance of research as a university mission, funding for research was scarce—indeed, such money often had to come from the pockets of the researchers themselves. The first line-item in a university budget specifically allocated to research appeared at the University of California in 1915. After that, funding specifically designated for research slowly increased, although it remained subject to the whims of donors.

Between 1890 and 1920 the assignment of the performance of basic research specifically to the research university slowly became more common. Support of research by the federal government also increased during this period, but this support, administered by a number of scientific bureaus, was supplied by a parsimonious Congress, anxious to see practical results from its investment. Agriculture and geological surveys thus were the largest beneficiaries of federal research support funds. With the federal government concentrating on the relatively mundane, and with academic scientists in some universities becoming free to pursue loftier problems, the professions of academic and government scientist began to diverge.

Another factor helped to solidify the university's hold on basic research. Between 1890 and 1900 in the industrial corridor of the Northeast, informal partnerships were formed between large industrial firms and certain universities. Unable to purchase the latest equipment and instruments, the universities concentrated on training students in the fundamentals of science and left their practical training to industry.[3] This division of labor formed the basis for the schism that eventually developed between academic and industrial science, a rift that persists to this day.

The growth of university research and its relation to business interests is the beginning of the story of technology transfer in universities. Before 1900, financial support for research came from wealthy individuals and was generally connected with a particular facility, such as an observatory, a piece of equipment, or a scientific collection. After 1900, as the cost of research rose, support came more often in the form of endowments from foundations.

THE 1920S

The First World War brought an unprecedented coordination, principally under the direction of the National Research Council, among the three partners in the research enterprise—universities, foundations, and the federal government. This coordination continued after the war in a consensus that research support ought to go to the "best science," that is, to the institutions and individuals most highly qualified to carry out research successfully. Despite recognition of the important boost that university research had given to the war effort, however, by 1920 most universities still did not have the money needed to conduct research.

Nonetheless, the 1920s saw the top American research universities accumulate considerable wealth. In the early part of the decade many foundations shifted their emphasis from general support of education to specific support of elements within higher education, including research. Although there was some concern at the time that the foundations might favor the establishment of private research institutions not associated with universities, by 1926 it was clear that this would not happen to any great extent. The university's position as the major performer of basic research had become secure. Fears that foundations would come to control the universities with their funds also failed to materialize as the academic profession solidified. The foundations settled into a practice of giving money to trusted individuals in a few institutions. Thanks in part to this increased support, American universities became roughly equal to their European counterparts, and the seriousness with which American students pursued higher education increased markedly.

Although there are no reliable data on the extent of direct support of university research by corporations until 1936, when the IRS began keeping statistics, there is some evidence that such support was commonplace but small and that it increased during the 1920s. Industries usually supported very specific types of research, and their support could not be depended upon. Even more important, it became clear that the purposes of the university (disinterested inquiry) and the corporation (a specific result) were often in conflict.

THE 1930S

By 1930 universities had learned that they could not depend on individual patrons for support of their research; funding sources were going to have to be diverse, and donor preferences for particular institutions and research subjects were going to prevail. Curiously, the 1930s, marked by major economic depression elsewhere, was a time of relative prosperity for university research. Universities had accumulated capital and facilities during the 1920s, and many of the patrons of university research were not hard hit by the depression. Furthermore, because jobs were scarce, academic researchers tended to stay in universities during this time, adding human capital to the already substantial physical capital possessed by the institutions in the form of equipment and facilities.

Once the Depression ended, additional research dollars returned to universities, and by the end of the 1930s American university research was in very good shape. In 1939, however, only 16 research universities existed, and the research they conducted was diverse because it was funded by diverse sponsors; there was no coordination among nongovernmental patrons of research. Furthermore, the federal government continued to confine its research support to a few narrow and applied fields, and there was no clear national or federal science policy.

WORLD WAR II AND AFTER

The situation changed after World War II. In 1940 Vannevar Bush of the Carnegie Institution (and formerly of M.I.T.) convinced President Franklin Roosevelt not only that scientists could produce the new military technology needed to defeat Germany but also that they could do so with little help or supervision as long as enough money was provided. The spectacular contributions of university science to the war effort in atomic energy, radar, aeronautics, and many other fields clearly called for the continuation of federal support of university research after the war, but the need for coordination became obvious.

Bush again spoke for the scientific community in *Science: The Endless Frontier* which basically described the terms of a tacit contract between the scientific community and the federal government. The result was the development of a clearly articulated federal science policy and the establishment of a number of government agencies whose task was to allocate research funds to institutions. The scientific community was given control of this funding to an unprecedented degree. The peer review system of evaluating funding proposals became the ultimate expression of the "best science" principle that had been established earlier, and the university's role as the primary provider of basic research was virtually guaranteed.

This new partnership between academic institutions and the federal government, founded on massive federal support, changed the face of American institutions of higher education and overwhelmed the comparatively feeble ties that had been formed between universities and industrial sponsors. Although the amount of industrial sponsorship of university research probably did not diminish, its relative importance declined greatly during this period.

THE 1970S

Links between industry and universities were weakest in the early 1970s. Several factors besides increased federal support contributed to this decline. First, university research diverged from industrial needs and became concentrated in pure science on the one hand and military projects on the other. Second, the growth of universities' enrollments and research capacity lessened graduates' interest in industrial careers. Many university graduates came to view the kind of research carried on by industry as too applied, and they looked down on industrial positions, especially while academic jobs were plentiful. In turn, industry reduced its expenditures for basic research and therefore was no longer an employer of top university graduates.[4]

In the mid-1970s, university-industry relationships began to improve for several reasons. First, some of the basic research that universities had been doing began to pay off commercially, first in electronics and then in biotechnology. Much of the expertise in these fields resided in universities, and newly interested industry set about gaining access to it. Second, general concern about the competitive position of the United States in the world economy increased. Despite a formidable lead in basic research, the United States appeared to be losing markets to foreign competitors who were more skillful at developing and marketing products. As a result of this concern, policymakers and the federal government turned their attention to the process by which basic research, largely performed in universities, was transformed into a marketable product. Third, the federal commitment to continued funding of university research began to erode as cuts in the federal budget and an economic downturn decreased the amount of government funds spent on research. This loss caused universities to turn to industry to broaden their base of support and gain funding for research in areas which the government was no longer willing to underwrite.

Federal policy also began to encourage interaction between universities and industry. Some government funding agencies, notably the National Science Foundation (NSF), began to "leverage" federal funds by requiring industrial matching funding for some research projects. The NSF "centers" program is the most important example of this trend. Rather than supporting individual researchers at different campuses on a project-by-project basis, NSF devoted some of its funds to creating "centers for excellence"— groups that would receive long-term funding for research in particular subject areas. To try to ensure that the research produced results that would enter the commercial world rapidly, NSF made continuation of funding for the centers conditional on their ability to attract corporate matching funds.

Another significant change in federal policy was the revision of the patent law, primarily through the passage of PL 96-516, signed in December of 1980, which gave universities the right to patent and license discoveries made under federal contracts. Patent protection also was extended to new categories of discoveries and inventions, including bacteria produced in laboratories with the new methods in biotechnology, and later genetically altered animals and even mathematical equations.

These and many other events infused the university research environment with a new interest in relations with industry; at the same time they raised serious concerns that the integrity and autonomy of the university were being threatened by "introducing to the very heart of the academic enterprise a new and powerful motive—the search for commercial utility and financial gain."[5] The partnership among universities, the government, and industry that had existed intermittently before and during World War II but had languished after the war was beginning to be reestablished, based on the perceived threat of economic decline.

The history of technology transfer in particular institutions must be viewed against the backdrop of the general trends in the history of university research and relations with industry that have just been described. However, a knowledge of the specific history of a university also is absolutely necessary for understanding that particular university's present culture and responses to changes accompanying the renewed interest in technology transfer. The following sections describe briefly those elements of the history and the foundations of the culture of each institution that influence present attitudes toward technology transfer.

THE 1980S

The pace of change related to technology transfer in universities accelerated dramatically after 1979. As we will see in later chapters, every measure of such change—university patent applications, university-industry agreements, and technology transfer real estate developments, to name a few—point to 1979 as the turning point. A description of the extent of the changes in the 1980s and the many areas in which they occurred and an analysis of their impact on the university are the main underlying purposes of this book.

M.I.T.

M.I.T. is considered first not only because it has a history rich in relations with industry and technology transfer activities but also because events here influenced developments in other universities. Most important, some of these events provide early examples of issues related to technology transfer and relations with industry that now face all major universities. This narrative concentrates on three issues that are important throughout the book: the way in which the role of the faculty at M.I.T. has been shaped by relations with industry, the M.I.T. experience in balancing research and pedagogy between the demand for practical training on the one hand and the ideals of academic inquiry on the other, and the way in which M.I.T. has organized itself for coordinating its relations with industry. The history of M.I.T. not only contains lessons in each of these areas for higher education administrators and others trying to chart a course for institutions of higher education today but also explains much of the shape of the present culture of the institute.

EARLY YEARS

M.I.T. was founded in 1861 by a group of Boston citizens, led by William Barton Rogers, partly in reaction to what they perceived as Harvard's lack of commitment to scientific and engineering education. These wealthy and aggressive Boston industrialists felt that Harvard was not moving quickly enough into the utilitarian era.[6] The new school they started was to have three elements: "a society of arts, a museum of arts, and a school of industrial science"; it was to be dedicated to "the advancement, development, and practical application of science in connection with arts, agriculture, manufactures and commerce."[7] The most successful of these three elements was the school of industrial science, which was founded to provide a "complete course of instruction and training, suited to the various practical professions" and to meet "the more limited aims of such as desire to secure a scientific preparation for special industrial pursuits . . . having their foundation in the exact sciences."[8] After thwarting an attempt by the Massachusetts governor to merge M.I.T. and Harvard under the Land Grant Act in 1862, Rogers succeeded in securing a portion of the state's land-grant funds, and M.I.T. became a land-grant institution in 1863. The governor's effort was the first of at least four serious attempts to make M.I.T. Harvard's engineering school. If any of these attempts had succeeded, it undoubtedly would have threatened the preservation of the tenets of close association with industry and practical application of knowledge upon which M.I.T. had been founded.[9]

M.I.T. remained an important but relatively small school well beyond the turn of the century, attaining nothing approaching the status of a university. Until well after 1900, the institute had very little outside support; over 90 percent of M.I.T.'s income came from tuition.[10] This situation began to change, however, in 1912 when an anonymous donor, later revealed to be George Eastman, donated $2.5 million to move M.I.T. to Cambridge and to build a new physical plant for the institute. This gift, the first of several from Eastman, was substantial enough to begin attracting additional support. By the 1920s M.I.T. had entered the ranks of wealthy private universities, with an endowment fund valued at $15 million. By 1930 M.I.T. had the fifth-largest endowment in the country—$33 million, over half of which had been donated by Eastman.[11] Eastman was dedicated to the idea of practical engineering training, and his lavish support of M.I.T. strongly reinforced the principles upon which the institute had been founded.

SHAPING FACULTY ROLES

These principles, however, were somewhat at odds with the ideals characterizing most higher education of the time. As M.I.T. tried to remain true to its founding principles and, at the same time, enter the top ranks of research universities, it began to show signs of an institutional identity crisis. This crisis, which began in 1919, continued through the 1920s, reaching a kind of resolution in the early years of Karl T. Compton's presidency, which began in 1930. The crisis manifested itself in several ways, beginning with conflicts related to the role of the faculty.

Despite the increase in the institute's prosperity that began in 1915, faculty salaries at M.I.T. remained well below those at other research universities. This was partly because M.I.T.'s donors were more interested in contributing to the physical plant than to operating costs, as shown by the institute's failure to attract support for endowed chairs during the period. Also, the kind of education offered at M.I.T. required a larger operating budget than was needed at most other universities of the time.

One of the consequences of low faculty salaries was M.I.T.'s adoption of a rather liberal consulting policy that allowed the underpaid faculty to supplement their income with outside work. This kind of activity was seen as consistent with the institute's founding principles because it helped faculty members stay in touch with industrial needs. However, it also meant that they had less time for the basic research that was beginning to become important to the reputation of a university. M.I.T.'s progress toward entering the ranks of the top research universities was slowed by this situation which had to be corrected by President Compton on his appointment in 1930. However, the early consulting policy also helped to establish M.I.T.'s present culture, in which, as in no other institution in this study, faculty members move easily between the academy and industry, integrating their industrial contributions with laboratory and classroom duties.

BASIC VERSUS APPLIED RESEARCH

Beginning about 1903, M.I.T. aspired not only to train engineers for industry but also to have a role in graduate education and basic research—in other words, to enter the academic mainstream. Leading M.I.T. in this new direction was Arthur Amos Noyes, a professor of chemistry and graduate of M.I.T. and Leipzig. Noyes had a broad view of the education of an engineer: "The engineer is trained to put in application existing methods; and it seems to me that what is wanted of the factory chemist in this country is rather the power of solving new problems and making improvements in processes—a power to be acquired far more by a good chemical training, which should include a large proportion of research and other work requiring independent thinking."[12] Noyes and his colleagues successfully lobbied for the establishment of the Research Laboratory of Physical Chemistry (RLPC), which would be independent of the department of chemistry and would devote its facilities and full-time staff to chemistry research. This laboratory opened in 1903 with funding from M.I.T., the Carnegie Institution, and Noyes himself, and it quickly attained an international reputation. M.I.T.'s first Ph.D.s did their work in the RLPC.

The founding of the RLPC and the philosophy behind it did not go unchallenged. Shortly after the RLPC opened its doors, an opposing faction, led by William H. Walker, formed in the chemistry department. Walker's philosophy was opposite that of Noyes: "Science by itself produces a very badly deformed man who becomes rounded out into a useful creative being only with great difficulty and large expenditure of time. . . . It is a much smaller matter to both teach and learn pure science than it is to intelligently apply this science to the solution of problems as they arise in daily life."[13]

The story of the struggle between these two opposing views and the consequences of the struggle is well told by John Servos in an article that has become a classic in the history of university technology transfer. In 1908 Walker founded a rival laboratory, the Research Laboratory for Applied

Chemistry (RLAC), the purpose of which was to apply scientific principles to problems brought to the laboratory by industrial concerns. The RLAC was to be supported by user fees. This laboratory was intended to form one of many links between the institute and industry. It was consistent with the aims of the institute and the attitude of its industrialist trustees, who, in addition to Eastman, included several Du Ponts. Walker was well known and well liked by these trustees, all of whom shared the goal of creating a more efficient industrial society.

The respect for Walker that these activities engendered among the trustees, combined with large enrollments in Walker's programs, put Noyes on the defensive. Acrimony between the two factions increased, leading to the departure for U.C. Berkeley, in 1912, of G. N. Lewis and several other professors of chemistry who had supported Noyes's position. Walker finally threatened to resign unless Noyes was removed from the chemistry department. In 1919, when he returned from Washington after involvement in the war effort, Noyes saw that his political base at the institute was completely eroded, so he resigned. He went on to play a leading role in the founding of M.I.T.'s archrival, the California Institute of Technology.

Problems with Walker's vision for the institute and the RLAC soon began to surface, however. Although popular with the trustees, the RLAC and several other projects headed by Walker engendered considerable opposition from the faculty. Part of this opposition grew out of the faculty's concern over the narrow and applied nature of the problems addressed by the RLAC. Funded almost entirely from user fees, the RLAC was forced to perform whatever tasks were brought to it. For the most part these were not the broadly based, industrywide questions that originally had been contemplated. When the RLAC did produce a significant breakthrough, the laboratory was often prohibited by the sponsor from publishing it. The RLAC had administrative problems as well. Funding was uncertain from one year to the next, and, because of budgetary problems, RLAC researchers' salaries were lower than those for the institute as a whole and two to three times lower than what industry was paying for the same kind of work.

By the time Compton became president in 1930, these factors, combined with the deepening economic turbulence of the period, clearly spelled the end for Walker's vision and the organizations founded on it. In 1934 the RLAC was terminated because it had little work to do.

ORGANIZATION OF RELATIONS WITH INDUSTRY

Walker was also instrumental in an early effort to centralize and coordinate M.I.T.'s relations with industry. In 1919 M.I.T. was in the middle of a fund-raising drive that appeared to be in trouble. Eastman had promised the institute $4 million for a building fund if his contribution could be matched by $3 million from other sources by the end of the year. By October the campaign was far from reaching its goal. In desperation, M.I.T. endorsed a hastily prepared "Technology Plan," authored by William Walker, that basically offered to sell the services of the institute to corporations in return for "a retainer of (blank) dollars, in five annual installments payable on the second day of January of each year from 1920 to 1924."[14] In return for this retainer, the institute agreed, among other things, to "make available its libraries and files and to arrange conferences with its technical staff on problems pertaining to the business of the Company," "maintain a record of qualifications, experience, and special knowledge of its alumni," and assist directly in the solution of company problems.[15] To coordinate all industrial research at M.I.T. and organize fund-raising under the plan, M.I.T. established the Division of Industrial Cooperation and Research under Walker. Within months, over $1 million in contracts had been signed under the plan, carrying M.I.T. over its goal. When M.I.T.'s president, Richard C. Maclaurin, died suddenly in 1920, Walker was named chairman of the administrative committee charged with overseeing the institute until a successor could be found.

Again Walker's personality and willfulness came into play. He exercised almost dictatorial control over the Division of Industrial Cooperation, and when the administrative committee voted to stop his aggressive sales tactics and exercise some selectivity in the choice of research problems, he resigned as director of the division. Thereafter the Division of Industrial Cooperation and the RLAC became, in effect, competitors, often contacting the same clients for funds.

At the end of its first five years of existence (1920–1925), the Technology Plan attracted virtually no renewals or new takers, and by 1930–1931 the Division of Industrial Cooperation administered only $30,000 in contracts.

COMPTON RESTORES THE BALANCE

In the late 1920s, as the importance of foundation support rose, M.I.T.'s reputation as merely an engineering school dedicated to practical training began to hurt. Several of the institute's proposals for funding were turned down by large foundations, notably the Rockefeller Foundation, on the grounds that M.I.T. should turn to its natural constituency, industry and private industrialists, for support.

Added to these budgetary and administrative problems was mounting evidence that M.I.T. was slipping in national prestige. The decline was clear not only from the institute's previously mentioned failure to secure significant foundation support but also from evidence that top students were not selecting M.I.T. From 1919 to 1930 only 8 National Research Fellows in Chemistry chose to attend M.I.T., as opposed to 16 who went to Cal Tech and 19 to U.C. Berkeley.[16]

Shortly after his inauguration, President Compton, intent on improving

M.I.T.'s image as a research university, initiated a new policy: if a faculty member wanted a promotion, he would have to agree to turn over 50 percent of all his consulting income to the institute for a "professor's fund" that would be used to pay for sabbatical leaves. Compton also set forth a list of faculty obligations that placed consulting last and decreed that consulting should be "of such important and dignified nature that its carrying on enhances the prestige of the institution, vitalizes teaching, and provides helpful contacts for students."[17]

In order to control the character of the research performed at the institute, Compton redefined the mission of the Division of Industrial Cooperation as monitoring the kinds of agreements entered into by the institute and its faculty. Projects inappropriate for university involvement no longer would be accepted, nor would any contract calling for secrecy. Once institute control over all research contracts had been centralized in the Division of Industrial Cooperation, separate units like the RLAC were eliminated.

With these changes, Compton worked a transformation on M.I.T., restoring it to prominence and ending its philosophical detour from the accepted standards of the academic world. For example, the physics department became one of the top three in the country by 1937. Foundation funds began to flow to M.I.T., and the climate for research improved. By 1937–1938 research at M.I.T. had surpassed its 1930 peak, and only a small proportion of support for that research came from industry.

LESSONS FROM THE M.I.T. EXPERIENCE

M.I.T.'s history in the 1920s and 1930s is rich with precedents that provide useful lessons for the present. First, industrial support proved too slight and too uneven to sustain alone the research required to maintain a high institutional reputation. Second, the nature of the research sponsored by industry proved to be incompatible with the goals of the university, pulling it away from those areas in which it had a competitive advantage (basic research) into a realm in which it could not hope to compete (industrial research). Third, faculty roles—that is, the ways the faculty allocated their time—proved to be pivotal in restoring academic standards to M.I.T. Fourth, Compton's use of policies on faculty consulting and research sponsorship illustrated the administrative potential for policy formulation and implementation in universities.

This period also set the stage for M.I.T.'s present relations with industry. The RLAC, and even more the pre-Compton Division of Industrial Cooperation, were early attempts at a centralized organization of relations with industry. Like the present Industrial Liaison Program at M.I.T., the early Division of Industrial Cooperation handled a range of university-industry interactions, including development efforts, continuing education, faculty consulting, access to university facilities, and contracts.

DURING AND AFTER WORLD WAR II

The success of Compton's efforts to restore M.I.T.'s scientific reputation, combined with the institute's long-standing reputation for service to industry, positioned it extremely well for taking advantage of the growth in research support stimulated by World War II. Its close ties with General Electric, Westinghouse, and others helped it become a full partner in the military-industrial complex. During the war it developed another organizational form relevant to this study, the semi-independent but associated federal laboratory. These labs, most significantly the Lincoln and Instrumentation Laboratories, allowed the institute to dissociate itself from an activity likely to be viewed as incompatible with the purposes of a university—in this case, classified research—while still maintaining some actual involvement with it. Its connection with these labs also allowed M.I.T. to develop a large cadre of nonfaculty researchers. This model of organizational definition and isolation of activity, although often judged inappropriate for universities, is one in which many universities wish to take part in some way. It is a model that is repeated in many forms throughout this book.

By the end of the war, M.I.T. was the largest university contractor of research in the nation, and it maintained this position throughout the years of the Cold War. In the mid-1960s M.I.T. had $47 million in defense contracts, three times the amount of the second-ranked university, and this sum did not include the $80 million in contracts administered by M.I.T.'s affiliated laboratories.[18] The institute's ties with industry, which Walker had seen as directly influencing its capacity to do research, had, by associating M.I.T. with one element of the military-industrial complex, contributed indirectly to a surge in its scientific reputation and its ability to attract research funds.

U.C. Berkeley

Like M.I.T., but unlike the other two universities in this study, U.C. Berkeley had its scientific reputation well established before World War II. If anything, that reputation increased during the war. Unlike M.I.T., however, U.C. Berkeley was set early on a course that stressed the traditional values of the academy. Indeed, much of the university's early history can be viewed as a struggle between the patricians of university education who founded it and external forces that attempted to "popularize" the university and make it more responsive to the immediate demands of the society that supported it. As made evident throughout the book, these two opposing forces are still at work.

EARLY YEARS

The year of the founding of the University of California, 1868, is more noted for the completion of the transcontinental railroad than for the beginning of a great university. The completion of the railroad caused considerable political and social unrest in California, precipitating what have been called "the terrible seventies." The state was flooded by goods, opportunists, and speculators brought by the new railroad, and its government was dominated by powerful and corrupt rail and land monopolies. California also was flooded by a large labor pool of Chinese workers who had been employed in building the railroad but were now mostly jobless, as well as by products of the increased migration encouraged by the the Burlingame Treaty of 1868. Economically depressed farmers and workingmen began to organize, blaming the Chinese, the railroads, and public institutions—including the university—for their unhappy state.[19]

During the 1870s U.C.'s autonomy survived a number of attacks by those wishing to force it to serve more practical ends. Finally, in 1878 a new state constitution decreed that the university "shall be entirely independent of all political or sectarian influence."[20] Of course, such independence could never be complete in an institution supported by state funds, and there were to be many subsequent situations in which the ideals of the university clashed with popular demands, but the principle of independence still operates at U.C. as perhaps in no other state university.

In large measure, U.C. owes its independence to the quality and vision of its leaders during this time, notably Daniel Coit Gilman. Gilman served for only three years, from 1872 to 1875, but his philosophy, which found greater institutional expression in his founding of the first graduate university in the country, Johns Hopkins University, clearly guided U.C. Berkeley through these turbulent times. This philosophy was supported by the faculty, including John LeConte, who succeeded Gilman as president. Gilman argued that the public's demands for a practical education were compatible with a comprehensive education and with both the traditional purposes of liberal education and the emerging goal of knowledge development. "The most practical service which the University can render to the State is to teach the principles of science, and their application to all the wants of man—and at the same time to teach all that language and history have handed down as the experience of humanity."[21]

RESEARCH AS COMPETITIVE STRATEGY

The 1880s brought prosperity to California and paved the way for the implementation of a process designed to place U.C. Berkeley among the country's top universities.[22] Unable to compete with the salaries and the proximity of cultural and educational resources offered by the wealthy private universities of the East, Berkeley embarked on a campaign to recruit the best of the young scientists coming out of these schools by offering support for their research. It was able to make this offer primarily because of a steady stream of large private research-related donations. With state appropriations covering basic instruction and the building and maintenance of its physical plant, U.C. could direct private contributions to research-related programs and facilities. For instance, in 1888 James Lick contributed money to build the 36-inch telescope at the Lick Observatory, thus propelling U.C. into preeminence in the field of astronomy. Phoebe Apperson Hearst donated funds to build a museum of anthropology, and other gifts established laboratories for the study of physiology and anatomy. U.C. also began to acquire research institutes. The Scripps Institution of Oceanographic Research at La Jolla was transferred to the university in 1901, and a donation of over $1 million in timber lands from George W. Hooper established the Foundation for Medical Research in 1913.

These gifts were only a part of U.C.'s strategy for attracting eminent research faculty. The university also instituted policies designed to be attractive to such people. For example, it kept teaching loads at Berkeley among the lowest of any for institutions classified as research universities at that time.

Other policies and practices at U.C. also clearly supported research. In 1915 the university established a separate fund for research that was to be controlled by the faculty, the first such fund in the country. By 1929 that fund had increased to over $100,000, making it by far the largest fund of its type in any state university. U.C. also established a policy that discouraged the hiring of its own graduates into faculty positions, thus avoiding the "inbreeding" that harmed some other institutions.

U.C.'s program of encouraging faculty research succeeded spectacularly in attracting not only top young scholars but also a few mature scholars, such as G. N. Lewis, who fled the temporarily research-averse M.I.T. to come to Berkeley and build its physics department. The period after World War I saw Berkeley rise unmistakably into the ranks of top research universities. Under President Wallace Campbell (1923–1930), himself a researcher and a longtime director of the Lick Observatory, the faculty gained unprecedented control of the university. The physics department became the jewel in Berkeley's crown. Foundation support for physics had produced a large crop of new, aggressive physicists, and Berkeley attracted many of them to its newly built (1923) physics lab, which offered amenities such as ample support for research, flexible teaching loads, graduate fellowships, and travel money. All of Berkeley's appointments during this period were former National Research Council (NRC) Fellows; in other words, they had been judged the best of the nation's Ph.D.s and had had experience at the nation's best laboratories. One of the early appointments of this era, Leonard Loeb, pulled off the greatest coup by attracting E. O. Lawrence from Yale in 1928. The department was further strengthened when J. Robert Oppenheimer accepted a joint appointment at Berkeley and Caltech in 1929.

Although the Depression temporarily reduced the pace of Berkeley's advancement, by 1940 conditions for research at the university and the reputation of Berkeley's faculty had never been better. Berkeley faculty on average spent less than half their time teaching, with the rest going to research, and Berkeley's expenditures for research ranked it among the top five universities in the country. Despite the popular demands that continually pushed at it, Berkeley had defended its "ivory tower" ideals and, by the end of World War II, had proved those ideals to be practical as well. The tension between these academic ideals and more worldly concerns persists, however, and is very important in determining the university's present stance toward technology transfer and involvement with private industry.

Penn State[23]

Like U.C., Penn State is a land-grant, public institution. There are undoubtedly many similarities in the history of these two institutions, but it is their differences that are important for this discussion.

EARLY YEARS

Unlike U.C., Penn State has not been able to gain the autonomy needed for the academic ideals of a top research university to assert themselves. At the same time, however, its identification with practical applications and its ability to serve its publics have created unusual opportunities to be effective in technology transfer. Penn State's history demonstrates a number of organizational arrangements that can serve as models for universities undertaking technology transfer activities.

Founded in 1855 by the Pennsylvania state legislature as the Farmer's High School, the new school was to "adopt a system of instruction which shall embrace, to the fullest extent possible, those departments of all sciences which have a practical or theoretical bearing upon agriculture." The legislature called for students "to be instructed and taught in all things necessary to be known as a farmer." Located purposely in a rural area in the geographical center of the state, far from Pennsylvania's industrialized cities, the school, soon renamed the Agricultural College of Pennsylvania, struggled to survive in its early years. Even its success in being named recipient of land-grant funds did not assure its survival.

Evan Pugh, the first president of the new school and a chemist with a European degree, saw the potential for the revival of Pennsylvania lands through the use of modern chemical fertilizers. He was an articulate spokesman and advocate for the college, and his sudden death of typhoid fever in 1864 was a serious blow to the institution.

He was followed by a series of short-term presidents, who, although they succeeded in fighting off several other Pennsylvania colleges that were vying for a piece of the Morrill Act pie, proved unable to expand the mission of the college beyond its original narrow agricultural focus.

James Calder, who was appointed president in 1871, realized that the scope of instruction would have to be broadened if the college were to survive. He instituted a curricular change that brought many traditional university subjects to the college, which was renamed Pennsylvania State College in 1874. Despite these changes, and partly because Calder did not institute engineering courses, the institution floundered, graduating only four students in 1876, seven in 1877, and three in 1878.

In late 1879 Calder resigned. He was followed by another series of short-term presidents and finally, in 1882, by George W. Atherton, who at last provided stability and leadership to the institution. Atherton immediately set about building a new undergraduate curriculum that included engineering. By 1900, thanks to his direction, Penn State had become tenth in the nation in number of engineering undergraduates. It had 209 such undergraduates, compared to U.C. Berkeley, which had 729, and M.I.T., which had 535.

RELATIONS WITH STATE GOVERNMENT

While this growth in undergraduate engineers indicates that Penn State had achieved viability by the turn of the century, it is also one of the reasons that the institution did not enter the ranks of major research universities. Penn State satisfied the demands of the state legislature by expanding its undergraduate programs, but these programs quickly became overcrowded and underfunded. In this atmosphere, faculty research could hardly avoid assuming a relatively low priority. Perhaps because of its rural location and concentration on the applied fields of agriculture and engineering, Penn State was unable to attract the kind of private or foundation support that U.C. and M.I.T. enjoyed. Indeed, the historical archives of the College of Engineering at Penn State are filled with applications by faculty members for research equipment and support, many of which apparently were ignored.

The events following the fire that burned down the main engineering building in November 1918 underlined the difficulties that Penn State faced. Valuable equipment and buildings were lost in the fire. Enrollments in the college had to be drastically reduced, dropping Penn State's ranking in number of enrollments from sixth place before World War I to tenth afterward. Instead of providing funds for rebuilding the college, which had supplied trained manpower to Pennsylvania industry for years (over half of Penn State's engineering graduates lived and worked in the state), the legislature and governor actually reduced the size of the institution's appropriations. When appeals to the legislature failed, the college turned to private sources with an emergency campaign to generate the $2 million it needed to rebuild. Initiated in 1922, three years after the fire, the campaign had netted only $1.2 million by June 1924, and much of this was in the form of pledges.

Again, Penn State had leadership problems. President Edwin E. Sparks had a nervous breakdown in 1919, and his successor, John Martin Thomas, frustrated by the lack of support from the legislature, resigned in June 1925 to become president of Rutgers University. It was not until the fiscal year

1927–1928, after President Ralph Dorn Hetzel was able to establish good relations with the governor and a surplus appeared in the state treasury, that Penn State finally received an appropriation for a new building— almost ten years after the old building had been destroyed.

This incident is typical of Penn State's early development. During a period of prosperity in the history of higher education that saw a few universities break away from the others by virtue of their commitment to academic ideals and high-level research, Penn State actually lost ground. Held captive by political idiosyncracy and diminished state financing and unable to attract private philanthropy, Penn State concentrated on the basis of its support—undergraduate teaching and (another element important to this story) public service through its agricultural and engineering extension services. These were important elements in Penn State's growth beyond its original campus, with extension centers forming the basis of a system of two-year colleges located throughout the state.

AFTER WORLD WAR II

The next relevant historical features of Penn State appeared in the mid-1940s. Like the other universities in this study, Penn State participated in the growth of federal support of research that occurred just after World War II. Some of this support was for classified research. In 1945 the Ordnance Research Laboratory (ORL) was moved from Harvard to Penn State, where it was administered by the School of Engineering. The ORL's scientists became Penn State faculty members and staff. Much of their work was classified. The laboratory was closely supervised by the Navy, which maintained 24-hour security on the research facility and supplied armed guards for some shipments into and out of it. Thus the kind of activity that on other campuses had been administratively and geographically isolated, often through separate buffer organizations, was taken to the heart of the institution at Penn State. This circumstance may have been the result of Penn State's marginal reputation and the compromises it was willing to make with the traditional values of a university. In this respect its history coincides with that of Stanford.

The role of the ORL in Penn State's history is tangentially related to the last bit of university history covered here. In 1956 Eric Walker, former dean of engineering, became Penn State's twelfth president. Walker retired in 1970, having presided over the most successful period in the university's history. Some of the organizational innovations made by Walker in connection with technology transfer are worth mentioning here.

In 1963 Walker removed the ORL from the College of Engineering, placed it under the direction of the vice president for research, and later changed its name to the Applied Research Laboratory. These changes were intended partly to encourage the growth of nonmilitary research and to reflect the increasingly intercollegiate and interdisciplinary research that was being conducted by the laboratory. In addition, Walker made this laboratory the chief technical component of the newly formed Commonwealth Industrial Research Corporation, which had been founded jointly by Penn State and the state of Pennsylvania to promote the development of research and development centers throughout the state. This joint enterprise to encourage economic development in the state presaged many future such arrangements and set a pattern for further relationships.

One of these future developments, founded only two years later, was the Pennsylvania Technical Assistance Program, or PENNTAP, perhaps the first industrial technical assistance program of its kind. Walker also was responsible for the foundation of a business incubator and for the formation of the university's Industrial Research Office, which was designed to help small businesses gain access to the university.

COMPARISON WITH OTHER UNIVERSITIES

Penn State's history presents an interesting contrast with the history of the other universities in this study. During most of its history, Penn State certainly could not be classified as a research university. Without a well-recognized research capability, it could not make the kind of contributions that M.I.T. and U.C. made to industry. On the other hand, Penn State's popular and practical orientation, combined with the organizational modes it used to serve farmers and industrial concerns, allowed it to make significant contributions to the economy of the state. In 1970, at the time of President Walker's retirement, Penn State had the infrastructure and organizational facilities to serve the coming increase in the demand that universities take part in the economic life of the country and the state. However, for a time it lacked the leadership and the will to be a full player in this new arena.

Stanford

Stanford entered the ranks of top research universities only after World War II, and the war marks a sharp disjunction in its history. The strategies it employed to achieve prominence were based partly on relations with industry.

EARLY YEARS

Stanford University was founded in 1885 by Mr. and Mrs. Leland Stanford as a living memorial to their son, Leland Stanford, Jr., who had died in Europe the year before. The new university was endowed with the Stanfords' entire estate, estimated at more than $20 million. Although the Stanfords' first thought was to found an engineering school, their ideas soon broadened. The charter of the new school mentioned agriculture and engineering (mechanics), but it stated that the school's main purposes were "to qualify students for personal success and direct usefulness in life" and to provide "the studies and exercises directed to the cultivation and enlargement of the mind."[24]

Two provisions of the original charter were to have a significant impact on the university. The first of these stated that the founders (Mr. and Mrs. Stanford) during their lives could modify the terms of the charter and exercise all of the functions, powers, and duties of trustees. The second provision prohibited the future sale of any of the 8,800 acres of land left to the university.[25]

Leland Stanford recruited David Starr Jordan, then president of Indiana University, to be Stanford's first president. Jordan quickly recruited a faculty, and the new university opened its doors in October 1891 to 559 students (Berkeley enrolled 520 in the same year). These students were attracted to the new school by free tuition, virtually no entrance requirements, an innovative curriculum, and the admission of special categories of students.[26]

Clouds gathered over this sunny beginning, however, when Mr. Stanford died in 1893, leaving control of the university in the hands of the sometimes eccentric Jane Stanford and plunging the estate into a legal battle that severely hampered funding of university operations until 1896. 1893 also saw a sharp drop in the value of the Southern Pacific stock that formed the basis of Stanford's endowment. President Jordan struggled against all these problems to keep the university on its original track. On the death of Jane Stanford in 1905, Jordan helped to establish a pattern of active trustee and presidential governance that was to be important to Stanford's later development and exists to this day. However, on his retirement in 1913, Jordan must have been disappointed that the promise of Stanford's early beginnings had not been realized.[27]

REGIONAL UNIVERSITY

After a brief transition period, Ray Lyman Wilbur assumed the presidency of Stanford in 1915. Under his administration, which lasted until 1942, Stanford gained stability and gradually rose to the status of a sound regional university that flirted with entry into the ranks of the top research universities. Appealing largely to the West's middle class, Stanford benefited from the "collegiate age" in the history of higher education, during which university social (rather than academic) life was the feature that attracted financial support. Symbolic of this period was the football stadium that Stanford built in 1921; on its expansion in 1927, this stadium became one of the largest in the country.

Taking advantage of its broad-based support, Stanford became one of the first universities to institute a professionally managed fund drive. This drive netted $1.48 million between 1919 and 1925, a modest sum measured by the standards set by the top eastern schools (during the same period Harvard raised almost $14 million, and Princeton raised almost $10 million) but significant nonetheless.[28] Almost half of this total came from Stanford alumni, and another 30 percent came from the general public. Even more significantly, Stanford began to attract foundation money, including money from the Carnegie Foundation in 1921, to found the Stanford Food Research Institute and over $1 million during the 1920s from Rockefeller-related foundations.[29]

Also during the 1920s, Stanford was reorganized from over 26 separate departments into 10 schools, including a new School of Business instituted in 1925 by alumnus Herbert Hoover. Although this reorganization eliminated duplication, it established a pattern that allowed extensive variation in the internal organization and governance of the schools and the preservation of departmental authority. This reorganization had a considerable effect on Stanford's academic culture and present organizational structure.

Conditions for research at Stanford during this period were marginal. Although graduate enrollments increased from 58 in 1916 to 1,200 in 1925 and over 1,800 in 1940, funding for research lagged. Funding allocated specifically for research was only $3,300 in 1927–1928, as compared to $112,000 at U.C. and over $200,000 at M.I.T.[30] Nevertheless, by the end of the 1930s Stanford had a solid reputation in physics and electrical engineering, based primarily on the reputations of William W. Hansen and Frederick Terman and the role that they and Stanford played in Russell Varian's invention of the klystron tube, which became important in microwave research. This reputation got a further boost when Felix Block, who later received the Nobel Prize, fled Germany in 1934 and became a Stanford faculty member.

WORLD WAR II AND "STEEPLES OF EXCELLENCE"

In spite of these advantages, by the beginning of World War II Stanford's reputation as a major research university was not yet secure. During the war most of Stanford's eminent faculty were recruited to the research centers of the East, Hansen to Sperry Corporation for further development of the klystron tube and Terman to head the Radio Research Laboratory in Cambridge which, at its peak in 1944, operated with a staff of 800, of whom 200 were professional researchers.[31] At the end of the war these faculty members returned to Stanford. Stanford's subsequent rise to prominence is one of the most dramatic success stories in the history of American universities and is a great tribute to the leadership the university had at the time.

Research administration appealed to Terman, and he saw in the organization of research during the war a pattern that he felt could be used in

Stanford's development. Correctly foreseeing the coming partnership between universities and the federal government and the tremendous increase in federal research support, Terman, Hansen, and J. E. Wallace Sterling, who assumed the presidency of Stanford in 1948, moved quickly to put Stanford in a position to capture a part of the federal research largesse. They called their strategy "steeples of excellence" to indicate that, instead of trying to compete head-on with more established universities by building a broadly based reputation in research, Stanford would concentrate on identifying and supporting research in a few areas of science that were not well addressed by other universities.

The foundation for one of the first "steeples" was the Microwave Laboratory, which was founded by Terman in 1945 with a federal contract snatched away from M.I.T. This laboratory was followed by the Electronics Research Laboratory (1947) and an offshoot of it, the Applied Electronics Laboratory. Stanford's strategy paid off as federal research contract volume climbed dramatically: from only $127,599 in 1947 to almost $7 million in 1957 and over $13 million in the early 1960s. The strategy reached a pinnacle in 1962 when the federal government built the Stanford Linear Accelerator at a cost of $114 million. At the time it was the largest single federal research facility ever built.

As at Penn State, much of the federally sponsored research done at Stanford was classified. Although Terman at first tried to keep classified research in separate administrative units, he eventually agreed to fold it into the regular administrative apparatus of the university. He recognized that giving up classified research would seriously impair Stanford's ability to compete on the leading edge of science.[32]

RELATIONS WITH INDUSTRY

Another part of Stanford's postwar development strategy was the close association of the research and teaching functions of the university with corporations and industrial interests. Again, Frederick Terman, who had become dean of engineering and later vice president and provost, took the lead. Building on his prewar experiences with the Sperry Gyroscope Company, its development of the klystron tube, and its arrangements for recovering royalties and using them for further research support, Terman, with the support of President Sterling, undertook a series of initiatives to foster relations with industry. These initiatives were rewarded by large donations for research support from the Varian brothers and Hewlett and Packard. The initiatives included use of university lands for a research park, installation of a part-time Master of Engineering degree program, and development of industrial affiliate programs.

Because of this strategy and his role as a spokesman for close university-industry relations, Terman is widely credited with the creation of Silicon Valley. Although his leadership and Stanford's research capabilities undoubtedly had some effect, recent questions have been raised about the extent of Stanford's role in the development of the valley. Part of this controversy is examined in chapter 11, which also compares Stanford's role in creating Silicon Valley with M.I.T.'s role in promoting the developments around Route 128 in Boston. More important to the history of Stanford is the postwar development at the university, under Terman's leadership, of a set of policies and organizational forms that supported relations with industry.

Thus, Stanford's postwar rise from a good regional university to a research university with a high national reputation was based on a strategy whereby the university picked its spots carefully, exploiting areas that the competition had overlooked or in which Stanford had a competitive advantage.[33] This "steeple-building" strategy had two main elements: the attraction of federal research support and the careful cultivation of corporate support. It is a strategy that still exerts a strong influence on Stanford's organizational structure, its academic culture, and its response to the changes brought about by technology transfer.

Conclusion

The history of university efforts at technology transfer goes back to the development of the research university as an identifiable entity. M.I.T. presents early examples of the problems faced by universities as they attempted to balance the demands for service to economic goals with the need to maintain the integrity of the goals upon which universities are based. The history of each of the four universities of this study provides illustrations of institutional adaptations to external forces—adaptations of policy, organizational form, and resource allocation.

This brief account of the history of the four universities in this study provides the foundation on which the rest of the examination is based. Understanding each university's history is indispensable to understanding the present institutional setting within which responses and changes related to technology transfer are taking place.

Notes

1. Except where noted, the information contained in this section is summarized from Roger Geiger, *To Advance Knowledge: The Growth of American Research Universities, 1900–1940* (New York: Oxford University Press, 1986).
2. Geiger, *To Advance Knowledge*, 69.
3. David F. Noble, *America by Design: Science, Technology, and the Rise of Corporate Capitalism* (New York: Oxford University Press, 1977), 27.
4. Denis J. Prager and Gilbert S. Omenn, "Research, Innovation, and University-Industry Linkages," *Science* 207 (January 1980): 379.

5. Derek C. Bok, "Business and the Academy," *Harvard Magazine* 83 (1981): 26.
6. Leonard S. Reich, *The Making of American Industrial Research* (Cambridge, England: Cambridge University Press, 1985), 17.
7. Karl L. Wildes and Nilo A. Lindgren, *A Century of Electrical Engineering and Computer Science at M.I.T., 1882–1982* (Cambridge, Mass.: The M.I.T. Press, 1985), 4, quoting "An Act to Incorporate the Massachusetts Institute of Technology, and to Grant Aid to said Institute and to the Boston Society of Natural History," enacted by the Massachusetts Senate and House of Representatives, 10 April 1861.
8. Reich, *The Making of American Industrial Research*, 17.
9. Wildes, *A Century of Electrical Engineering*, 378–390.
10. Geiger, *To Advance Knowledge*, 13.
11. Ibid., 179.
12. John W. Servos, "The Industrial Relations of Science: Chemical Engineering at M.I.T., 1900–1939," *Isis* 71 (1980): 533–534. The material contained in this section is summarized from this reference and Roger Geiger, *To Advance Knowledge*.
13. Servos, "The Industrial Relations of Science," 535.
14. William H. Walker, "The Technology Plan," *Chemical and Metallurgical Engineering* 22 (March 1920): 464.
15. Ibid.
16. Geiger, *To Advance Knowledge*, 180.
17. Ibid., 182.
18. Stuart W. Leslie, "Playing the Education Game to Win: The Military and Interdisciplinary Research at Stanford," *Historical Studies in the Physical and Biological Sciences* 18 (1987): 55.
19. Kathleen Rockhill, *Academic Excellence and Public Service: A History of University Extension in California* (New Brunswick, N.J.: Transaction Books, 1983), 8–9.
20. Ibid., 9.
21. Ibid., 13.
22. Much of the material in this section is summarized from Geiger, *To Advance Knowledge*.
23. The factual material in this section is from Michael Bezilla, *Engineering Education at Penn State* (University Park: Pennsylvania State University Press, 1981).
24. Frank A. Madeiros, *The Sterling Years at Stanford: A Study in the Dynamics of Institutional Change*, (Ph.D. dissertation, Stanford University, 1979), 40.
25. Ibid., 41.
26. Ibid., 43.
27. Ibid., 46.
28. Geiger, *To Advance Knowledge*, 127.
29. Ibid., 153, 162, 170.
30. Stuart W. Leslie and Bruce Hevly, "Steeple Building at Stanford: Electrical Engineering, Physics, and Microwave Research," *Proceedings of the Institute of Electrical and Electronics Engineers* 73 (July 1985): 1173.
31. Leslie, "Playing the Education Game," 71.
32. Leslie and Hevly, "Steeple Building at Stanford," 1176.
33. Leslie, "Playing the Education Game," 56.

Bibliography

Bok, Derek C. "Business and the Academy." *Harvard Magazine* 83 (1981): 23–35.

Bezilla, Michael. *Engineering Education at Penn State.* University Park: Pennsylvania State University Press, 1981.

Geiger, Roger. *To Advance Knowledge: The Growth of American Research Universities, 1900–1940.* New York: Oxford University Press, 1986.

Leslie, Stuart W. "Playing the Education Game to Win: The Military and Interdisciplinary Research at Stanford." *Historical Studies in the Physical and Biological Sciences* 18 (1987): 55–88.

———— and Bruce Hevly. "Steeple Building at Stanford: Electrical Engineering, Physics, and Microwave Research." *Proceedings of the Institute of Electrical and Electronics Engineers* 73 (July 1985): 1169–1180.

Madeiros, Frank A. *The Sterling Years at Stanford: A Study in the Dynamics of Institutional Change.* Ph.D. diss., Stanford University, 1979.

Noble, David F. *America by Design: Science, Technology, and the Rise of Corporate Capitalism.* New York: Oxford University Press, 1977.

Prager, Denis J., and Gilbert S. Omenn. "Research, Innovation, and University-Industry Linkages." *Science* 207 (1980): 379–384.

Reich, Leonard S. *The Making of American Industrial Research.* Cambridge, England: Cambridge University Press, 1985.

Rockhill, Kathleen. *Academic Excellence and Public Service: A History of University Extension in California.* New Brunswick, N.J.: Transaction Books, 1983.

Servos, John W. "The Industrial Relations of Science: Chemical Engineering at M.I.T., 1900–1939." *Isis* 71 (1980): 531–549.

Walker, William H. "The Technology Plan." *Chemical and Metallurgical Engineering* 22 (March 1920): 463–464.

Wildes, Karl L., and Nilo A. Lindgren. *A Century of Electrical Engineering and Computer Science at M.I.T., 1882–1982.* Cambridge, Mass.: M.I.T. Press, 1985.

3

Present Institutional Setting for and Policies Related to Technology Transfer in Four Universities

---

There were many precursors of present developments related to technology transfer in the four universities described in this book and the present relations between those universities and industry. However, a "new era" for technology transfer began in the late 1970s, and the early 1980s ushered in a series of important developments and incidents. These events include the following:

• Developments in and controversy over university ownership of intellectual property, including the passage of amendments to the patent law that allowed universities to patent inventions arising from government-sponsored research. There were a series of disputes over the ownership of intellectual property in which universities sued or were sued by corporations.

• The prospect of tremendous developments of high commercial value in the field of biotechnology. Because of the prominence of university people in these developments, commercial exploitation concentrated for a time on university intellectual property and the services of university researchers.

These developments also brought on increased university interest in the ownership of research-based companies, resulting in (among other things) a plan by Harvard, ultimately dropped, for the founding of a company jointly with a Harvard faculty member.

- A number of widely publicized large-scale agreements between universities and corporations for cooperative research, including agreements between Monsanto and Harvard, Monsanto and Washington University, and Hoescht and Massachusetts General Hospital (Harvard). This last agreement prompted a congressional inquiry in 1981 and 1982 into the details of such agreements. These agreements also prompted a top-level conference between eleven corporate leaders and the presidents of five major universities (Havard, the University of California, Stanford, M.I.T., and California Institute of Technology) at Pajaro Dunes, California, in March 1982. Because of the secrecy surrounding this meeting, it created considerable controversy.

- A number of highly publicized disputes, also brought on by developments in biotechnology, between university researchers or between researchers and their own universities.

These developments and others combined to focus public attention on university research and its administration and exploitation.

The Present Institutional Setting for Technology Transfer

The four universities in this study are responding in different ways to the increased attention being paid to their roles in research and technology transfer. Some of these differences can be explained by the institutions' different histories, but another source of the differences comes from a complex of additional elements: current events, organizational goals, competitive issues, and—hardest to define—differing academic cultures. This section is an examination of elements of the current setting for technology transfer at each of the four universities.

M.I.T.

M.I.T. today ranks third among American research universities in total federal research funding and is among the highest in volume and percentage of research funding from industry. Physicist Alvin Weinberg, who coined the term *big science*, spoke in the early 1960s of M.I.T.'s organizational structure: "[it is hard] to tell whether the Massachusetts Institute of Technology is a university with many government research laboratories appended to it or a cluster of government research laboratories with a very good educational institution attached to it."[1] This organizational characteristic carries over into the internal structure of M.I.T. Over two-thirds of the institution's research is carried out by extradepartmental units—centers, consortia, institutes—and M.I.T. has developed a unique and innovative complex of organizational subunits and an administrative structure to control and monitor this research.

M.I.T.'s organizational flexibility was partly responsible for the controversy over the Whitehead Institute, which became one of the highly publicized events relating to biotechnology. In 1981, after failing to persuade Duke University to agree to the terms of his gift, Edwin Whitehead donated $20 million to M.I.T. to build an institute dedicated to biological science, $5 million annually to support its operations, and $7.5 million in unrestricted funds. The controversial aspect of this gift was the right of the board of the Whitehead Institute, which was dominated by non-M.I.T. people, to appoint up to 20 members of the M.I.T. biology department, which at the time numbered 40. After some opposition from its faculty members and considerable national attention, M.I.T. accepted Whitehead's offer, and the Whitehead Institute was built.[2] It serves as a recent example of M.I.T.'s willingness to accommodate academic principles and to adopt new organizational forms to take advantage of opportunities.

M.I.T.'s organizational complexity and innovativeness is partly the result of an academic culture that embraces faculty entrepreneurship, both within the academic community and in building bridges between that community and the commercial world. The historical basis of M.I.T.'s culture was described above, as were the institute's early attempts to organize and centralize relations with industry. Both of these elements are featured in two incidents from the 1980s, which also illustrate how a prevailing academic culture works to bring the organization of a university into accordance with it.

The first incident was the reorganization of M.I.T.'s patent office in 1985. The patent office was widely criticized during the late 1970s and early 1980s by the institute's faculty who saw it as too unaggressive and legalistic in its licensing activities. In 1985 almost the entire patent office staff was fired and replaced by license officers who were not attorneys. The name of the office also was changed to the Technology Licensing Office.

The second incident was the development in the early 1980s of a new, or rather adapted, organizational form to coordinate relations with industry. The Industrial Liaison Program (ILP) had been formed in 1948 to serve as a single point of industry contact with M.I.T. and the research produced by the institute. Through a single person in the ILP a company could gain "facilitated access" to the institute and its faculty. The ILP also served development purposes and provided the vehicle for corporate contributions and research support. However, as individual faculty members began to have close ties with industry, the services of the ILP came to seem bureaucratic and burdensome. A number of research consortia and "mini-ILPs" therefore sprang up to foster these closer ties.

Both these examples reflect the prevailing academic culture of M.I.T. Today, as in no other university, M.I.T. faculty members move easily from the classroom to the boardroom, integrating their research and teaching duties with consulting, advising, and other ways relating to and serving industry. To be sure, an almost constant tension exists between the ideals of the university and the traditions of service to industry upon which M.I.T. was founded; in other words, the institute deals daily with the same issues and arguments that were present in the debate between Noyes and Walker in the first decade of the century. However, M.I.T. is determined to manage this tension and conflict and to walk the line between academic virtue and the values of the marketplace. It is this balancing act that makes M.I.T. such an interesting case for study.

U.C. BERKELEY

Of the four universities studied, U.C. Berkeley appears to be the least affected by the issues surrounding technology transfer. With a combination of state support, private philanthropy, and its share of federal research funding, Berkeley has steadily added to its national reputation while being shielded from some of the exigencies that compelled other universities to bend traditional university values. For instance, it has never had to accept classified research on its campus. Today, however, U.C. Berkeley, like the other universities in this study, is sorting through a number of issues related to technology transfer and faces perhaps the most difficult task in making adjustments. Possessing an academic culture unsupportive of (even hostile to) the commercialization of research, unaccustomed to dealing with industrial concerns, and lacking an organizational infrastructure available for mediating relations with industry, Berkeley is struggling through a minor identity crisis over technology transfer.

This struggle is complicated by two features of Berkeley's present organizational setting—features that provide good examples of the difficulties Berkeley faces. First, as part of a nine-campus statewide system, the Berkeley campus is not entirely the master of its own fate. Although many administrative tasks were decentralized from the systemwide administration to the campuses years ago, authority for some aspects of patent administration was not transferred until early 1990. The former inability of the centralized university patent office to serve the campus adequately has created problems for Berkeley's administration, which is under pressure from faculty inventors to do something to correct the situation but at the same time must circumscribe its actions to fit systemwide policy. This impasse, aspects of which persist today, represents both a practical and a symbolic failure of the university to be effective at technology transfer.

The second fact of life that Berkeley has to face is that it is a public institution and its faculty members have been defined as public servants. For political reasons, the historical roots of which were examined earlier, the university must carefully monitor the activities of faculty members, protecting them and the university from charges of unjust enrichment or conflict of interest. This necessity has led the university, including the Berkeley campus, to require faculty researchers to prepare an annual disclosure statement listing all financial ties of a value in excess of $250, including consulting arrangements, to any company supporting research in the university. Although perhaps reasonable from a public policy point of view, this requirement has the effect of dampening faculty entrepreneurial activity and discouraging relations with industry.

A current example of the interaction of Berkeley's academic culture and the university's efforts to be involved with technology transfer is the proposed development of a "research campus" on university lands about seven miles from the main campus. The planners of this development rightly concluded that their first step should be to secure the backing of the faculty, and they therefore set up an internal public relations campaign. The perception that the faculty are so important to the effort is an indication that consideration of faculty concerns is essential to developments relating to technology transfer on the Berkeley campus.

At present, then, U.C. Berkeley is steadily digesting the issues surrounding technology transfer. It is groping for some unifying or rationalizing principle that will guide its action, but it still does not fully understand the dimensions of the problems that crop up or the degree of their interconnectedness.

PENN STATE

If U.C. Berkeley is undergoing a minor identity crisis, Penn State is facing a major one. Although the programs fostering relations with industry that were put in place by President Walker continued after his retirement in 1970, and although the volume of research at Penn State increased, the atmosphere surrounding research on campus deteriorated during the 1970s. This decline occurred partly because Walker's successor, Jack Oswald, became preoccupied with student concerns and campus unrest and failed to recognize the importance of mobilizing the university for the support of the economic initiatives that were being sponsored by the state government.

Facing severe economic depression and realignment as its economic base declined in the extractive and "smokestack" industries and agriculture, Pennsylvania mounted a series of programs that were designed to create jobs and bring prosperity back to the state. Key among these programs was the Ben Franklin Partnership, a comprehensive program of business incentives and institutional assignments. The partnership included a large number of smaller programs that could be classified as technology transfer programs. Penn State, with its statewide infrastructure, was bound to play a major role in this program. However, since the partnership provided very little support for basic research, the faculty and administration of the university were not enthusiastic about it. When Penn State's foot dragging became apparent, then Pennsylvania Governor Thornberg threatened the university with budgetary cuts if it did not cooperate with the state.

Penn State hastily organized its share of the program, in the process holding informational meetings and asking support from a still-reluctant faculty. These meetings were clearly the start of a major shift in Penn State's stance toward technology transfer and economic development. From this experience it was obvious that Oswald's successor would need to be someone who was in tune with the state's program.

That successor was Bryce Jordan, who came to the university in July 1983. Jordan quickly took steps to assure the governor and the legislature that Penn State was now going to be an active partner in Pennsylvania's economic development plans. These steps culminated in a set of "economic development program initiatives" that were adopted by the board of trustees in May 1987. The trustees' resolution is quoted here at length because it is a model of explicitness about what the university should do, yet it also incorporates functions that seem at odds with traditional university missions.

Now, Therefore, Be It Resolved, that the Board of Trustees authorizes and directs that the officers of the University undertake program initiatives to increase and strengthen the contributions of The Pennsylvania State University to the economic development of the Commonwealth of Pennsylvania; and

Be It Further Resolved, that, in connection therewith, the officers of the University develop, for review and approval by the Board of Trustees, specific programs as follows:

1. A program to strengthen advisory and consultative services to government to assist in the formulation and implementation of necessary public policies and programs to foster economic growth, job creation, maintenance of adequate infrastructure and preservation and restoration of environmental quality.

2. A program for the orderly use of University lands for the development of research-office parks, business incubators, conference centers and other potential uses.

3. A program to promote a business sensitive environment within the University which facilitates access and supportive relationships between the academic and industrial communities.

4. A program to assess potential venture capital and equity investments by the University to promote economic development.

5. A program to establish, maintain and strengthen state-of-the-art University facilities for research and technology transfer.

6. A program to coordinate existing programs and activities which relate to economic development in order to maximize their effectiveness.

7. A program to encourage and assist faculty, staff and students to engage in entrepreneurial activities related to University research activities.

8. A program to build recognition for and support of an active role for the University in the economic development of Pennsylvania.[3]

Many faculty members found this resolution contradictory to the avowed goal of Penn State to become one of the top ten research universities in the country, particularly point seven, which suggested steering the faculty toward applied research and business development activity. The liberal arts faculty, concerned that the initiatives were misdirecting the university into noneducational and nonscholarly activities, issued a report on the possible negative effects of the program.[4]

In addition to this controversy over mission, technology transfer created organizational changes, itself becoming an organizing principle as a number of units having some role in relations with industry and technology transfer were gathered together under a new vice president. This reorganization caused the resignation of a patent administrator and may have occasioned several higher-level resignations as well.

Despite the turmoil caused by these attempts to change the culture of Penn State, faculty morale and conditions for research appear to be improving. A significant number of faculty members approve of the new initiatives and are becoming active and open participants in businesses and enterpreneurial activities, presenting quite a contrast to the "public servant" role thrust on the faculty in the U.C. system. Although Penn State is in a trying period of adjustment, and although it may be making some beginner's mistakes in its programs, its direction now has been clearly established by its board and its president.

STANFORD

Perhaps the most striking thing about the present setting for technology transfer at Stanford is the authority and independence of its departments. Unlike the other private university in the study, M.I.T., which has about half of its faculty in engineering fields, Stanford is truly a "multiversity" with a full complement of schools and colleges. Although, like M.I.T., Stanford is known for its interaction with industry, it has never made a serious attempt at centrally organizing these relationships. Symbolic of this attitude is the almost complete lack of interaction or coordination among the 34 industrial affiliate programs at Stanford today. This "hands-off" attitude by the central administration has allowed individual faculty members and departments to form liaisons with industry without help or intervention, thus encouraging faculty entrepreneurship. The same attitude has carried over into relations between individual faculty members and the institution. Alone among the major research universities, Stanford allows its faculty members to retain title to inventions and discoveries made with university funds and facilities.

This entrepreneurial spirit has played a part in the development of several organizational subunits in addition to the affiliate programs already mentioned. The Office of Technology Licensing (OTL) at Stanford has the highest reputation in the country even though faculty cooperation with the Office is voluntary. The necessity for the OTL to "sell" its services to faculty members is a significant factor in its success.

While this entrepreneurial spirit and autonomy on the science and engineering side rarely comes into direct conflict with the humanities and social science elements of the university, one current issue is indirectly pitting the two sides against one another. Stanford's research contract overhead rate, that is, the amount of money that Stanford charges on research contracts and grants to cover the operating (overhead) costs indirectly associated with the performance of those contracts, is in excess of 70 percent—among the highest in the country. U.C. Berkeley's overhead rate, for comparison, is 49 percent. Because Stanford is a comprehensive private university, its costs of operation are higher than those of most research institutions. This high rate places Stanford researchers at a competitive disadvantage when they apply for research funding, and they resent it. Although most of this resentment is directed toward the administration, some researchers feel that the university is asking them to subsidize other disciplines that cannot bring in contracts. This conflict has the potential for disturbing the atmosphere around research at Stanford.

Although Stanford today is secure in its position as a top-ranked research university and is comfortable with its technology transfer activities and relations with industry, it still has some issues to resolve. Certainly the overhead rate matter must be dealt with. In addition, there are strong indications that Stanford must take steps to coordinate and control some of the activities that have so far enjoyed almost complete independence.

As this narrative continues, many of the differences among the four universities that have been mentioned here will come into sharper focus. The influence of the universities' history and present culture appear in diverse forms as their individual technology transfer mechanisms are examined in later chapters.

Policies Related to Technology Transfer

A final important element of the institutional background for technology transfer is the administrative policies related to this subject. In this section, policies of the four universities are described and their differences are highlighted. A knowledge of institutional policies is important because they serve as guides to action, so they are in some ways predictive of behavior. They also reflect current values. Official changes in policies often signal a shift in the academic culture, and the adoption of new policies can reflect new pressures or concerns affecting the university.

Although there is a great deal of overlap and a considerable blurring of distinctions among the names given to policies on different campuses, those related to technology transfer can be placed roughly in six categories: intellectual property, conflict of interest, conflict of duties, consulting, improper use of university facilities, and use of proprietary information. Each of these categories contains several subcategories, all of which are examined below.

INTELLECTUAL PROPERTY

Policies related to intellectual property are vital to technology transfer—so much so that they are treated at length in a separate chapter in this book, chapter 5. These policies set forth the ownership rights to intellectual property (especially those of the university and the faculty originator of the property), the obligation of university personnel to disclose inventions, and the rights of sponsors of research to the intellectual property resulting from such research. Because sponsors of industrial research often request that the results of research be withheld from the public for a time in order to secure protection of intellectual property, policies related to delay in publication are also treated in chapter 5. Another issue falling in this category concerns the extent to which the university is willing to develop its own intellectual property by investing in start-up companies designed to develop and market university inventions. This issue is treated in chapter 8.

Two elements of the detailed treatment that is to come should be mentioned here. First, universities usually assert ownership rights over patentable intellectual property, although most universities allow faculty members to retain full ownership of property that can be copyrighted. U.C. Berkeley, Penn State, and M.I.T. follow this pattern, although, like many universities, they have developed separate policies for copyrightable computer software. Only Stanford allows its faculty members to retain ownership of patentable property. Second, alone among the universities in this study but by no means alone in the country, M.I.T. is aggressively pursuing the development of start-up companies based on the research of its faculty. Both these policy differences are symbolic of the unique academic cultures of the respective universities.

CONFLICT OF INTEREST

Conflict of interest policies cover a multitude of issues, but they fall into two main categories. The first relates to the obligations of faculty and arises when a faculty member's commitments to the university are not met because of involvement with outside activities. Issues in this category are examined in the section of this chapter called "conflict of duties." The second category arises when a faculty member uses his or her position in the university inappropriately to advance personal gain. As the value of university intellectual property increases and as universities more vigorously assert their ownership rights to this property, conflict of interest issues become more important.

Of the many possibilities for conflict of interest in technology transfer, two may serve as examples for this description. The first occurs when a faculty member transfers university intellectual property to a company though a consulting contract. Rather than disclosing a valuable invention to the university, the faculty member takes the invention to an outside company and "sells" it by executing a lucrative consulting agreement with the company. The second example is less blatant. It arises when a university faculty member who has an equity stake in a company causes that company to support further research in the professor's university laboratory. While this may seem on the surface to benefit the university, it is fraught with possibilities for conflict of interest. The professor may be tempted to use university facilities inappropriately, may distort the work of graduate students in favor of company research interests, or may shuffle intellectual property to the company.

The first line of defense against these kinds of abuse is always, of course, the personal ethics of faculty members. The second line of defense, and the focus for most university policy in this area, is the disclosure of ties to outside firms. It is in disclosure requirements, rather than in actual statements of policy, that universities differ significantly. In a study of 46 research universities conducted in 1984 by the Association of American Universities, 27, including U.C. Berkeley and M.I.T., required faculty members to disclose whether they had any financial ties with outside firms sponsoring their research. Nineteen others, including Penn State and Stanford, required such disclosure only in cases where a faculty member had an equity interest in an outside firm.[5]

U.C. Berkeley has perhaps the most restrictive disclosure regulations, partly because its faculty members fall under the Fair Political Practices Act of the state. Under the terms of this legislation, whenever a faculty member accepts research support from a nongovernmental sponsor, he or she must disclose any financial interest in the sponsor in excess of $250 and describe the nature of that interest. Financial interest includes consulting agreements, loans from the company, an equity ownership in the company by the faculty member or any member of the faculty member's family, or any position in the company in which some management authority might be exercised. This disclosure is made to a centralized, systemwide (rather than campus) office.[6] In the first year of this requirement, 4,340 disclosures were filed, 210 of which were found to indicate potential conflicts of interest. Three contracts were modified or canceled as a result of disclosures.[7]

The other three universities in this study have less restrictive requirements. Stanford, for instance, requires faculty members to disclose only when asked to do so. Penn State faculty members must inform the department head of the type and extent of their outside activities so that the head may determine whether the activities are appropriate. M.I.T. comes closest to U.C., requiring disclosure to "designated officials" (usually department heads) of the number of days spent in outside activity and the extent and nature of financial involvement in outside firms.

At first glance, this pairing of U.C. Berkeley and M.I.T., and of Penn State and Stanford, may seem odd. However, after one considers the historical information presented earlier in this chapter, the pairing appears more logical. Restrictive measures have been imposed at Berkeley and M.I.T. for different reasons. At U.C., restrictions are the result of legal requirements and the pressure to be publicly accountable; at M.I.T., they are an attempt by the institute to "manage" the conflicts that arise out of a great deal of activity. Stanford's less restrictive requirements are consistent with its "hands off" policies, while Penn State's lack of restrictions is probably further evidence of its desire to facilitate faculty contacts with industry.

CONFLICT OF DUTIES

Conflict of duties represents a special case of conflict of interest and is based on the premise that faculty members owe their primary allegiance to the university. Almost every university has a policy that echoes that of M.I.T.: "The primary loyalty of a full-time member of the faculty must at all times be to the Institute. This obligation underlies all others."[8] At M.I.T. and U.C. Berkeley, again paired on this issue, this means that faculty members may not serve as operating officers of outside companies (although they may serve on boards of directors). Both Penn State and Stanford allow faculty members to be officers of companies, however, and Stanford goes even further in making provision for extensive management involvement by permitting half-time tenured professorships.

A conflict-of-duties issue that many universities have not faced could occur when a faculty member, whether or not an operating officer, becomes a founder of an outside business. Founding a business usually requires extensive time and effort of kinds not normally associated with a faculty member's research—arranging financing, hiring staff, developing business plans, and so on. The limitations set by consulting policies, examined in the next section, cannot possibly be complied with if a faculty member is involved with the founding of a business. Nonetheless, powerful forces, such as those present now at Penn State, may favor such involvement. The subject of faculty enterpreneurs is ripe for further policy development.

CONSULTING

Policies on consulting are designed to control the extent and the nature of faculty consulting. Almost by tradition, most universities allow faculty members to consult one day a week. Not only does this practice allow faculty members to supplement their salaries, but it also allows them, theoretically at least, to stay in touch with the practical aspects of their fields and with important external constituencies. However, consulting policies are notoriously subject to interpretation, sometimes admitting large-scale, even abusive, faculty involvement in outside activities. For instance, does one day a week mean one week day, or can it also mean one week day combined with two weekend days? If a faculty member works on university business on Saturday, can he or she spend Thursday and Friday consulting? What about the summer months? The possibilities for abuse and its rationalization are endless. Furthermore, consulting policies are often as difficult to enforce as they are to interpret.

Despite these policy issues and the possible abuses related to conflict of interest and conflict of duties, consulting is a traditional and extremely effective method of university technology transfer. It also can protect the university's research agenda from inappropriate projects. As noted by a Berkeley faculty member interviewed for this study: "If a sponsor wants me to do basic research, I ask him to sign a university contract or make a gift to my lab. If he wants me to do applied research, I ask for a consulting contract."

Because all the universities in this study have about the same stated policies concerning consulting, and because enforcement of such policies generally is localized in individual departments and therefore is subject to many interpretations, it is difficult to discover differences among the four universities in this area. In all four cases, however, consulting policies are presently inadequate to deal with the current range of outside activities, including entrepreneurial ventures, engaged in by faculty members.

IMPROPER USE OF UNIVERSITY FACILITIES

Other policies are designed to make sure that university property and facilities are not diverted from the primary tasks of teaching and research to which they are dedicated. Improper use includes not only use of the facilities by university employees for personal gain but also use for the mundane tasks of testing and analysis if those tasks are part of outside projects. William Walker's Research Laboratory for Applied Chemistry at M.I.T. in the early 1900s is an extreme example of using university equipment to serve outside interests.

While most universities agree on the principle upon which their very similar policies are based, their practices sometimes differ. The story of the Varian brothers' use of Stanford facilities to make their first klystron tube is often cited as an example of how universities can affect the technology transfer process in a positive way. Penn State has given early support to a number of business ventures in the form of laboratory space and use of equipment. On the other hand, as was the case with Stanley Kaplan, a faculty member at UCLA, professors have been fired for using university facilities for their personal gain. The ambiguity in this policy area is another symptom of the underlying lack of coherence in the stance of universities toward technology transfer and the appropriate role for the university in the process of economic development.

USE OF PROPRIETARY INFORMATION

As university-industry collaborations increase, so do opportunities for universities to gain access to knowledge and equipment in industrial firms, and for universities to use problems encountered by industrial firms as subjects for research. These opportunities sometimes involve tangible or intellectual property that a firm owns and wishes to keep confidential. Such potential conflicts present two kinds of difficulties for universities. First, the idea of proprietary information seems to contradict the dicta of science and the university, which are based on the free exchange of ideas and the opportunity to replicate experimental findings. Many of the same arguments used against doing classified research in universities are used against doing research that involves proprietary information.

The second policy issue is a logistical one. Few universities can guarantee the secrecy of data or knowledge; they possess neither the will nor the administrative apparatus for assuring that information remains secure. U.C. Berkeley rarely accepts research involving proprietary information, but in exceptional cases it will agree to the sort of terms now generally accepted by other universities, including the other three in this study. Under these terms, the university agrees to use its "best efforts" to keep information secret, though the primary responsibility for doing so remains vested in the involved faculty members. However, most universities also insist that the involvement of proprietary information should in no way prevent publication of the results of research.

Conclusion

Many changes in university policies related to technology transfer have occurred recently. A study of university policies and ethical issues based on a survey in the spring of 1988 by Louis, Swazey, and Anderson showed that 59 percent of the 86 research universities and 57 percent of the 268 universities studied had modified old policies or developed new policies related to patent ownership and royalties during the last three years. For changes in or development of new policies related to university-industry relations, the respective percentages were 46 and 43 percent; for policies related to faculty involvement in firms whose products are based on their research, the percentages were 41 and 39 percent.[9] The same study reported that 47 percent of the 268 graduate deans polled considered university-industry research relations to be one of the four most critical issues facing universities, and 26 percent considered patent ownership and royalties to be among the top four critical issues.[10] The results of this study reaffirm the central hypothesis of this book, namely that a significant change is occurring in colleges and universities regarding the issues involved in technology transfer.

This chapter and the last have set the stage for the more detailed examination of technology transfer mechanisms that follows. The history of the four institutions, their present condition in regard to technology transfer, and their policies form a background that must be understood if one is to make sense of the developments that are occurring so rapidly on these and other campuses today.

This chapter has focused on those forces within an institution that are summarized under the term "academic culture"—the norms and collective attitudes of faculty and university leadership that serve constantly as a check on activities inappropriate to the university. The discussion also has illustrated the importance of the policy dimension of this analysis. Policies both determine and reflect the behavior and attitudes of those within universities. The subtleties of the influence and meaning of policies and policy changes are examined throughout the book.

The next chapter begins the discussion of specific technology transfer mechanisms, starting with the most central engine of university technology transfer, patent and licensing policy and administration.

Notes

1. Stuart W. Leslie, "Playing the Education Game to Win: The Military and Interdisciplinary Research at Stanford," *Historical Studies in the Physical and Biological Sciences* 18 (1987): 56.
2. Nicholas Wade, *The Science Business: Report of the Twentieth Century Fund Task Force on the Commercialization of Scientific Research* (New York: Priority Press, 1984), 49–50.
3. Thomas D. Larson, "Towards Closer University-to-Industry Linkages: The Penn State Experience" (Paper delivered at the Higher Education International Conference, Braga, Portugal, April 1988), 32–33.
4. Raymond Hogler et al., "The Implications of the Penn State Economic Development Plan for the College of Liberal Arts" (University Park: Liberal Arts Faculty Affairs Committee, Pennsylvania State University, 1988).
5. Association of American Universities, *University Policies on Conflict of Interest and Delay of Publication* (Washington, D.C.: Association of American Universities, 1985), pp. E.
6. University of California, "Form 730-U, Principal Investigator's Statement of Economic Interests," January 1983.
7. Drew Digby, "UC Faculty Conflicts Extensive, Study Says," *Daily Californian*, 24 May 1983, 1.
8. Massachusetts Institute of Technology, *Policies and Procedures* (Cambridge: Massachusetts Institute of Technology, 1979), 24.
9. Karen Seashore Louis, Judith P. Swazey, and Melissa S. Anderson, "University Policies and Ethical Issues in Graduate Research and Education, Results of a Survey of Graduate School Deans" (Bar Harbor, Maine: Arcadia Institute, 1989), Appendix, Table 12.
10. Ibid., pp., table 18.

Bibliography

Association of American Universities. *University Policies on Conflict of Interest and Delay of Publication*. Washington, D.C.: Association of American Universities, 1985.

DIGBY, DREW. "UC Faculty Conflicts Extensive, Study Says." *Daily Californian*, 24 May 1983, 1.

HOGLER, RAYMOND, et al. "The Implications of the Penn State Economic Development Plan for the College of Liberal Arts." University Park: Liberal Arts Faculty Affairs Committee, Pennsylvania State University, 1988.

LARSON, THOMAS D. "Towards Closer University-to-Industry Linkages: The Penn State Experience." Paper presented at the Higher Education International Conference, Braga, Portugal, April 1988.

LESLIE, STUART W. "Playing the Education Game to Win: The Military and Interdisciplinary Research at Stanford." *Historical Studies in the Physical and Biological Sciences* 18 (1987): 55–88.

LOUIS, KAREN SEASHORE, JUDITH P. SWAZEY, and MELISSA S. ANDERSON. "University Policies and Ethical Issues in Graduate Research and Education, Results of a Survey of Graduate School Deans." Bar Harbor, Maine: Arcadia Institute, 1989.

MASSACHUSETTS INSTITUTE OF TECHNOLOGY. *Policies and Procedures*. Cambridge: Massachusetts Institute of Technology, 1979.

UNIVERSITY OF CALIFORNIA. "Form 730-U, Principal Investigator's Statement of Economic Interests." January 1983.

WADE, NICHOLAS. *The Science Business: Report of the Twentieth Century Fund Task Force on the Commercialization of Scientific Research*. New York: Priority Press, 1984.

4

A Brief History of Patent Policy and Administration

Patent ownership and patent licensing are clearly the most important and visible forms of technology transfer in universities. Patents and their commercialization are the issues that make many universities face for the first time the question of how deeply they can become involved in the commercial world without jeopardizing their traditional role, freedoms, and privileges. The special place universities enjoy in society is based in part on their maintenance of objectivity, impartiality, and aloofness from commercial concerns. However, universities are also expected to serve society, and recently this has meant engaging in activities designed to promote economic development or to improve the competitive position of the country. Furthermore, universities today are facing higher costs for research and greater competition for resources than ever before, problems which mean that they must constantly be alert for new sources of funding. University-owned patents embody both the visible value of university research and the university's willingness to sell and profit from this value.

This and the next three chapters show that the development of patent policy and the degree of sophistication of patent administration are clear indications of a university's commitment to and effectiveness in carrying out technology transfer.

The chapter begins with a brief description of the history of patent policy and administration at each of the four universities, three of which—

56

U.C., M.I.T., and Stanford—are estimated to issue from one-fourth to one-third of all the patent licenses granted by American universities and to garner a like proportion of the total royalty income received annually by American universities. A knowledge of the history of university patent policy and practice is necessary to understand what is happening today. Most of the current issues and controversies have antecedents, and universities have long memories, especially of problems or of incidents that were embarrassing. The history of patents in universities is especially important here, in that the study's purpose is to define what is new and to identify and project trends.

Patents and Intellectual Property

Before the history of universities and patents is examined, the role of patents, and particularly the term *patent* as used in this study, should be placed in a proper context and perspective. First, it should be noted that patents are only one form of commercializable intellectual property. Copyrights are another obvious form, and with the extension of copyright law to computer software, they have become very important in technology transfer. Tangible research property, such as cell lines, plasmids, electronic schematics, mechanical structural drawings, detailed mechanical specifications, and descriptions of laboratory procedures, has also become important and valuable in the commercial world. "Know-how" (sometimes synonymous with "trade secrets"), the transferable ability to perform or make something of value, is also intellectual property that can have commercial value. Finally, trademarks, logos, and other names and symbols also can be included in this category, even though they are not, strictly speaking, intellectual property.

Second, the reader should be aware that patents have only a limited use in the arena of technology transfer. William Ouchi, a frequent commentator on the differences between the commercial development processes of the United States and Japan, categorizes intellectual property as "private," "public," and "leaky."[1] Private property is that which can be appropriated to private benefit and protected. Patents fall into this category. Public property cannot be owned by any single person and cannot be protected even for a short time. "Leaky" property can be owned and protected only for a short time. Patents take a relatively long time to obtain and, once obtained, are available for examination by everyone, including potential infringers and those who seek information about them in order to "invent around" them. Much of the intellectual property that is important in technology transfer falls into this category, especially in industries and areas of science that are undergoing rapid change.

A third qualifying factor in using the term *patent* is that this term tends to reduce all patentable intellectual property to the same category and definition, when in fact such property may be quite diverse. Patentable inventions differ widely in their distance from the market. Some may need extensive further development and testing (for example, most pharmaceuticals), while others may be immediately marketable (as with some plant or seed varieties). Patentable material may also differ in the extent to which it is "new" or possesses the potential for creating new markets and the degree to which the university or the inventor needs to be involved in further development. Finally, patentable inventions differ in the extent to which their development may benefit or harm society and the environment.

Because of these factors and because patents are in a category with other forms of intellectual property that can be legally protected and commercialized through a license for use, some have suggested that the term *Technology Licensing Office* be substituted for *Patent Office*, and that focusing on patents when examining issues related to technology transfer can be misleading. However, patents are the earliest and most visible form of commercializable intellectual property, and much of the current debate about such property still centers on patent policy and administration. Policies and administrative practices concerning other forms of commercializable intellectual property have followed the pattern set by those for patents. Therefore, this study concentrates on patents as reasonable examples of developments involving copyrightable property and tangible research property. It refers to other forms of intellectual property when that seems appropriate or when practices concerning them differ from those concerning patents.

This survey of the history of patent policy and administration begins with the early days of patent awareness in universities—the period before 1930—and proceeds, by decades, to the present.

Before 1930

The early history of university involvement with patents was dominated by philosophical questions concerning the ethics of universities that asserted ownership of inventions created on campus. The term *intellectual property* seemed a contradiction when associated with a university because universities were not supposed to "own" the knowledge created within them. During this time, many attempts were made to protect the university from the supposedly negative effects of owning and profiting from important inventions by university scientists.

EVENTS AT U.C.

One of the earliest illustrations of the prevailing attitudes toward university ownership of patents was the invention, in 1907 at the University of California by F. G. Cottrell, of an electrostatic precipitator designed to reduce air pollution from industrial smokestacks. Cottrell obtained a patent and set up a company to exploit his invention, but early on, in the tradition of scientists such as Curie, Pasteur, and (more closely analagous) Professor Charles E. Munroe (inventor of smokeless powder) and Dr. Marion Dorset (who developed hog cholera serum), he sought to give away profits from his invention. His first thought was for his university: "I wanted nothing in the way of financial profit from it myself but did want a fair share of any profits which resulted from the invention to come directly to the University."[2] However, on further reflection, Cottrell and his associates decided not to involve their institution, believing that it was inappropriate for universities to compete in the marketplace and that concerns for profit might lead to control of university research by industry.[3] But Cottrell was dedicated to the idea of directing the profit from his invention back into scientific research. After trying to sell this idea to the American Chemical Society, the Bureau of Mines, and the Smithsonian Institution, he finally convinced a group of businessmen to charter a corporation to hold patents on scientifc inventions and to channel the proceeds from such inventions back into university research.[4] This organization, the Research Corporation, exists to this day and has fulfilled Cottrell's hopes. It figures in later discussions in this book, since Penn State still contracts with a successor to Research Corporation for the patenting and marketing of the university's inventions.

A decade later, again at U.C., Professor T. Brailsford Robertson discovered and patented "tethelin," a substance that apparently promoted the growth of human tissue and had been used effectively to speed the healing of wounds. Potentially this discovery had wide use and high commercial value. Amid considerable public attention, including two articles in *Science*, Robertson donated his patent to the university and directed that the expected profits from it be used to endow a university-controlled foundation for medical research. This may have been the first patent ever owned by the University of California.[5] Robertson sought to avoid situations in which "the investigator is placed in a relationship of direct or indirect dependence upon his patron" and wanted to create instead a mechanism "whereby a proportion (and a very small proportion would be sufficient) of the values created by scientific investigation would flow back to provide the material foundations of further discoveries."[6] The October 19, 1917, issue of *Science* presented a statement by Robertson that the magazine had solicited because it believed the arrangement proposed by Robertson was "a new development in the relationship of science to the industries, and of scientific investigators to the institutions employing them."[7] In this issue Robertson presented a strong argument in favor of universities' holding and administering patents (under carefully controlled conditions) and cited the success of Cottrell and the Research Corporation.[8] Unfortunately, tethelin did not prove to be the great commercial success its inventor expected it to be. By 1923 the invention had netted the University only $242.47, and no research foundation was ever established.[9]

INSULIN AND VITAMIN D

In December, 1921, F. G. Banting and C. H. Best, of the University of Toronto, reported discovering a way to produce insulin. Banting and J. B. Collup received a patent for the process in 1922 and assigned it before issue to the university. The University of Toronto, in turn, gave the process away without strings and entered into several cooperative arrangements for the production and distribution of insulin. The wide public acclaim for the discovery and recognition of the generosity of the inventors in giving up their rights to it led to the establishment of the Banting and Best Research Fund. This fund received contributions from the Canadian government and many other agencies and individuals and used the money to support university research. Unlike the foundation proposed by Robertson, the fund did not depend upon the returns from commercialization of research.[10] It is an early illustration of a theme that appears throughout this book, that patent administration and policy are parts of the much larger subject of the relationship between the university and industry and of the generation of financial support for university research.

In 1924 Professor Harry Steenbock of the University of Wisconsin published research showing that vitamin D could be activated by irradiating food. His discovery led to the virtual eradication of rickets as a childhood disease in the United States and helped to launch the first and most successful university-affiliated research foundation in the country. Spurred by desires to have his university share in the benefits of his work, to guard the public against improper use of his findings, and to protect Wisconsin dairy interests against the makers of oleomargarine, Steenbock offered his patents to the University of Wisconsin.[11] The Board of Regents turned down his offer, saying that the university could not "be expected to allot money for a patent application when it is not certain that it will not [sic] receive something for such an expenditure."[12] Also, the state's Attorney General had ruled that the university did not have the legal capacity to defend a patent, so it appeared that the patent would have no value.[13] Finally, there were serious questions (even from the dairy interests, which stood to gain the most from the restrictive use of the patents) about the propriety of a university's owning and controlling something that had been invented with public funds.[14]

The alumni of the University of Wisconsin, troubled at the prospect of losing potential benefit to the university and further angered by the regents' 1925 restriction on private gifts (inspired by progressive Republican objections to a gift to the university from John D. Rockefeller), established the Wisconsin Alumni Research Foundation (WARF) on November 14, 1925.[15] WARF was created specifically to receive gifts to the university, and the Steenbock patents were its first assets and the source of its funding for the first ten years of its existence. These patents ultimately returned over $14 million to WARF.[16] WARF was the first entity established to administer patents based on university research that was completely separate from the university. It became the model for many similar foundations.

CONCLUSIONS FROM EARLY EVENTS

These early ventures into the commercial world, especially the case of WARF, illustrate many of the problems associated with modern patent administration. Steenbock's desire to protect the dairy industry, which seemed to run contrary to the nonpartisan ideals of science and universities, eventually resulted in a 1943 federal antitrust suit because of WARF's refusal to license even one oleomargarine manufacturer.[17] WARF had to invest heavily to defend the patents it held, and during the controversy the distinction between the foundation and the university was not always recognized, at least by the public. The failure of Robertson's tethelin points out the inherent risk involved in investing in patents. Another recurring problem was highlighted by a controversy that arose in the mid-1920s when Herbert Evans of U.C. Berkeley obtained a patent on a discovery made with the financial help of the Rockefeller Foundation and the National Research Council Committee on Sex Research. Although neither sponsor had a policy on patenting inventions, the Rockefeller Foundation was so angry at the prospect of a researcher's reaping a personal benefit from the research that it cut off funding for Evans until he relinquished his stake in the patents.[18] This conflict over the rights of sponsors to the results of sponsored research continues today.

During the late 1920s and early 1930s a few research universities, recognizing the value of their research, began to develop patent policies. Generally these policies were adopted as a reaction to a particular problem associated with patent ownership or because of the prospect of a particularly lucrative invention. By the end of the 1920s a few universities were beginning to feel more comfortable about owning patents on university inventions, but there was still considerable philosophic resistance in many quarters.

The 1930s

Some of the opposition to university ownership of patents was apparent at an International Research Council meeting in 1935 that was called to consider the financial condition of universities and included a reexamination of the ethics of patenting. The WARF arrangements were roundly criticized as bringing the university too close to commercial concerns. At this time the University of Wisconsin was obtaining funding not only from WARF but also from a number of food producers, who began featuring the name of the university in their advertising.[19] This practice was offensive to many university scientists, but the deepening Depression brought with it the incentive to look for new sources of research support, and those scientists had to admit that the WARF experience, however repugnant it might seem in some respects, had nevertheless brought the university a significant amount of research funding.

The record at M.I.T. illustrates the ambivalence that major universities showed toward patent ownership in the 1930s. In April 1932, the Executive Committee of the M.I.T. Corporation, with the approval of the Faculty Council, adopted the institute's first patent policy. This policy asserted M.I.T. ownership of all inventions resulting from research funded by the institute and pledged that the institute "shall hold and administer these rights for the ultimate benefit of the public."[20] Within a few years M.I.T.'s administration became concerned that, although much of the research being done at the institute had potential commercial value, very little of that potential was actually being developed. In March 1937, after considerable study, M.I.T. entered into an agreement with Research Corporation "whereby this organization will handle all legal and commercial aspects of inventions assigned to it by Institute inventors."[21] The official statement of policies and procedures issued in 1940 contains considerable discussion and justification of M.I.T.'s policy and the arrangement with Research Corporation. In the years between the adoption of the policy and the Research Corporation agreement, the statement maintained "there have been no unresolved misunderstandings, and the inventions which have inevitably arisen from the research programs in the Institute laboratories have been handled smoothly and in a manner which has neither unduly distracted the research worker from his more important objectives nor brought criticism of the policies pursued."[22] The statement also addressed the concern that the adoption of a patent policy would have a negative public relations effect and "render individuals less willing to support Institute programs of research. This has certainly not occurred. . . . Not one of the contributors has criticized the presence of the system."[23]

This tentative awakening to the potential financial return and public benefit from the development of university research was echoed in a small number of other universities, but there was by no means a consensus on the issue. Penn State, at about the same time (1935), tried to solve its patent problems by establishing the Pennsylvania Research Corporation, which was to receive and evaluate invention disclosures and develop promising inventions. Although the University of California began a "patent program" in the 1920s, it did not adopt a formal patent policy until 1943. By 1949 only 50 universities had patent policies, and half of these had developed their policies within the previous ten years.[24] In some quarters, ethical and philosophical objections to university ownership of patents persisted well into the 1970s; Harvard did not change its policy of not filing for patents on medical inventions until 1975.

During the 1930s, then, a few universities saw both the potential value of university research and the potential problems associated with its development. These universities, spurred by the financial exigencies of the Great Depression and presented with several examples of inventions with big payoffs, began to organize their invention disclosure and development processes and to adjust their policies to admit patent ownership. Nonetheless, the ethics of university patent ownership was still a serious issue, and as universities moved to assert ownership rights, the rights of research sponsors and inventors also became issues. For most universities, furthermore, the amount of research and the level of research funding were so low that patents and patent ownership were not particularly important.

The 1940s and World War II

In the early 1940s it became clear that the patent ownership situation was beginning to change. At the University of California, Davis, two scientists formulated calcium pantathenate, a chicken food additive and later a common ingredient in vitamin pills, that had considerable commercial potential. The scientists assigned their rights to this discovery to the university, an act which prompted the regents to adopt their first formal patent policy. But it was at M.I.T., which had led the field of patent policy development and patent administration, that the direction of change was clearest. The following excerpts are from a letter dated October 19, 1942, from Howard Poillon, president of Research Corporation, to N. McL. Sage, head of the Division of Industrial Cooperation at M.I.T. Sage was seeking to capture a share of royalty income on inventions produced by researchers from his division; Poillon's reply outlines the institute's original thinking on income distribution and gives a hint of some possible changes:

> In my previous discussions with you I find that I have let myself drift away from the original course that I think was established when the Institute first decided upon a patent policy and entered into its cooperative agreement with Research Corporation. Briefly it was the consensus of opinion of those in charge of the program that the duty of the Institute was to educate and to increase the frontiers of knowledge.... As I recall, it was rather distinctly understood that any patents would be considered as entirely collateral to the main object of the Institute, and any income would not be used in a major way to increase the appropriation to the department in which the discovery was made, lest such procedure place undue emphasis on the importance of making discoveries....
>
> It is realized that certain conditions today are quite different than were those which then existed, and I think that it might be well to determine whether or not certain fundamental concepts have changed in the meantime, in light of certain incomes which are now on the horizon, and if your department is concerned with the possible income from patents that result from discoveries made by men working on projects financed by Institutional appropriations....
>
> From a purely administrative standpoint, it would appear best to route all income from all patents through the same channel, which would lead directly

to and into general funds of the Institute, where their identity would be lost. . . . All of this may have some influence in the presentation of the reasons why M.I.T. wants a royalty from the Government, and it might be the better procedure to use the whole patent program and all the out-of- pocket expenditures . . . as the reason for the royalty, supported with the intimation that any royalties received would be used for further research, than to select the specific patent and establish a definite use for its royalties.[25]

This letter has been quoted at length because it contains evidence of several trends that characterized the 1940s and were brought into sharper focus by World War II. First, it indicates a subtle shift from high-minded goals for patenting university research, which placed income generation low on the priority list, to a more financially self-interested attitude. Second, it shows that allocation of royalty income was an issue at M.I.T. and that some departments, at least, were not satisfied with allowing such income to vanish into the general fund. Finally, it demonstrates that the federal government was a party of interest in the distribution of patent royalty income. M.I.T. was seeking a royalty from the federal government on several inventions and was in the process of trying to justify that royalty on the basis of the cost of patenting.

The war made it clear to federal policymakers that the federal government could not do the scientific research it needed in government laboratories alone; it had to contract with universities and private laboratories. As federal support of university research therefore mounted, and as the number of important inventions that came out of federally supported research increased, three-way disputes among inventors, institutions, and the federal government became common.[26] While the war continued and government-supported research was primarily dedicated to developing war-related rather than consumer or commercial goods, these disputes could be muted through appeals to patriotism and the ambiguous rules relating to governmental rights to inventions developed with government funds. After the war, however, when it became clear that federal support of university research would continue under the auspices of the Office of Naval Research (1946), the National Institutes of Health (1946), and later the National Science Foundation (1950), these issues could no longer be sidestepped.[27]

The 1950s

In the 1950s a pattern of federal support for university research was established that continues to this day. As federal agencies became active contractors for university research, the issue of ownership of rights to inventions surfaced in virtually every contract. Each agency developed its own guidelines and policies, and by the late 1970s there were at least 26 separate policies and many more regulations.[28] The goals of federal policy were to make the results of federally sponsored research widely available and to protect the public from the wrongful appropriation to private interests of the results of research supported by public funds. The development of federal patent policies compelled universities to adopt more comprehensive patent policies and procedures of their own to define ownership rights and, more important for agency purposes, to spell out how royalty income would be used.

At the the University of California, for example, the University Patent Fund was established in 1952 "to invest the accumulated earnings of University-owned inventions in the General Endowment Pool and to provide income to finance patent expenses and research activities."[29] In 1956 the Regents of U.C. first addressed the issue of the rights of sponsors of research to any patentable inventions that might result from the research they paid for.[30]

Federal policies developed in the 1950s had additional important impacts. Federal agencies followed one of two approaches to the ownership of inventions. The federal "title agencies" acquired full title to all inventions, even when such inventions were only incidental to sponsored projects. In other words, these agencies denied universities the right to apply for or own patents on government-sponsored inventions. These agencies then dedicated the patents to the public, although sometimes they would grant a nonexclusive, royalty-free license on request. The federal "license agencies" would permit universities to apply for, hold, and obtain royalty income from government-sponsored inventions but would reserve a royalty-free license for the government to use the invention for governmental purposes.[31]

A common provision of license agencies was to require that a contractor possess a "licensing capability" in order to obtain rights to an invention. This requirement was loosely defined, but meant that a university aspiring to hold title to a government-funded invention had to demonstrate the organizational capacity and willingness to pursue patent applications and to market inventions. A number of universities, including Stanford and Penn State, seeking a quick and inexpensive way to satisfy the requirement, contracted with licensing corporations such as Research Corporation in the 1950s. Universities that did not make such arrangements had to depend largely on patent boards composed of university scientists and outside patent attorneys, supported by modest clerical staffs.

By the end of the 1950s, most major research universities found that they needed not only to develop a patent policy but also to establish procedures to deal with patentable intellectual property and with the income such patents might generate. Much of the debate over the ethics of patent ownership by universities that had dominated earlier times had by now evolved into discussion of more narrowly focused issues such as the appropriate distribution of income and the rights of sponsors.

The 1960s

The 1960s saw the continuing development of policies reflecting federal requirements and of procedures for invention disclosure and patent administration. For Penn State and Stanford the 1960s were relatively stable. Both universities continued to rely on Research Corporation for the bulk of their patent administration, and the volume of patent activity at both was very small. At M.I.T. and U.C., however, the decade brought some significant changes.

EVENTS AT U.C.

For U.C. inventors, submitting inventions through the university's "patent program" was optional until 1963, when the Board of Regents mandated that inventions be assigned to the university. In order to take the sting out of this mandate and to encourage invention disclosures, the inventor share of royalties was increased from the previous sliding scale of 25 percent to 15 percent of gross royalties to a flat 50 percent of net royalties. These changes had the desired effect. The number of inventions disclosed by inventors to the university increased from between 20 and 30 per year to between 200 to 300 per year. The administration decided to handle this increase in-house rather than contract with an external patent development corporation. A patent board composed primarily of academics was established to develop policy and procedures related to patents. The day-to-day operations of patent administration were carried out by an attorney in the office of the General Counsel.[32]

Because of the relatively large number of disclosures and the small staff, U.C. was quite conservative during these years in applying for patents, selecting for processing only those inventions that appeared to have clear commercial potential and in regard to which the university was reasonably assured of achieving a defensible patent position. Of the 200 to 300 disclosures handled by the patent administrator annually, an average of only 25 resulted in patents, and only two or three of these ever produced any income.[33]

What happened at U.C. was fairly typical of events at other universities that decided to perform patent administration functions internally. These activities were generally carried on at a low level and were concerned with the legal aspect of patenting rather than with marketing. They were designed primarily to meet the minimum federal requirements of "licensing capability."

EVENTS AT M.I.T.

At M.I.T., however, something less typical was happening. M.I.T. scored a "big hit." In the late 1940s, J. Forrester of the Digital Computer Laboratory, working under an Office of Naval Research contract, had developed magnetic core memory, an invention that was to serve as the basis for the computer industry for over two decades. Both the Navy and Research Corporation, in the words of Research Corporation's president, thought at first that, since the invention was "in the narrow, specialized field of computers, it can never be expected to earn much."[34] Nonetheless, Forrester and M.I.T. insisted that a patent be filed, and a patent was issued on February 28, 1956.

As the computer industry and the technology surrounding it matured, Forrester's patent became very important. Research Corporation, however, continued to mis-estimate the market for the invention and set a royalty figure so high that no segment of the industry was willing to take out a license. At the same time, RCA instituted interference procedings against M.I.T., claiming that one of its employees had discovered magnetic core memory prior to Forrester. When Research Corporation reached an impasse over licensing fees with IBM, the prime candidate for a license, M.I.T. sought and obtained permission from Research Corporation to take over the negotiations. During a break in these negotiations, Research Corporation, without consulting M.I.T. or giving the institute advance warning, brought suit against IBM over issues related to the licensing negotiations. For this reason and because M.I.T. felt that there was simply no way that any of the matters at issue could be resolved under the assumptions that Research Corporation was following, M.I.T. canceled its contract with Research Corporation.[35]

There is much more to this story: how, after a great deal of litigation, M.I.T. prevailed over RCA in the RCA infringement litigation; how M.I.T. finally came to licensing terms with IBM and the rest of the computer industry; and how the Forrester core memory eventually returned many millions of dollars to the M.I.T. Corporation. But even this brief description highlights several recurrent themes in the history of patent administration. First, it shows that the prospect of a "big hit" is likely to precipitate a change in patent policy, patent administration, or both. Second, the use of an outside patent management firm has its disadvantages, especially in new areas of science or in areas in which the name of the university and the reputation of its scientists are at stake. Third, the patenting of a "big hit" increases dramatically the chances that the university will be sued or that it will have to sue to protect itself. Universities are never comfortable in court, especially when the opposition has a stake in attacking the reputation of the institution and the integrity of its faculty.

With Research Corporation out of the picture, M.I.T. established its own, in-house patent office. No doubt because of its experiences with the Forrester patent, it staffed the office with patent attorneys. For many years the office operated very conservatively, seeking patents on relatively few inventions.

Despite the activity at U.C. and M.I.T., interest in patenting and its potential benefits was by no means universal in institutions of higher learning. In a 1967 survey, 7 institutions of the 31 surveyed still did not have a patent policy.[36] Still, the 1960s saw an acceleration of interest and activity in this area that carried into and increased further during the 1970s.

The 1970s

By the early 1970s, the full commercial impact of scientific breakthroughs from university laboratories in electrical engineering, computers, and several other fields was apparent. The potential economic impact of major universities, as exemplified by Silicon Valley and Route 128, had also been demonstrated and had been seized upon by politicians and policymakers as the "answer" to a host of economic ills. The economic downturn of the early 1970s and the demonstrated success of the Japanese in American markets also made it clear that the superiority of the United States in basic research was not sufficient to ensure continued American domination of the world economy. If our basic research establishment (predominantly universities) remained sound, these people concluded, the problem must reside in the technology transfer function. Federal action to address this problem, which continued through the decade, was to have a significant impact on patent policy and administration in universities.

FEDERAL ACTIONS

By the mid-1960s the dampening effect of the previously mentioned complex of federal policies on the commercialization of government-ponsored inventions was beginning to worry people. Even when the government had the right to waive title to inventions, it rarely did. Furthermore, agencies extended federal patent regulations to inventions developed with only a small amount of federal funding. Advocates of the government's right and obligation to retain ownership of inventions developed with federal funds were powerful and quite vocal. However, they were opposed by those who saw an anomaly in government ownership of patents. Patent law, after all, had been designed to give inventors a protected monopoly on inventions in order to encourage economic growth, but government agency regulations and, later, the Freedom of Information Act effectively excluded most university inventors from the protection of patent law.

A presidential memorandum issued in 1963 set up guidelines for a more uniform and equitable federal patent policy. It resulted in revision of the regulations of several federal agencies, including the Department of Health, Education, and Welfare (DHEW) and the National Science Foundation (NSF), two of the largest. These agencies developed Institutional Patent Agreements (IPAs), which allowed universities to take title to inventions under certain circumstances. Even this change, however, had only a marginal effect on the problem. Each agency still had its own rules, and these were so complex that only the more sophisticated universities could take advantage of the new provisions. There remained a tangled web of policies that effectively meant "publish *and* perish" for most university inventions.[37] This issue continued to affect universities throughout the 1970s, culminating in the passage of PL 96–517, the Patents and Trademarks Amendments Act, in 1980.

RESEARCH PARTNERSHIPS

In the 1970s, too, industrial patrons of university research began to have a significant effect on university patent policy and administration. The 1970s saw the advent of the giant university-industry research partnership. One of the earliest and largest of these "mega-agreements" was between Harvard and Monsanto. Negotiated in secret in 1974, this agreement provided Harvard with $23 million in research grants and other support over ten years in return for the assignment by Harvard to Monsanto of the rights to discoveries involving the tumor angiogenesis factor (TAF) blocking agent, which had the prospect of curing certain types of cancer. As a result of this agreement, Harvard formally adopted a significant change in its patent policy in November, 1975. The original policy, adopted in 1934, read: "No patents primarily concerned with therapeutics or public health may be taken out by any member of the University, except with the consent of the President and Fellows; nor will such patents be taken out by the University itself except for dedication to the public."[38] Monsanto would not have been willing to provide such a large research fund without some right to the results of research. The new policy required faculty to disclose patentable inventions and admitted university ownership of patents.

This agreement was followed by several similar ones in the late 1970s and early 1980s between Monsanto and Washington University, Exxon and M.I.T., and Hoescht and Massachusetts General Hospital (which is affiliated with the Harvard Medical School), to name a few that made the biggest news. In most cases these agreements not only created changes in university patent policies and procedures but also focused public attention on the issues surrounding patent ownership by universities.

Although such "mega-agreements" affected only a few large universities, other factors pushed many universities and corporate sponsors of research to the bargaining table to consider patent rights. Beginning in 1971 and accelerating through the 1970s to about 118 agreements by 1981, for example, NSF developed a series of programs designed to encourage university-industry partnerships.[39] Factors such as these agreements forced universities to adopt more extensive, sophisticated, and flexible patent policies and administrative procedures.

At two of the four universities in this study, U.C. and M.I.T., significant changes in patent administration took place during the 1970s. These changes reflected the universities' growing concern with patent matters and the increasing allocation of resources that this concern warranted.

EVENTS AT U.C.

In 1973, patent administration at the University of California was removed from the Office of the General Counsel, which reported directly to the regents, and was established as a separate office under the president of the university. This change in administration recognized the increasing volume of patent matters and the changing nature of patent administration, from a process involving primarily legal concerns to one that was more closely associated with the concerns of the faculty and the administration on the various campuses. The new position of Patent Administrator reported to the Vice President-Financial and Business Management. Though the patent board continued to be appointed by the regents, it was now chaired by a representative from the president's office.[40] The board was charged with the responsibility of establishing policy and procedures for patent matters, while the Patent Administrator handled the day-to-day functions.

During the early and mid-1970s, U.C. revised several aspects of its patent policy, including policy relating to the rights of sponsors of research, procedures for reporting inventions, and extension of the coverage of the policy to nonfaculty employees and students. In 1975, for the first time, an official copyright policy was adopted. During this same period the volume of patent activity increased greatly: the number of patents issued rose from 18 in 1973 to 59 in 1979.[41]

Despite the organizational rearrangement, patent administration at U.C. continued much as it had prior to 1973. Mark Owens, who had been patent counsel in the General Counsel's office, simply moved over to the Patent Administrator position. He was supported by a clerical staff of three plus outside counsel for applying for patents. Under his direction the office remained conservative, seeking patents only on those inventions that appeared licensable. When Owens left the position in late 1977, he wrote that "the program is expected to be continued with little or no change in the immediate future."[42]

A significant change, however, was in the wind. In late 1978 the university recruited Roger Ditzel from Iowa State University to be the Patent Administrator and established a new Patent, Trademark, and Copyright Office (PTCO) under his directorship. Although the office remained small through 1979, when it consisted primarily of Ditzel and clerical help of one or two, it grew to a staff of ten in 1980.[43] In 1978, U.C.'s royalty income was close to $1 million per year; by 1982 it had climbed to $1.3 million. Despite these changes, U.C. patent administration was far from having achieved organizational stability.

EVENTS AT STANFORD

In 1970 Stanford began a process that would take it from almost no patent activity to a position of well-recognized preeminence among universities in the field of patent administration and technology licensing by the end of the decade. On April 14, 1970, the university's board of trustees adopted the following resolution: "Except in cases where other arrangements are required by contracts and grants for sponsored research or where other arrangements have been specifically agreed upon in writing, it shall be the policy of the University to permit employees of the University, both faculty and staff, and students to retain all rights to inventions made by them."[44] Stanford may be the only major university with such a liberal policy. In reality, since most research is funded under agreements with sponsors, and since most Stanford inventors choose to assign their inventions to the university, Stanford's policy may not result in practices radically different from those at other universities. Nonetheless, the board, in adopting this policy, took an important step in establishing a culture that encouraged invention disclosure and technology transfer.

At about the same time this policy was adopted, another development was occurring that would help the university realize favorable results from the policy. In 1968, Niels Reimers, a contract officer in the Stanford Office of Sponsored Projects, was asked to review an invention disclosure form submitted by a Stanford faculty member. At that time it was standard practice to record the submission of the form in the Office of Sponsored Projects and then send it to Research Corporation for review. In accordance with standard practice, Reimers was instructed to send the form to the Research Corporation for evaluation. Reimers became interested in the way Stanford handled inventions and did some investigation. At the time, Stanford was processing approximately 30 invention disclosures per year. But in the period from 1954 to 1967, Reimers learned, Stanford had realized less than $5,000 in royalty income.[45]

Not satisfied with this record, Reimers requested and received permission to try a different approach. Rather than seek a patent and then attempt to license it, Reimers tried to sell particular technologies before they were patented, contacting companies that seemed to be involved in each technology at issue. By 1970 this pilot program had proven successful, and Reimers became the director of the newly established Office of Technology Licensing (OTL). Given a mere $125,000 in start-up funding, this office grew steadily in staff, licenses issued, and stature through the 1970s. It was the first, and for a long time the only university patent administration office that had marketing as its primary purpose. By 1980, after ten years of operation, the OTL had licensed and/or patented some 170 inventions, which produced about $2 million in annual royalties.[46]

At Stanford, then, the conjunction of a very liberal intellectual property policy and an effective, market-oriented technology licensing office began to show results. These two factors combined to produce stability in the organizational structure of patent administration and a technology transfer-oriented culture that carried the university smoothly and profitably through the many changes that the 1980s held in store.

EVENTS AT M.I.T. AND PENN STATE

At both M.I.T. and Penn State the 1970s saw relatively few changes in patent administration. Penn State continued with Research Corporation. Although Penn State had adopted its first patent policy in 1935 and had supplemented this brief policy statement with a number of procedures for handling invention disclosures and for submitting these disclosures to Research Corporation, the board of trustees did not adopt a comprehensive patent policy until 1975. This policy served as the basis of "Inventor's Handbooks" issued in 1977 and again in 1979 by the university patent counsel, Robert Custard. The handbooks were designed to encourage researchers to disclose inventions. At this time Custard and his secretary were the only university staff assigned to patent administration, and even they did not have this as a full-time job. Invention disclosures, estimated at fewer than 40 per year, were handled by Research Corporation, which sought patents on an average of 10 percent of the disclosures received.

M.I.T. had a detailed patent policy and a separate patent licensing office in 1970. In 1973, apparently to encourage more faculty disclosures, the policy governing the amount an inventor would receive was changed from a flat 12 percent of gross royalty income to a sliding scale, ranging from 35 percent down to 15 percent of gross royalties. M.I.T. also issued a more fully articulated policy on the obligation of employees, including faculty, to disclose inventions.

By the late 1970s, the conservative and legally oriented posture of the M.I.T. patent office was beginning to annoy faculty members and members of the institute administration. The office filed patents on an extremely high percentage of disclosures. For the year 1978–1979, for example, 137 disclosures were submitted and 81 patents were filed.[47] By comparison, Research Corporation reviewed about 500 disclosures in 1978 and decided to file on about 35.[48] This low rate of filing reflects the conservative posture of Research Corporation: patent filing costs are high. In spite of M.I.T.'s high patent filing volume, few licenses were issued, usually 12 to 15 per year. Royalty income was around $1.2 million, but it came from a very few inventions. This low "throughput" and some administrative problems combined to raise the issue of patent administration to the surface in 1977, when the M.I.T. Development Foundation, Incorporated (MITDF), created in 1972 to develop M.I.T. inventions, went out of business. At that time the administration considered the recommendation that several former staff members of MITDF set up a new patent office to supplant the existing one. This plan was not implemented, however, and dissatisfaction with the patent office's ineffectivness in the marketing of M.I.T. technology continued.[49]

As we have seen, the 1970s saw the development of a full-service patent and technology licensing office at Stanford, the further development and separation of the Patent, Trademark, and Copyright Office at U.C., and some policy revisions but few changes in patent administration at M.I.T. and Penn State. The example set by Stanford was to have significant impact on other universities that decided to "go it alone" rather than depending on a professional patent management company.

The 1980s

The 1980s began with an avalanche of events that have had a significant impact on university patent administration—an impact that continues to this day. These events set the stage for the recent changes in patent administration at the four selected universities that are described in chapter 7.

THE PROMISE OF BIOTECHNOLOGY

On January 16, 1980, a "major announcement" in molecular biology was made at a news conference in Boston. The conference was called by Charles Weissman of the University of Zurich and Walter Gilbert of Harvard on behalf of Biogen, a gene-splicing company whose board of directors included both scientists. The announcement had to do with the cloning by Weissman of a human leukocyte interferon that might be very important one day in cancer treatment. The making of such an announcement by two eminent scientists caused consternation in some sectors of the scientific community. Similar work had already been reported by others, and full development and testing of the product was many years away. This situation prompted one observer to term the scientist's behavior "gene cloning by press conference."[50] Ten months later, when Genentech, another biotechnology company (founded by Herbert Boyer of the University of California at San Francisco), went public for the first time, its stock went from an opening price of $35 per share to $89 per share in 20 minutes, despite the fact that at the time Genentech had not a single product to sell.[51] These and related events in the fields of the "new biology" reflected and reinforced in the public eye the direct connection between university research and commercial possibilities.

THE NEW PATENT LAW AND FEDERAL INVOLVEMENT

At the same time, less attention-grabbing but even more significant and wide-ranging changes were occurring in federal policy. Congress, seeking to promote the development of commercial products in the hope of improving the United States' international competitive status, passed the Patent and Trademarks Amendments Act (PL 965–17). President Carter signed it into law on December 18, 1980. This law allowed nonprofit organizations (including universities) and small businesses to retain title to inventions made in the course of government-funded research. It is commonly referred to as the Uniform Federal Patent Legislation because it finally clarified federal policy and abolished the maze of regulations that had previously governed the ownership of inventions produced by universities with federal funds. It clearly set forth, for the first time, the intent of Congress to use the patent system to transfer government-financed inventions to the public courage technology transfer by fostering the interaction between universities and commercial concerns.[52]

A number of specific provisions of this law, in addition to the general incentive of investing ownership of inventions with universities, had a significant effect on university patent policy and administration. The act requires that nonprofit organizations share royalty income with inventors and use the balance of royalties to pay for the costs of patent administration and to fund further scientific research and education. It also limits the period of exclusivity on licenses granted on federally funded inventions. Furthermore, in a clear revelation of the underlying motive of Congress for passing the act, it requires that domestic (U.S.) licensees be required to manufacture products resulting from such patents "substantially" in the United States.[53] Through its "march-in" provisions, which provide that the government can reclaim title to an invention if it feels that licensing is not being pursued aggressively enough or that the patent system is being abused by allowing the withholding from the market of an invention, the act continues the requirement that a university maintain the "licensing capability" alluded to earlier as one of the important provisions of the "license agencies."

In a 1986 survey of 42 research universities, respondents were split on the significance of the new law.[54] While some credited it with providing an incentive for the creation of aggressive patent licensing programs, others felt it had no effect. This divergence highlights a more basic split in universities' views of patent operations that existed at the beginning of the decade. Institutions that had patent programs had already worked out arrangements with the government, pushing the limits of federal procedures close to the final provisions made by PL 96–517. These universities saw the act as part of a general trend—nothing new, in other words—while those that had not been aggressive saw the act as a causal factor in the development of patent administration.

A number of other changes in federal law or policy taking place in the 1980s also reflect an increased interest on the part of the federal government in encouraging commercial development of university research. Below are a few of the more noteworthy changes:

• March 31, 1981: first patent issued on a genetically engineered microorganism. In 1972 Ananda M. Chakrabarty applied for a patent on a bacterium he had bred that would clean up oil spills. The U.S. Supreme Court, in a 5-to-4 ruling, upheld his application in 1980, and he was issued a patent in March, 1981.[55]

• June 8, 1981: congressional hearings on the commercialization of biomedical research. Spurred by a "mega" research partnership between the Hoescht Corporation of Germany and Massachusetts General Hospital (Harvard), which raised the specter of a foreign corporation "cashing in" on federally sponsored research, Senator Albert Gore chaired a two-day hearing designed to explore the possible effects of commercial involvement by universities on university research agendas, freedom of inquiry, and especially on the issue of who was entitled to the financial return on discoveries made with federal money. Congress feared that a corporation, through an agreement with a university, might be able to appropriate for itself the results of years of federal sponsorship of research. These hearings and another two-day set of hearings in June, 1982, in which university officials were called to testify, clearly underlined the continuing federal concern about patent policies and practices in universities.[56]

- October 12, 1984: PL 98-620 became effective. This law, spurred in part by developments in biotechnology, continued the reform of federal policy begun by PL 96-517 by including plant varieties in the term "subject invention," eliminating the exclusive license limitation, lengthening the time allowed for universities to elect to take title to federally funded inventions while shortening the time allowed to file patents, and strengthening the confidentiality of disclosures and applications.[57]

- April 10, 1987: Executive Order: Facilitation of Access to Science and Technology issued. Although directed primarily at government laboratories and concerned primarily with encouraging the development of inventions produced in those labs, this executive order, signed by President Reagan, articulates the government's and the administration's commitment to economic development through technology transfer. It refers to universities as being a part of several "partnership" programs, including the research center programs of the federal government.[58]

- April, 1988: first patent on animals issued. Harvard received a patent on genetically altered mammals (the first one being a mouse) that can be used to detect cancer-causing substances. Harvard had applied for the patent in June, 1984.[59] This action added animals to the growing list of categories that the Patent Office accepted as patentable.

- March, 1989: University of Utah held a press conference to announce that two of its researchers, Stanley Pons and Martin Fleishmann, had succeeded in producing what came to be called "cold fusion," a process that promised to solve the world's energy problems. The great excitement over this announcement, coupled with the university's desire to protect its patent rights by withholding information about the experiment from other scientists, focused public and scientific attention on the increased readiness of universities to seek financial returns from their research. As the experiment was discredited during the ensuing months, its overpromotion became indelibly associated with the University of Utah, an example of what can go wrong when accepted scientific practices are not followed.

These external events had significant effects on the four universities being studied here. In all four, new or revised policies were adopted to govern the rights of nongovernmental research sponsors to the intellectual property developed from sponsored research, actions that reflected public and congressional interest in such agreements. As university researchers, particularly in the biological sciences, became more visible and valuable in commercial enterprises, concern about conflict of interest also increased. Most universities revised their conflict-of-interest policies to cover more explicitly the involvement of faculty in outside ventures. However, by far the most important development in universities associated with patent administration during the 1980s has been the sharp growth in the volume and complexity of patent activity, with concomitant increases in both risk and potential reward.

Conclusion

This brief history of patent policy development and patent administration at the four selected universities traces a significant turnabout in university attitudes. From a reluctance to be involved in patents and the ownership of intellectual property in the early decades of the century, universities have come to be concerned over how to manage patents to maximize royalty income. Serious university patent administration can be seen as a relatively recent activity in most universities, beginning after 1970.

However, pursuit of profits is not unrestrained. At every university the increase in patent activity has resulted in a growing concern over the operations of patent administration and a search for measures of effectiveness in dealing with intellectual property. Chapters 6 and 7 demonstrate that the manner in which a university approaches a review of patent administration has proved to be a good index of the degree to which it has managed to incorporate technology transfer into its mission.

Notes

1. William Ouchi and Michele Kremen Bolton, "The Logic of Joint Research and Development," *California Management Review* 30 (Spring 1988): 11.
2. Frank Cameron, *Cottrell: Samaritan of Science* (Garden City, N.Y.: Doubleday, 1952), 150.
3. Ibid., 153.
4. Ibid., 150–160.
5. Charles Weiner, "Academic Science and Industrial Links: Historical Perspectives on Current Problems," transcript of a seminar presented at the Center for Studies in Higher Education, University of California, Berkeley, 13 January 1983.
6. T. Brailsford Robertson, "The Utilization of Patents for the Promotion of Research," *Science* 44 (1917): 372.
7. Ibid., 371.
8. Ibid., 372–379.
9. Weiner, "Academic Science and Industrial Links."
10. Charles Weiner, "Science in the Marketplace: Historical Precedents and Problems," in *From Genetic Engineering to Biotechnology—The Critical Transition*, ed. W. J. Whelen and Sandra Black (New York: Wiley, 1982), 126.
11. David Blumenthal, Sherrie Epstein, and James Maxwell, "Commercializing

University Research: Lessons from the Experience of the Wisconsin Alumni Research Foundation," *New England Journal of Medicine* 314 (June 1986): 1621.

12. Howard W. Bremer, "University Technology Transfer—Publish and Perish," in *Patent Policy*, ed. Willard Marcy (Washington, D.C.: American Chemical Society, 1978), 55–56.
13. Ibid., 56.
14. Weiner, "Science in the Marketplace," 126.
15. Bremer, "University Technology Transfer," 56.
16. Blumenthal, Epstein, and Maxwell, "Commercializing University Research," 1623.
17. Ibid.
18. Weiner, "Academic Science and Industrial Links," 21.
19. Ibid., 17.
20. Massachusetts Institute of Technology, *Policy and Procedures,* 1940, 32.
21. Ibid., 34.
22. Ibid., 33.
23. Ibid., 37.
24. Archie Palmer, *University Research and Patent Problems* (Washington, D.C.: National Research Council, 1949), 24.
25. Howard A. Poillon, letter to N. McL. Sage, 19 October 1942, in Leroy Foster, *Sponsored Research at MIT, 1900–1960*, vol. 1 (MIT Archives).
26. Weiner, "Science in the Marketplace," 129.
27. Willard Marcy, "Patent Policies at Educational and Nonprofit Scientific Institutions," in *Patent Policy*, ed. Willard Marcy (Washington, D.C.: American Chemical Society, 1978), 79.
28. Arthur A. Smith, Jr., "Legislative Developments: Implications of Uniform Patent Legislation to Colleges and Universities," *Journal of College and University Law* 8 (1981–82): 83.
29. Roger G. Ditzel, letter to Senior Vice President Brady and Associate Vice President Pastrone, 12 October 1986.
30. University of California, "A Summary of Sponsor Rights Applicable to Funding Agreements with Industrial (for Profit) Sponsors of Research," March 1984.
31. Bremer, "University Technology Transfer," 59.
32. Mark Owens, Jr., "Patent Program of the University of California," in *Patent Policy*, ed. Willard Marcy (Washington, D.C.: American Chemical Society, 1978): 65.
33. Ibid., 90.
34. Robert J. Horn, Jr., and Melvyn R. Jenney, letter to Paul V. Cusick, 2 May 1974. This letter is in the MIT Archives, Office of the President and Chancellor, 1971–80 (Weisner/Gray) (80–27), Box 30, "Patent Policy."
35. Ibid.
36. O. J. Wilson, *Patent and Copyright Policies in Forty-five Colleges and Universities* (Bowling Green: Office of Institutional Research, Western Kentucky University, 1967).
37. Bremer, "University Technology Transfer," 60–61.
38. Barbara J. Culliton, "Biomedical Research Enters the Marketplace," *New England Journal of Medicine* 304 (14 May 1981): 1198.
39. National Science Foundation, *University-Industry Research Relationships:*

Myths, Realities, and Potentials (Washington, D.C.: National Science Foundation, 1982), 3.

40. Owens, "Patent Program," 65.
41. Peat, Marwick, *University of California Patent, Trademark and Copyright Office, Review of Operations, Final Report,* Regents of the University of California, February 1988, Appendix C.
42. Owens, "Patent Program," 68.
43. Roger Ditzel, interview with author, Berkeley, Calif., 9 November 1980.
44. Stanford University, "Inventions, Patents, and Licensing: Guide Memo 75," 15 November 1980.
45. Niels Reimers, interview with author, Palo Alto, Calif., February 1988.
46. Hugh McCann, "University's Inventions Provide Hefty Income," *Detroit Michigan News,* 29 December 1980.
47. Massachusetts Institute of Technology, *Reports to the President, 1978–79,* M.I.T. Archives.
48. Marcy, "Patent Policies," 90.
49. Richard S. Morris, letter to Jerome B. Weisner, 14 July 1976, in M.I.T. Archives, 80-33, Box 10, "M.I.T. Development Foundation, Inc."
50. Culliton, "Biomedical Research," 1195.
51. Ibid., 1195–1196.
52. Smith, "Legislative Developments," 86.
53. Ibid., 92.
54. Association of American Universities, *Trends in Technology Transfer at Universities* (Washington, D.C.: Association of American Universities, 1986).
55. David L. Wheeler, "Grant Patents on Animals? An Ethical and Legal Battle Looms," *Chronicle of Higher Education,* 25 March 1987, 1, 8.
56. U.S. Congress House Committee on Science and Technology, Subcommittee on Investigations and Oversight and Subcommittee on Science, Research, and Technology, *Hearings on the Commercialization of Academic Biomedical Research,* 97th Cong., 1st sess., June 8 and 9, 1981 (Washington, D.C.: U.S. Government Printing Office, 1981).
57. U.S. Congress House Committee on Science and Technology, Subcommittee on Investigations and Oversight, *Hearings on University/Industry Cooperation in Biotechnology,* 97th Cong., 1st sess., June 16 and 17, 1982 (Washington, D.C.: U.S. Government Printing Office, 1982).
58. U.S. President, *Executive Order: Facilitating Access to Science and Technology,* 10 April 1987.
59. David L. Wheeler, "Harvard U. Receives First U.S. Patent Issued on Animals," *Chronicle of Higher Education,* 20 April 1988, A1, A8.

Bibliography

ASSOCIATION OF AMERICAN UNIVERSITIES. *Trends in Technology Transfer at Universities: Report of the Clearinghouse on University-Industry Relations.* Washington, D.C.: Association of American Universities, 1986.

BLUMENTHAL, DAVID, SHERRIE EPSTEIN, and JAMES MAXWELL. "Commercializing University Research: Lessons from the Experience of the Wisconsin Alumni Research Foundation." *New England Journal of Medicine* 314 (1986): 1621–1626.

BREMER, HOWARD W. "University Technology Transfer—Publish and Perish." In *Patent Policy*, edited by Willard Marcy. Washington, D.C.: American Chemical Society, 1978.

CAMERON, FRANK. *Cottrell: Samaritan of Science*. Garden City, N.Y.: Doubleday, 1952.

CULLITON, BARBARA J. "Biomedical Research Enters the Marketplace." *New England Journal of Medicine* 304 (1981): 1195–1202.

McCANN, HUGH. "University's Inventions Provide Hefty Income." *Detroit Michigan News*, 29 December 1980.

MARCY, WILLARD. "Patent Policies at Educational and Nonprofit Scientific Institutions." In *Patent Policy*, edited by Willard Marcy. Washington, D.C.: American Chemical Society, 1978.

MASSACHUSETTS INSTITUTE OF TECHNOLOGY. *Policies and Procedures*. 1940 edition, available in the M.I.T. Archives.

———. *Reports to the President, 1978–79*. M.I.T. Archives.

NATIONAL SCIENCE FOUNDATION. *University-Industry Research Relationships: Myths, Realities, and Potentials*. Washington, D.C.: National Science Foundation: 1982.

"New Public Law 98–620." *Disclosure Newsletter*, January-February, 1987.

OUCHI, WILLIAM G., and MICHELE KREMEN BOLTON. "The Logic of Joint Research and Development." *California Management Review* 30 (1988): 9–33.

OWENS, MARK, Jr. "Patent Program of the University of California." In *Patent Policy*, edited by Willard Marcy. Washington, D.C.: American Chemical Society, 1978.

PALMER, ARCHIE. *University Research and Patent Problems*. Washington, D.C.: National Research Council, 1949.

PEAT, MARWICK. *University of California Patent, Trademark, and Copyright Office, Review of Operations, Final Report*. Berkeley: Regents of the University of California, 1988.

PENNSYLVANIA STATE UNIVERSITY. *University Inventor's Handbook*. University Park: Office of the Vice President for Research and Graduate Studies, Pennsylvania State University, 1979.

"Professor Robertson's Gift to the University of California." *Science* 46 (1917): 371–379.

ROBERTSON, T. BRAILSFORD. "The Utilization of Patents for the Promotion of Research." *Science* 44 (1917): 371–379.

SMITH, ARTHUR A., Jr. "Legislative Developments: Implications of Uniform Patent Legislation to Colleges and Universities." *Journal of College and University Law* 8 (1981–82): 82–96.

SMITH, R. JEFFREY. "Patent Policy Changes Stir Concern." *Science* 199 (1978): 1190.

STANFORD UNIVERSITY. "Inventions, Patents, and Licensing Guide Memo 75." 15 November 1980.

UNITED STATES CONGRESS. House Committee on Science and Technology, Subcommittee on Investigations and Oversight and Subcommittee on Science, Research, and Technology. *Hearings on the Commercialization of Academic Biomedical*

Research. 97th Congress, 1st sess., June 8 and 9, 1981. Washington, D.C.: U.S. Government Printing Office, 1981.

————. House Committee on Science and Technology, Subcommittee on Investigations and Oversight. *Hearings on University/Industry Cooperation in Biotechnology.* 97th Congress, 1st sess., June 16 and 17, 1982. Washington, D.C.: U.S. Government Printing Office, 1982.

United States President. *Executive Order: Facilitating Access to Science and Technology,* 10 April 1987.

University of California. "A Summary of Sponsor Patent Rights Applicable to Funding Agreements With Industrial (For Profit) Sponsors of Research." March 1984.

————. "Item for Action #506 for (Regents') Meeting of 18 November 1982." Office of the President.

Weiner, Charles. "Science in the Marketplace: Historical Precedents and Problems." In *From Genetic Engineering to Biotechnology—The Critical Transition,* edited by W. J. Whelen and Sandra Black. New York: Wiley, 1982.

————. "Academic Science and Industrial Links: Historical Perspectives on Current Problems." Unpublished transcript of a discussion at the Center for the Studies in Higher Education, University of California, Berkeley, 13 January 1983.

Wheeler, David L. "Grant Patents on Animals? An Ethical and Legal Battle Looms." *Chronicle of Higher Education,* 25 March 1987, 1, 8.

————. "Harvard U. Receives First U.S. Patent Issued on Animals." *Chronicle of Higher Education,* 20 April 1988, A1, A8.

Wilson, O. J. *Patent and Copyright Policies in Forty-five Colleges and Universities.* Bowling Green: Office of Instructional Research, Western Kentucky University, 1967.

5

Patent Policy at Four Universities

University patent policies both reflect and determine the attitude of university administration and faculty toward the commercialization of research. Formal, written policies are only the skeleton, but an examination of the formal patent and related policies of a particular university can reveal the degree of sophistication that university has developed in dealing with the ownership of intellectual property and the degree to which it recognizes the importance of technology transfer. Because the distinction between formal policy and less formalized practice is never clear, such an examination must include related practices and procedures. Universities adjust their policies and practices as new opportunities or situations arise. Thus, current policies are partial encapsulations of their history in dealing with their intellectual property.

Purpose and Elements of Patent Policy

The goal of university patent policies is to strike a balance between the needs of inventors, the institution, research sponsors, invention developers, and the general public. Policies should encourage commercialization of research and protect the intellectual property produced in the university while at the same time guiding the university and its inventors and protecting them from conflicts of interest or the appearance of improper conduct. Also, policies should comply with regulatory and contractual requirements imposed by the federal government and other sponsors of research.

Early questioning of the ethical legitimacy of university ownership of patents has currently given way to the opposite view, that university owner-ship of patents on inventions is a good way to "clean up" possible compet-ing claims and place ownership in the hands of a socially responsible institu-tion that, at least theoretically, has both the capacity and the incentive to provide sound patent administration for the ultimate good of the public. This change has meant an increase in the scope of patent policies.

Early statements of institutional patent policy usually were relatively brief and were concerned mainly with sorting out the conditions under which the university would assert ownership of the inventions of university researchers. Later policy statements were much more complex. For the four universities in this study, many changes and additions have come in the last ten years.

University patent policy statements typically include the following six elements:

1. a preamble, or statement of the purposes the policy is to serve

2. a statement telling to whom the policy applies

3. administrative arrangements

4. a statement of inventor obligation or commitment

5. a statement of the rights of the parties

6. income-sharing arrangements

Each of these elements is discussed in this chapter.

Of the four universities studied, Penn State has the most complete pre-amble in its policy statement. That preamble incorporates most of the pur-poses listed for a typical university patent policy:

Objectives—The principal objectives of this Policy are:

(a) to facilitate the transfer of University-developed technology to com-merce and industry and to encourage the broadest utilization of the findings of scientific research to provide the maximum benefit to the public;

(b) to encourage creative research, innovative scholarship, and a spirit of inquiry in order to generate new knowledge;

(c) to provide a procedure to determine the economic significance of dis-coveries so that commercially valuable inventions may be brought to the point of public utilization;

(d) to establish principles to determine the rights and obligations of the University, inventors, and research sponsors with respect to University inven-tions and to define the equitable disposition of interests in University inven-tions among the University, the inventors, and research sponsors;

(e) to provide incentives to inventors in the form of personal development, professional recognition, continuing research support, and direct financial compensation;

(f) to fulfill the terms of research grants and contracts;

(g) to safeguard the intellectual property represented by worthwhile inven-tions until appropriate patent protection is achieved; and

(h) to facilitate invention and patent management agreements with external organizations.[1]

U.C. has a one-sentence purpose statement that includes many of the same elements found in the longer Penn State preamble. Curiously, except for a statement confirming university commitment to the transfer of inventions to the public (Penn State's purpose [a]), neither Stanford nor M.I.T. has a purpose statement in its patent policies. Similar differences are found throughout this analysis: M.I.T. and Stanford, which are private universities, have policies that appear to be more concerned with internal issues than are those of Penn State and U.C., whose policies, at least in part, are directed at satisfying external constituencies.

University patent policies usually apply to anyone associated with the university who may invent or discover something. Students usually are included (as at U.C., Stanford, and Penn State, although the Penn State policy mentions only graduate students and the U.C. policy applies only to students who are employed at the university), but sometimes they are placed in a special category, as at M.I.T., that exempts student inventions except when they were made with external funding. In recent years, patent policies have been extended to apply to visitors, including visiting industrial scientists.

Descriptions of arrangements for patent administration and the distribution of authority over patents are usually included in patent policy statements. In this book a distinction is made between *patent policy* and *patent administration*, although, as pointed out earlier, the line between them is not always clear. Decisions made and consistently followed by patent administrators often become de facto policy, even when they are not mentioned in an official policy statement. Patent administration at the four universities is described in chapter 7.

In this chapter the remaining elements of university patent policy are discussed under five headings:

1. ownership, assignment, and disclosure

2. income sharing

3. sponsors' rights

4. delay in publication

5. other elements

The patent policies of the four universities are described and compared in each of these areas.

Ownership, Assignment, and Disclosure

The first element of patent policy establishes the right of ownership to inventions and imposes on the inventor the obligation to disclose any invention or discovery. In the four universities studied there is considerable variation in the expression of this element of policy.

U.C. and Stanford represent opposite extremes. U.C. requires that all employees, as a condition of employment, agree to assign their rights to all inventions to the university. Stanford, on the other hand, states that all inventions belong to the inventor unless third parties have gained rights by virtue of research sponsorship. The policies of the other two universities attempt to recognize the relative contributions of the parties involved in the invention and to apportion the rights between them. M.I.T. asserts ownership of inventions "subject to the terms of a sponsored research or other agreement" or those "involving the significant use of funds or facilities administered by M.I.T." Penn State makes the finest distinctions, categorizing inventions as "personal," "releasable," "university-sponsored," "externally sponsored, nonfederal," and "externally sponsored, federal government related."

Universities may assert ownership or rights to inventions in several ways. First, they may require the execution of an assignment agreement as a condition of employment (as at U.C.). This agreement requires that, if an invention is made, the inventor will disclose it to the university and then, in another document, assign all the inventor's rights and potential rights to the university. Second, universities may set up employment contracts that require compliance with university policy, including patent policy, which contains a description of invention ownership rights. That is the way M.I.T. does it. A third approach, taken by Stanford, requires patent agreements to be signed by those who accept research funding from outside sources. Penn State combines two of the preceding approaches by requiring employee compliance with its policy, which in turn requires university employees to make assignments after an invention is created.

One difficulty faced by universities in asserting ownership of inventions is that such ownership claims often are not consistent with the institutions' treatment of copyrightable intellectual property. Most universities traditionally have allowed ownership of copyrightable material to remain with the author, even when "substantial" university resources, including author salaries, were used in the production of the property. Only recently, with the inclusion of computer software in the copyrightable category, have copyright policies been altered. But the rationale for treating one type of intellectual property differently from another remains unclear for many university researchers.

Despite the seemingly wide divergence in the four universities' policies regarding ownership of inventions, their actual practices in this area prove to be quite similar. There are several reasons for this. First, each of the policies makes provisions for third-party agreements to take precedence over university policy. Since little research these days is done without an outside sponsor, and since the overwhelming bulk of research support comes from the federal government, most university researchers are in fact bound by similar policies, usually federal ones.

Second, all universities agree that inventions made on an inventor's own time ("personal" inventions) belong to the inventor (although Penn State invades this territory a bit by asserting ownership of all inventions except those that fall "outside the inventors' normal assigned activities and employment responsibilities"). Third, interpretation of obligations to disclose and assign patentable inventions depends entirely on the often unschooled judgments of the researcher, whose primary goal is usually to publish findings. Once research findings are published, obtaining patents on them is quite difficult, and much of the potential commercial value of the property is lost. Even so, universities are unlikely to discipline a researcher for publishing completed research too soon.

Finally, the perceived efficiency of university patent administration is a major factor in determining the degree of researchers' compliance with institutional policy. At Stanford, where policies are fairly permissive and researchers are not compelled to use the university office to handle inventions, there are relatively more invention disclosures than at Penn State, where policies are specific and detailed. In part this is because university researchers see the Office of Technology Licensing at Stanford as effective and helpful, whereas at Penn State patent administration procedures are viewed as relatively ineffective, primarily because very few inventions are patented and because the time between submission of an invention disclosure and its evaluation is so long. Researchers find it easier to violate or ignore university policy when they can blame the administration for not being effective.

It is the administration of patents and licensing, not official policy statements, that marks the underlying differences among the universities. Nevertheless, broad statements of university ownership policy and obligations by researchers to disclose inventions carry symbolic significance and set the tone, not only for patent administration, but for technology transfer activities in general.

Income Sharing

Policies governing the way royalty or other income received from a university's intellectual property is to be distributed are important indicators of an institution's overall attitudes toward university-produced inventions. Such policies reflect the relative values placed by the university on the contributions of the different parties involved—inventors, research sponsors, the academic departments of inventors, the university, and, for those universities using them, external patent management firms. As indicated in the last chapter, federal policy (PL 96-517) requires universities to share income with inventors, allows such income to be applied to cover the costs of patent

TABLE 5-1   Number of Universities Having Specific Percentages for Income Distribution to Inventors

TYPE OF DISTRIBUTION	MAXIMUM PERCENTAGES					
	1–15	16–25	26–35	36–50	51–75	76–100
Net income, sliding scale			4	14	8	3
Net income, fixed	1	5	8	15	1	
Gross income, sliding scale		1	3			
Gross income, fixed	11	3	1	3		

Source: K. W. Heathington, Betty S. Heathington, and Ann J. Roberson. "Commercializing Intellectual Properties at Major Research Universities: Income Distribution," *SRA Journal* 17 (1986): 34.

administration, and encourages universities to use the remaining income to support further scientific research. Most universities follow these guidelines. Considerable variation exists, however, primarily because the issues are so complex.

Although everyone agrees that an inventor should be entitled to some return, the real value of an idea or discovery is often very difficult to establish. Ideas vary greatly in their distance from the market and in the amount of development resources necessary to bring them to market. Thus the relative contribution of the inventor to the success of an invention and the relative reward the inventor should receive differ significantly from one invention to another. Yet the impulse is to adopt a policy that provides the same sharing agreement for all university inventors. It also should be noted that the inventor's share is awarded primarily for work already performed (and paid for), while the other players are rewarded for work directly associated with creating market value.

Universities try to incorporate all of these factors by varying the percentage of sharing with inventors and by using different bases for computing percentages. Income-sharing formulae are based on either gross income or net income (income less the costs associated with obtaining, processing, licensing, and defending patents) and are distributed on either a fixed or a sliding percentage. Using net income as a base is preferable because it recognizes that the costs associated with patents may vary considerably and that inventor shares should be reduced when patent-related costs are large.

A study conducted in 1985 confirms that universities vary considerably in income-sharing policies.[2] Table 5-1 shows the results of part of the study. Of the 78 universities surveyed, 69 based their policies on net income and 22 on gross income. Some universities offered inventors a choice or had more than one policy.[3]

U.C.

U.C. has not changed its stated policy since 1963, when patent assignment became mandatory for all university employees. Offering inventors 50 percent of net income, U.C. has the most liberal income-sharing policy of any of the universities in this study. The policy does not specify how the other 50 percent will be distributed, although it does instruct that "in the disposition of any net income accruing to The Regents from patents, first consideration shall be given to the support of research."[4] In contrast to official policy, actual practice of income distribution at U.C. has varied over the years and is still changing. In addition to distributing the inventor's share, the university is obligated to refund to the state of California an amount representing the state's estimated contribution to the production of patentable inventions. In the past few years this share has amounted to 22 percent to 24 percent of net royalties. Current practice is to distribute to the campuses any income left after payment of patent administration costs (including the cost of the operation of the Patent, Trademark, and Copyright Office), the inventor's share, and the state's share. This income is distributed in the proportion that the inventor shares to each campus bear to the total inventor shares to all campuses. This distribution pattern is a change from the one that prevailed just a few years ago when any excess income went into a patent fund, which supported graduate students on each campus and then was allocated to specific research projects by the president. In fact, no distribution to campuses was made in fiscal 1986–1987; there was a deficit in the patent fund of over $1 million. Figure 5-1 shows how gross income for 1987–1988 was distributed. The reasons for these changes in the context of some of the broader changes taking place in the U.C. system are examined in the next chapter.

M.I.T. AND STANFORD

M.I.T and Stanford have policies similar to each other, since M.I.T. modeled its present policies on those of Stanford. Both universities deduct from gross income out-of-pocket costs (costs associated with applying for and protecting patents) and a 15 percent administrative fee to fund the operations of the technology licensing office and then distribute the remainder as follows: one-third to the inventor, one-third to the inventor's department, and one-third to the university general fund. M.I.T. has changed its income-sharing policy from time to time as it searched for the right distribution and struggled with patent administration procedures. The earliest published rates at M.I.T., adopted in 1952, provided the inventor with 12 percent of gross royalties and a 50-50 split of the net after deducting the 12 percent. This distribution was changed in 1975 to a sliding scale of from 35 percent of the first $50,000 of gross royalties down to 15 percent of gross royalties over $100,000. In 1985 the sliding scale was increased to 35 percent of the first $100,000 of gross royalties, and for the first time the policy provided a share of income to the inventor's department. M.I.T.'s recent switch to a formula based on net royalties is similar to the shift that has occurred at many universities.

FIGURE 5-1   Distribution of Income from Patent Licensing Agreements, University of California, 1987–1988

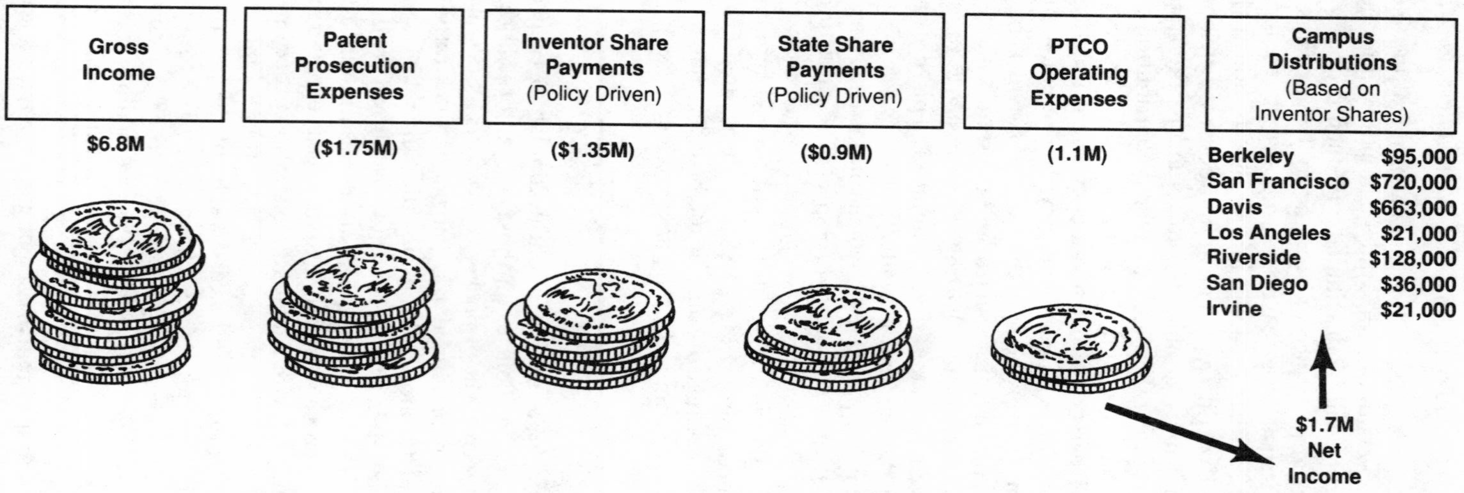

Gross Income	Patent Prosecution Expenses	Inventor Share Payments (Policy Driven)	State Share Payments (Policy Driven)	PTCO Operating Expenses	Campus Distributions (Based on Inventor Shares)
$6.8M	($1.75M)	($1.35M)	($0.9M)	(1.1M)	

Campus Distributions:

Berkeley	$95,000
San Francisco	$720,000
Davis	$663,000
Los Angeles	$21,000
Riverside	$128,000
San Diego	$36,000
Irvine	$21,000

$1.7M Net Income

PENN STATE

Because Penn State receives only a "net" share from Research Corporation, its income-sharing policy is not complicated with provisions relating to the costs of patent administration. However, there is considerable dissatisfaction at Penn State with the relatively low acceptance rate of inventions for patenting by Research Corporation and the small amounts returned to inventors and the university from royalty income (although most of the share distributed to the Pennsylvania Research Corporation eventually comes back to the university in the form of research grants). Because so little income is generated, this aspect of patent policy has not received much attention. It should be noted that Penn State is the only one of the four universities that provides a bonus ($1,000) for the acceptance of an invention for patenting.

COMPARISONS AND CONCLUSIONS

Comparing the effects of the income-sharing policies of these four universities is difficult. Obtaining comparable numbers for royalty income and inventor shares is hard because of accounting and timing differences. Inventor shares often are distributed in the year following the earning of the related income. In recent periods, when income has been increasing so rapidly, this practice has distorted the figures. Also, where policies have changed, as at M.I.T., the effects of the changes are not immediately reflected in inventor share figures. Finally, because most royalty income is derived from relatively few inventions and because litigation costs can be substantial in a single year when there is a problem of patent defense, inventor shares can vary widely from year to year. The data in Table 5-2, which show inventor shares related to total income for specified years for the four universities, should be interpreted with these caveats in mind. Despite the problems in interpreting the numbers in this table, the figures indicate a wide variation among universities. For instance, inventor shares at U.C. range from about 20 percent to 32 percent of total income, whereas at Penn State they are in the 15 percent range. It also appears that income-sharing policies have a significant effect on the general relationship of inventor share to total income.

Although wide variation in university income-sharing policies still exists, there is evidence that policies are beginning to converge for several reasons. First, most universities are required to fund patent administration from royalty income. Adequate funding of patent administration at income and activity levels achievable by most major American research universities is probably between 15 percent and 30 percent of total income. Less than adequate funding of this function means diminishing returns. With this level of funding going for administration and with other contributors to the patenting, developing, and marketing of inventions taken into account, inventor shares should come to somewhere near one-third of net royalties. Stanford and M.I.T. have adopted this percentage arrangement, and although

TABLE 5-2   Summary of Royalty Income and Inventor Shares for Selected
             Years for Research Universities (dollars in thousands)

	U.C.	M.I.T.	PENN STATE	STANFORD
	1987–1988	1986–1987	1985–1986	1987–1988
Royalty income	$6,822	$3,100	$456	$9,179
Inventor share	$1,342	$400	$68	$1,809
Percentage of inventor share	19.7%	12.9%[a]	14.9%	19.7%
Range of percentage of inventor share	19.7–32.9%	13%[a]	NA	15.2–20.8%
No. of years included in calculation	7	2	NA	4

Sources: U.C.: David Pierpont Gardner, "Annual Report Pertaining to University Patent and Other Intellectual Property Matters" (Oakland, Ca.: University of California, Office of the President, May 1989). M.I.T.: John T. Preston, interview with author, 3 March 1987. Stanford: Sally Hines, letter to author, 13 September 1989.

[a]Inventor shares at M.I.T. may be artificially low because M.I.T. inventors who license their own inventions from the Institute are not paid inventor shares of royalty income.

it is still too early to judge the situation at M.I.T., the arrangement appears to be working. At U.C., because of the problems associated with funding patent costs, there have been discussions about lowering the inventor share; this is a future possible policy change.

A second trend is the inclusion of the inventor's department in income-sharing policy. M.I.T., Stanford, and Penn State are doing this, and U.C. is moving in the same direction. Although inclusion of the department or administrative unit in the royalty-sharing arrangements puts yet another hand in the till, it has some strong advantages. First, it places the sometimes powerful engine of departmental interest behind the invention disclosure and development process. Second, it moderates potential intradepartmental jealousies and contention over ownership and responsibility for inventions. Colleagues are not so quick to quibble over invention ownership when they see that the common as well as the individual good is being served. This attitude is very important in fields where cooperative or team research is becoming more common. As shown in chapter 12, the faculty interviews revealed few instances of damage to collegial relations as a result of commercial success and many cases in which inventors dedicated their royalties to their departments.

It is easy to overemphasize the importance of income-sharing policy because it is the part of patent policy that most clearly sets forth the university's view of the relative value and rights of inventors and others, including the university itself. Such policy affects very few university researchers, however. Furthermore, although it is increasing, royalty income at most universities remains very small, and only a slight percentage of university patents ever produce net income. In many cases where inventor shares are potentially significant, inventors assign their shares to the university or the department in which they work. Nevertheless, income-sharing arrangements will continue to be important, and they will continue to change as new situations arise and the overall level of university patent activity changes.

Sponsors' Rights

Although the overwhelming portion of research sponsorship comes from the federal government, whose rights to the results of research are covered by federal law and policy, an increasing portion is coming from corporations, many of which seek rights to use or commercialize what comes out of the research they sponsor. Most universities (Stanford is the exception) have stated policies describing the rights they are willing to grant in research contracts and the conditions under which they will grant those rights. Such policies have been formulated with several objectives. First, universities want to select licensees and grant licenses under terms and conditions that they believe will best carry out the transfer of technology and the commercialization of research. Second, most policies try to grant more favorable terms to sponsors who are willing to contribute more toward research. Third, universities set policies to minimize up-front and patent maintenance costs. Finally, university policies are designed to provide royalty income to the university, although this aspect has been greatly overemphasized.

In negotiating research contracts, universities and research sponsors face several important issues. A number of options, discussed below, are available for addressing them.

PRE- OR POSTINVENTION AGREEMENT

Since most sponsored research does not result in a patentable invention, most universities try to avoid the inclusion of a sponsors' rights clause in a contract, preferring to enter negotiations over licenses only after an invention has been created. This simplifies contract negotiations and, when an invention does result, allows subsequent negotiation to be more focused and relevant to the situation. In most circumstances, universities are unlikely to sell an invention sponsored by one company to another company, so the sponsoring company will not be denied the right to a license.

RIGHT TO LICENSE, OPTION TO LICENSE, AND RIGHT OF FIRST REFUSAL

Research contracts may provide the absolute right to license an invention to the research sponsor, or they may be less specific. In the latter case they may provide only the option to license or the right of first refusal to license under terms to be determined later.

EXCLUSIVE LICENSE, NONEXCLUSIVE LICENSE, AND TERM OF LICENSE

Licenses to inventions may be granted on an exclusive or nonexclusive basis and for varying periods of time. Universities generally prefer to grant nonexclusive licenses because they believe such licenses are more consistent with the underlying goal of promoting the public good through the broadest dissemination of university products. However, most universities now incorporate into their policies some provision for exclusive licenses, recognizing that some technologies will not be developed unless the developing company can be assured of some period of exclusive use in which to profit from its initial investment. Universities also recognize the "field of use" exclusive license, which limits exclusivity to a particular field and allows development of the technology in fields unrelated to the business of the licensees.

ROYALTY-BEARING LICENSE, ROYALTY-FREE LICENSE, AND ROYALTY-SHARING AGREEMENT

Universities have a broad array of royalty arrangements available to them in negotiating research contracts, and agreements often include an initial license fee or provision for the licensee to pay for patent costs. Royalty-bearing licenses may require a fixed minimum annual royalty, and they usually provide for a running royalty based on annual sales of associated products. Royalty-free arrangements often are used to grant sponsors the right to use the technology for internal purposes only, but they may be extended to commercial use. Royalty-sharing arrangements, in which the university agrees to share any royalties with the sponsor, are a useful option when the sponsor is unable or unlikely to be able to commercialize the product itself but nevertheless wants to reap some reward from the sponsored research.

DUE DILIGENCE AND MARCH-IN RIGHTS

Most university policies require that a licensee use "due diligence" in attempting to commercialize a patent and that the patent not be used for "defensive" reasons, that is, the holding of a patent to keep an invention off the market so that it will not compete with a current product. A failure to use due diligence often triggers "march-in" rights, which allow the university to void the license and regain control of the rights to the patent.

SPONSORS' RIGHTS AT FOUR UNIVERSITIES

The four institutions studied, like most universities, do not grant outright ownership of inventions to sponsors. Rather, they grant the right to license, the option to license, or the right of first refusal to license the invention. Licenses may be exclusive or nonexclusive and they may be royalty free, royalty bearing, or royalty sharing.

U.C. has the most restrictive policy, limiting sponsors' rights to right of first refusal to royalty-bearing licenses and not admitting royalty-free or royalty-sharing agreements. However, by special action of the Board of Patents in 1982, the granting of royalty-free, nonexclusive licenses is allowed when the sponsor funds all of the subject research. This change was necessary because the university's insistence on royalty-bearing licenses effectively prohibited it from contracting with IBM, most oil companies, and research and development partnerships.[5]

At the other extreme, Stanford's policy is to place responsibility for the negotiation of contract clauses relating to sponsors' rights with the director of the Office of Technology Licensing. In these negotiations the director is instructed first to protect the "basic purposes" of the university and to make sure that patent considerations do not unduly influence university research. The second guiding principle is encouragement of the development of university research for public use. This principle is interpreted broadly and admits a range of contractual arrangements, including exclusive licenses, royalty- free licenses, and royalty-sharing arrangements.

M.I.T. offers sponsors who fund total costs of research four possibilities, including a "royalty-free" option (which requires the payment of an annual fee), a six-month option to negotiate exclusivity, and a royalty-sharing arrangement. This policy was adopted in March 1988. Before that time, M.I.T. had a much simpler policy that gave sponsors a royalty-free, nonexclusive license.

Penn State adopted its policy in 1981 in order to accommodate industrial sponsors with terms that were more flexible than before. It goes further than the other universities in providing extra rights to research sponsors willing to pay more money "upfront." Reflecting Penn State's close ties with the state economy, it is also the only university that has a special policy for small industries.

PROBLEMS

Because sponsors' rights policies in all four universities have tended more toward promotion of technology transfer than toward generation of royalty revenues, university administrators have had to protect themselves and their institutions from the charge of "giving it away." This has become particularly important in recent years because competition for research support is increasing and a number of university constituencies, including donors, are becoming aware of the value of university research.

University sponsors' rights policies so far have not addressed several other recent problems. One problem has to do with equities among sponsors under consortia or joint research sponsorship agreements. What happens when an invention arises out of research sponsored jointly by several companies? Are exclusive licenses possible under such arrangements? Can research so sponsored result in licenses to companies not involved in the research sponsorship?

Another problem, discussed in chapter 8, is called "pipelining." Pipelining is the granting of licenses to favored companies or, to put it another way, the unfair exclusion of some companies from access to university technology. This last was illustrated by the government antitrust action against WARF for failure to license the Steenbock patents to oleomargarine manufacturers. Theoretically a university should search for the company best able to commercialize a technology and/or benefit the public. Financial return, particularly financial return associated only indirectly with the exploitation of the technology (increase in equity value as opposed to royalty income), should be a secondary factor in choosing a licensee.

Both of these problems were illustrated recently at Penn State when a part of the university's endowment funds was invested in a new company formed to exploit a diamond technology produced by the Materials Research Laboratory. The new company was granted a license to the technology. This action disturbed several of the companies that belonged to the lab's liaison/affiliate program, which had sponsored diamond research at the laboratory for a number of years. The problem was resolved by making the new company a part of the research consortium, but the impression was left that the university was more interested in speculating on stock of an an unproven company than in recognizing the years of research support that had been provided by established companies.

CONCLUSIONS

Over the last ten years, as the percentage of university research sponsorship provided by industrial sponsors has increased, a great deal of attention has been devoted to institutional policies concerning the rights of research sponsors to the results of sponsored research. University policy in this area is still changing, and policies vary considerably from one university to another. Indeed, there seem to be too many variables and too many different situations for any statement of policy to handle effectively. Recent policy changes at many universities have been in the direction of increased flexibility and promotion of technology transfer. They are moving toward Stanford's model which includes no stated policy but merely provides broad guidelines and assigns responsibility for the negotiation of specific agreements to an appropriate individual. Universities have retreated a bit from the idea that licenses are directed primarily at producing royalty income. Instead they are adopting the more comprehensive view that an enlightened and flexible policy regarding preinvention license agreements can produce greater research support and a wider array of research patrons. So few inventions ever produce income that universities have become more willing to trade possible future royalty income for actual current research support.

Delay in Publication

The university is based on the free exchange of ideas and the rapid communication of research results to the intellectual community, usually accomplished through scholarly publications. University policies covering conditions under which a delay in the publication of research may occur, although not formally a part of patent policy, are so closely related to patenting and the protection of intellectual property that they cannot be left out of this discussion.

In a study conducted in 1985 by the Association of American Universities (AAU), 32 of 49 universities had a written policy covering delay in publication. Nineteen of these listed patent review as the only reason for delaying publication, and 22 listed patent review and sponsor review for confidential information as the only two reasons.[6] The length of permitted delay is rarely stated in university policy, but in practice it varies from 30 days to one year. Most universities grant either a 60- or a 90-day delay. Delays for patent review are tolerated under the theory that patents are publications and that they enhance the dissemination of research rather than impede it.

Delays are often necessary to protect intellectual property prior to the filing of a U.S. patent. Foreign patents represent another important reason for delays because most foreign governments will not allow a patent to be filed once an invention has been disclosed in a publication.

University researchers are not particularly concerned with this issue, however, as the normal lag between submission of an article and its publication usually exceeds any delay that would be imposed for patent-related reasons. Furthermore, since the individual researcher determines whether or not to publish and when, researchers may voluntarily withhold publication if they deem it desirable to provide a sponsor with more time for a review.

Unlike many of the other policies related to university patent administration, delay-in-publication policies are reasonably uniform in American universities. The four universities in this study all permit reasonable delays for patent review. Although a good deal of attention is given to this issue, publication delay provisions are rarely sticking points in contract negotiations. They present problems only with unsophisticated sponsors who are not familiar with university contract practices.

Other Policy Elements

University patent policies may contain other elements, the most important of which deal with the taking of equity positions in lieu of royalty payments and the conditions under which inventions will be released back to the inventors. The equity issue is present in only a few universities. It opens a Pandora's box of problems, which are considered in chapter 8.

The four universities in this study all permit the inventor to take full ownership of an invention when and if the university feels that the invention is not worth further development and when such repossession is permitted by research sponsors. Stanford goes one step further, offering to help inventors reclaim rights to their inventions from sponsors. This element is a good escape hatch for universities, since it allows the university to extricate itself from situations in which there is a difference of opinion between the inventor and the university concerning the value of a particular invention. It is also used sometimes in a pro forma way to separate the university from an invention to which it has only a tenuous relationship in the first place.

One problem associated with releasing rights to an invention back to the inventor is that the university then has no control over the invention or its use and often has a difficult time dissociating itself from subsequent efforts to develop or commercialize the product.

Faculty Attitudes toward Patent Policies

A total of 90 faculty members from the four universities in this study were questioned about their attitudes toward patent administration on their campuses (see chapter 12). Most faculty had to be reminded what the patent policies at their institutions were, and many equated patent policy with inventor income-sharing arrangements. They also had difficulty in making the distinction between patent policy and patent administration. Perhaps partly because of this confusion, faculty at M.I.T. and Stanford, which have relatively effective patent administration offices, were positive about policies, while faculty at U.C. Berkeley and Penn State, where administration is perceived as ineffective, were negative. When asked how policies should change, most faculty members responded with suggestions about improved administrative procedures such as more rapid response on invention disclosures or greater marketing capability.

Conclusion

Early university patent policies were developed in response to government requirements, "big hits," or problems, but recent changes in or additions to such policies appear to be driven more by internal considerations and by a desire to promote technology transfer, or, less charitably, to increase royalty income. Some elements of policy, such as delay in publication and release to inventor, are similar in all four universities in this study. In the other main elements of patent policy, however—ownership, assignment and disclosure, royalty sharing, and rights of sponsors—the four universities show some important differences.

One of the premises of this chapter is that a review of university patent policy can indicate the degree of an institution's sophistication and commitment to technology transfer. This is only part of the story, however. An analysis of patent administration is necessary to complete the picture. Nonetheless, an examination of differences in the patent policies of the four universities in this study can help to show how technology transfer is viewed and how important it is on each campus.

U.C.

Patent policies at the University of California are somewhat confused, contradictory, and unsophisticated. They seem to be governed mainly by concern for avoiding the appearance of conflict of interest or violation of public trust—a concern quite justified in a public university in a highly politicized state. An example of the policy confusion can be seen by comparing U.C.'s disclosure and ownership policies with its inventor share provisions. Policies at U.C. are among the most restrictive, comprehensive, and procedurally enforceable concerning the ownership, disclosure, and assignment of inventions to the university, but they are among the most liberal concerning inventor share of royalties (although this liberality may be more the result of historical happenstance and bureaucratic inertia than design). The source of some of this confusion is the sheer size and complexity represented by a nine-campus "multiversity" and an annual research budget of over $800 million. Also, U.C.'s position as a public university imposes some restrictions on it that private universities can avoid.

As with most universities, U.C.'s patent administration must be funded from royalty income. But with such a large share (50 percent of net) going back to the inventor and with the requirement that the state of California be reimbursed for some costs, gaining enough funds for competent and comprehensive patent administration is difficult at U.C. For instance, in fiscal 1985–1986, patent income totaled $3.42 million. Patent prosecution and operating costs were $2.34 million, inventor shares were $948,000, and the state share was $260,000, leaving a deficit of $133,000. This difficulty will persist until inventor shares are reduced.

U.C. also has the most restrictive and inflexible policy relating to the rights of sponsors. Again, this may be because the university wants to avoid the charge that it is selling out to commercial interests and/or because it wants to promote the broadest dissemination of research results. By strongly favoring nonexclusive, royalty-bearing licenses, however, the university actually may be discouraging effective technology transfer, which must be based, in part, on providing sufficient incentive for outside firms to invest in product development. As indicated earlier, U.C. policies concerning both inventor share and rights of sponsors may be revised soon.

U.C. patent policies, when compared with those of M.I.T. and Stanford, do not indicate any strong commitment to technology transfer, but recent changes and the indication of more change to come provide evidence that technology transfer is climbing in the priorities of the university. This combination of lack of commitment with growing concern is also reflected in U.C. patent administration (see chapter 6).

PENN STATE

The other public university in the study, Penn State, indicates in its policies a greater commitment to technology transfer than U.C. seems to have, but these policies also show considerable confusion. Penn State, spurred by Pennsylvania's legislature and governor, has stated clearly a commitment to creating jobs in the state, to supporting economic development, and to promoting technology transfer. Its patent policies and administration have not yet caught up to this commitment, however. Penn State does have a reasonably flexible sponsors' rights policy, which is consistent with both its interest in technology transfer and its history of attracting industrial sponsorship for research. Still, its arrangement with Research Corporation for patent administration restricts inventor shares and means that relatively few inventions are patented and licensed, thus weakening the incentive for invention disclosure and research leading to inventions. The desire of a public university to avoid controversy related to patent administration is an important reason behind Penn State's agreement with Research Corporation. Research Corporation serves as a buffer between Penn State and the faculty and also between Penn State and the commercial world. However, this buffering also dampens Penn State's potential for technology transfer through the licensing of technology.

M.I.T.

M.I.T.'s patent policies, especially the recent changes in them, reflect a new aggressiveness in technology transfer. M.I.T. asserts ownership of inventions and backs this policy with active patent administration. Its policies relating to inventor share and royalty distribution to the inventor's department appear to be well balanced and to reflect accurately the relative contributions of the various parties to the process of an invention's development. The many options it offers to research sponsors are an indication of an underlying flexibility in this policy area. Perhaps as a result, M.I.T., like Penn State, has a high percentage of industrial research sponsorship.

M.I.T.'s sponsors' rights policy also shows an attempt to provide flexibility by covering a great many possible situations in the policy statement. This approach can be contrasted with Stanford's, which, instead of spelling out details in a policy statement, authorizes the director of the Office of Technology Licensing to make determinations in individual cases.

STANFORD

Stanford has less written patent policy than the other three universities. The cornerstone of Stanford's policy is that inventions are the property of university researchers. This basic stance symbolizes Stanford's willingness to stand aside from the development process. (At the same time, by providing an effective Office of Technology Licensing, Stanford has demonstrated that technology transfer is important to it.) Stanford's "Guide Memo 75: Inventions, Patents, and Licensing" is less a statement of policy than a source of useful information to University inventors. Most of what in other universities is a matter of written policy, at Stanford is left to the discretion of the director of the Office of Technology Licensing. This position allows considerable flexibility in making agreements and responding to new situations. In reality, the situation at Stanford may not be too different from that at M.I.T., where the director of technology licensing also enjoys considerable latitude in negotiating agreements.

The four universities in this study offer examples of different approaches to patent policy, ranging from the relatively restrictive written policies of U.C. to the more flexible policies of Stanford. Examination of these universities' policies illustrates several points. First, recent changes in patent policies indicate that universities are altering their attitudes toward intellectual property, becoming more aggressive and market-oriented. Second, universities are altering patent policies to encourage technology transfer and to relate these policies to the research mission of the university. For instance, at three of the four universities, policies related to royalty share distribution have recently changed or are expected to change to establish a more appropriate balance between the desire to reward inventors and campus departments and the need to provide adequate patent administration funding. Finally, policies concerning research sponsors' rights to inventions have been liberalized in order to encourage greater support from corporations and to speed technology transfer activities.

Policies usually offer official reactions to new situations. This chapter has viewed present patent policies as summaries of universities' official responses to some of the pressures of the new interest in the commercial value of their research. It has also viewed changes in patent policies, particularly recent or anticipated changes, as indications of trends in the universities' changing attitudes toward technology transfer. The number of changes and the direction of those changes are indications that universities are actively engaged in considering issues central to technology transfer. However, as has been pointed out repeatedly, university patent policy and patent administration are related and one must view them together in order to gain a complete understanding of the university's role in technology transfer. The next two chapters complete the description begun here, with an examination of patent administration, especially its recent history (since 1980), in the four universities being studied.

Notes

1. Pennsylvania State University, *University Patent Policy* (University Park: Pennsylvania State University, 1985).

2. K. W. Heathington, Betty S. Heathington, and Ann J. Roberson, "Commercializing Intellectual Properties at Major Research Universities: Income Distribution," *SRA Journal* 17 (1986): 34.
3. Ibid., 10.
4. University of California, "University Policy Regarding Patents," 1 April 1980.
5. August Manza. Interview with author. Berkeley, Calif., 6 January 1988.
6. Association of American Universities, *University Policies on Conflict of Interest and Delay of Publication* (Washington, D.C.: Association of American Universities, 1985).

Bibliography

ASSOCIATION OF AMERICAN UNIVERSITIES. *Trends in Technology Transfer at Universities*. Washington, D.C.: Association of American Universities, 1986.

———. *University Policies on Conflict of Interest and Delay in Publication*. Washington, D.C.: Association of American Universities, 1985.

COUNCIL ON GOVERNMENTAL RELATIONS. "Patents at Colleges and Universities: Guidelines for the Development of Policies." Washington, D.C.: Council on Governmental Relations, 1985.

CUNNINGHAM, R. G. "Examples of Conflicts of Interest and Publication Rights Problems in Sponsored Research at The Pennsylvania State University, A Report Prepared for the AAU Advisory Committee to Clearinghouse for Information on University-Industry Relations." 1984.

GARDNER, DAVID PIERPONT. "Annual Report Pertaining to University Patent and Other Intellectual Property Matters." Oakland: University of California, Office of the President, May 1989.

HEATHINGTON, K. W., BETTY S. HEATHINGTON, and ANN J. ROBERSON. "Commercializing Intellectual Properties at Major Research Universities: Income Distribution." *SRA Journal* 17 (1986): 27–38. (The *SRA Journal* is a publication of the Society of Research Administrators.)

LACHS, PHYLLIS S. "University Patent Policy." *Journal of College and University Law* 10 (1983–1984): 263–292.

MARCY, WILLARD. "Patent Policies at Educational and Nonprofit Scientific Institutions." In *Patent Policy*, edited by Willard Marcy. Washington, D.C.: American Chemical Society, 1978.

MASSACHUSETTS INSTITUTE OF TECHNOLOGY. *Guide to the Ownership, Distribution, and Commercial Development of M.I.T. Technology*. Cambridge: Massachusetts Institute of Technology, 1987.

PALMER, ARCHIE. *University Research and Patent Problems*. Washington, D.C.: National Research Council, 1949.

PENNSYLVANIA STATE UNIVERSITY. *University Patent Policy*. University Park: Pennsylvania State University, 1985.

STANFORD UNIVERSITY. *Copyrightable Materials and Other Intellectual Property: Guide Memo 76*. Stanford, Calif.: Stanford University, 1983.

———. *Inventions, Patents, and Licensing: Guide Memo 75*. Stanford, Calif.: Stanford University, 1980.

————. *Tangible Research Property: Guide Memo 77.* Stanford, Calif.: Stanford University, 1983.

UNIVERSITY OF CALIFORNIA. "Summary of Sponsor Patent Rights Applicable to Funding Agreements with Industrial (For Profit) Sponsors of Research." Office of the President, March 1984.

————. "University Policy Regarding Patents." 1 April 1980.

WILSON, O. J. *Patent and Copyright Policies in Forty-five Colleges and Universities.* Bowling Green: Office of Institutional Research, Western Kentucky University, June 1967.

6

Forms and Functions of Patent Administration

Early patent administration concentrated on the logistics of obtaining patents on inventions of high potential value. Today, what used to be known as patent administration has expanded to include many functions beyond the simple obtaining of patents, such as the administration of copyrights and other intellectual property rights. There also has been a major shift in attitude in recent years—a move toward a more market-oriented approach. This new emphasis is reflected in many universities (including M.I.T. and Stanford) by a change of the name of the patent office to the technology licensing office. However, in this chapter and the next, which concentrates on patents, the term *patent administration* is used in place of the more inclusive but rhetorically awkward *technology licensing administration*.

This chapter describes the organizational forms that patent administration takes in universities. It also examines the functions normally included in patent administration and considers how an institution's effectiveness in carrying out each of these functions may be evaluated.

Organizational Forms of Patent Administration

University patents are usually administered in one of three ways: through an outside patent management firm, through a university-affiliated research foundation or development corporation, or "in-house" through an office of technology licensing. There is some overlap in these three forms of organization; elements of the three may be combined, and each may vary in the
102

degree to which it becomes involved (beyond patenting and licensing) in the development and commercialization of inventions.

OUTSIDE PATENT MANAGEMENT FIRMS

The use of outside patent management firms was an early answer to the requirements of university patent administration. Since the rewards of patenting were so uncertain and universities had little expertise in patent administration, many institutions turned for help to organizations such as Research Corporation (now succeeded by a spin-off corporation to handle patents, called Research Corporation Technologies or RCT), University Patents, Inc., and Battelle Development Corporation. Stanford and M.I.T. at one time used such outside firms but then dropped them, although at Stanford the use of outside firms is an option that remains available to university inventors. A number of universities, including Penn State, continue to use outside firms. RCT, for instance, represents about 300 colleges, universities, and nonprofit organizations, although in 1984 it reported royalty income returns to only 40 such institutions.[1]

HOW THEY WORK. In a typical arrangement between a university and an outside management firm, disclosures of inventions created at the university are first routed through a small university patent administration office or through a university patent committee composed of scientists and administrators. The disclosures are then submitted to the firm for evaluation, and the firm decides whether to patent and market each invention at its own expense. In return for this evaluation process, for assuming the upfront costs of applying for patents, and for defending and maintaining patents, the firm retains a percentage of any royalty income received (typically 40 to 50 percent). It remits the remainder to the university, which usually keeps part of it and gives part to the inventor.

ADVANTAGES. Contracting with outside patent firms offers a number of advantages. The most obvious one is that these firms assume much of the risk involved in the patenting process, advancing funding for patent evaluation, applications, marketing, defense, and maintenance. In most universities, funding the administrative costs associated with patents is a significant problem, and use of outside patent firms avoids it. Another advantage is that outside firms, by pooling invention disclosures from many organizations, can assemble economically the "critical mass" of expertise necessary to make administration effective, especially in marketing, which requires people who understand both the underlying science involved in various inventions and the potential market for these inventions. This "critical mass" is very difficult to develop and maintain in any but the largest research university.

Outside organizations also can serve as important buffers between the university and those involved with patents. The patent process is inherently complex, with many factors besides scientific value influencing the economic and legal decision about whether to proceed with a patent application. Few university inventors understand these factors well, and egos may be bruised when inventions are turned down. It is often convenient for universities to let the outside firm "take the heat" from rejected inventors.

More important, outside firms can shield universities from certain kinds of liability, litigation, and related awkward situations. It is axiomatic that patent administration involves suing and being sued. When universities go into court on patent issues they are particularly vulnerable to criticism, since they do not appear to be acting in accordance with the public view that they should remain aloof from commercial concerns. For instance, one of the reasons advanced for Penn State's agreement with RCT is that it would be awkward for Penn State to sue a Pennsylvania corporation for patent infringement. Because many patent cases involve disputes over priority of discovery, the reputation of university faculty (and thus the university) is at issue in almost every court appearance. Although this problem cannot entirely be avoided by use of an outside firm, negative effects can be minimized if the university is not a primary party in a suit.

DISADVANTAGES. Despite these considerable advantages, there is some evidence that the use of outside firms by major research universities is on the decline and that these outside firms are having to modify their contractual arrangements to keep or attract universities into affiliation agreements.[2] This situation has developed because using outside firms also has some significant disadvantages. First, the university that contracts with such a firm relinquishes control over its inventions and assumes a passive role. Such a stance is counter to the growing notion that the university should take an active role in the technology transfer process.

Second, the fiscal conservatism of outside firms can be very frustrating to university inventors. Outside patent firms accept relatively few inventions for patenting. For instance, from its inception in 1912 through 1984, Research Corporation (now RCT) received about 15,000 invention disclosures, applied for and received patents on 1,500, and eventually issued licenses on only 300.[3] This 10 percent acceptance rate continued through 1987 and is expected to remain stable.[4] It can be compared with 34 percent for the combination of U.C., Stanford, and M.I.T. for fiscal 1986–1987. To most university researchers, low disclosure acceptance rates indicate a lack of responsiveness and perhaps a lack of understanding and competence, a perception which discourages invention disclosure and dampens enthusiasm for technology transfer.

Another disadvantage is the perception that outside firms take too high a percentage of royalties or that they "cherry pick"—select only inventions that seem likely to be "big hits" and let other good ideas wither. Although this same criticism is sometimes directed at in-house patent administration, the high acceptance rate by in-house offices is a clear indication that this is not a significant problem. In general, the relatively great distance (real and psychological) between the university researcher and an outside firm tends to intensify bitter feelings.

This matter of remoteness contributes to two further disadvantages. First, the geographical distance between universities and outside firms means that researchers have less contact with patent administration and therefore are less likely to make invention disclosures. Second, occasionally there may be also be philosophical differences between the firm and the university. This is well illustrated by M.I.T.'s disagreement with Research Corporation over the handling of the magnetic core memory patents when M.I.T. was sued for patent infringement by RCA and had to assume control of the litigation from Research Corporation (see chapter 4).

TRENDS IN OUTSIDE MANAGEMENT AGREEMENTS. To counteract some of these disadvantages, outside management firms have begun to alter their arrangements, a trend illustrated in a recent agreement between RCT and the University of Illinois. Illinois switched from University Patents, Inc. (UPI) to RCT partly because it expected RCT to be "more aggressive" than UPI.[5] To counter the charge that it was getting too large a share of patent royalties, RCT agreed to accept a low 40 percent of gross royalty income for its services. In addition, RCT agreed to pay the university $100 for each $1 million in research funding. To encourage invention disclosures, RCT promised to pay inventors, through the university, $5,000 for every disclosure accepted for patenting, even though this offer placed a disincentive on RCT to accept inventions for patenting. Finally, the university insisted on a nonexclusive arrangement whereby, at its option, it could patent and develop an invention on its own. In other words, the university was not obligated to send potentially lucrative inventions to RCT.[6]

This last provision indicates a trend in relationships between outside patent organizations and major research universities that also will be evident in the Penn State example covered later in this chapter. Seeking more control over the patenting and licensing process and anxious to garner a higher return from obvious (or seemingly obvious) winners, yet not willing to make the necessary investment for full-scale patent administration, or not possessing the volume of commercially viable inventions needed to make such administration affordable, some universities are moving toward a hybrid of the outside management firm and in-house patent administration. For instance, Stanford offers help to faculty inventors in securing the services of an outside patent administration organization. Another institution, the University of Illinois, has reserved for itself the right to patent and market inventions; therefore, it must have some kind of administrative apparatus to do preliminary evaluations and to go through the patent application and marketing process.

UNIVERSITY-RELATED FOUNDATIONS

Many universities, following the model established by the University of Wisconsin, have formed (or have had formed for them) separate foundations or corporate entities that assume some or all of the tasks of patent administration. This model has many variations. University-related foundations may be profit or nonprofit, entirely independent from or closely allied with the university, and confined narrowly to patent administration or heavily involved in the formation of new enterprises and economic development. Sometimes such foundations appear to university researchers to be similar to outside patent management firms, while in other cases they appear more like in-house patent administration offices or even venture capital firms.

In a typical arrangement with such a foundation, the university submits invention disclosures to the foundation and the foundation evaluates them. Inventions judged to be worthwhile are patented by the foundation in the name of the university, and the university in turn assigns the patent to the foundation. The foundation then undertakes the marketing of the patent, returning a share of any royalty income to the inventor and the university. It may also provide a fund for business start-up activity or technical assistance to businesses that use university technology.

EXAMPLES OF FOUNDATIONS.   In study conducted by the American Association of Universities in 1986, ten of the 39 universities surveyed had established nonprofit foundations and five had established for-profit entities.[7] The Wisconsin Alumni Research Foundation (WARF) is an example of an independent, nonprofit foundation that confines its activities to patent administration and passive investing. By policy, its board of directors does not include University of Wisconsin faculty, regents, or administrators. The income it receives from royalties is shared with inventors and the university, but the bulk of its present income derives from its endowment, which has been built up over the years from the royalties received from several highly successful inventions, including the Steenbock patents related to vitamin D production. WARF traditionally does not undertake business development activities, but it has gone beyond passive investing by occasionally taking equity in an inventor's company when the company is the licensee of the invention.[8]

Nonprofit foundations may take other forms. The Brown University Research Foundation is a nonprofit entity that is separate from the university, but its board of directors is composed entirely of university administrators. It goes beyond traditional patent administration in that it has helped to form new ventures and accepts equity in lieu of royalties. The Pittsburgh Foundation for Applied Science and Technology is a wholly owned, nonprofit subsidiary of the University of Pittsburgh that acts as a broker between the university, inventors, entrepreneurs, and venture capitalists to develop new businesses. The Washington Research Foundation is a similar organization designed to serve several universities in Washington state.[9]

Examples of for-profit external entities include the Michigan Research Foundation, the University Research Corporation (Colorado), and wholly owned subsidiaries of universities such as the Washington University Technology Associates and University Technology Incorporated (Case Western Reserve).[10] These for-profit entities are all heavily involved in technology transfer and business development activities. Patent administration is usually a secondary purpose for them.

ADVANTAGES. In establishing separate entities for patent administration, universities seek to gain some of the advantages of using outside patent management firms while at the same time retaining some control over the patenting process and a larger share of royalty and related income. Such entities can shield the universities from liability and can allow universities to engage in certain business activities without jeopardizing their tax-exempt status. Another significant advantage is that through a separate entity, universities can hire and compensate patent administrative staff, particularly technology license marketing staff, at appropriate levels. Universities, particularly public universities, rarely have the flexibility to offer the sort of bonus or incentive compensation plans that make powerful motivation instruments for technology marketing and development.

DISADVANTAGES. Nonetheless, there are disadvantages to having university-related foundations or other entities handle university patent administration. First, some control is inevitably relinquished to the external entity, although the degree of such surrender can vary widely depending on the legal form and pattern of governance of the entity. Second, although the university may be shielded from legal liability, it can rarely dissociate itself from the external organization in terms of public relations. This became clear in the WARF-oleomargarine antitrust actions and the University of Wisconsin product endorsement controversy cited in chapter 4.

A more significant disadvantage occurs when these separate entities are directed at business development, investment in new businesses, and management of endowment and venture capital portfolios. With these distractions, they tend to be deflected from the important and complex tasks of patent administration. As with universities using outside patent management firms, universities with related foundations or subsidiaries often have to establish some version of an in-house patent administration office to pre-screen invention disclosures or to handle inventions or discoveries that are not submitted to the external entity.

None of the four universities in this study uses an external foundation or entity for patent administration, although Penn State does use the separate, virtually dormant, Pennsylvania Research Corporation to act as a rather passive intermediary between the university and RCT. U.C., Stanford, and especially M.I.T., however, have experimented with external entities focused on specific technologies. The history of these experiments is discussed in the next chapter, which covers universities' equity participation in companies based on university technology.

IN-HOUSE PATENT ADMINISTRATION

Stanford, U.C., and M.I.T. all have full-scale in-house patent or technology licensing offices. These offices perform most of the tasks of patent administration, although usually the bulk of the detailed legal work of preparing and filing patent applications is placed in the hands of outside patent attorneys. Sometimes, as in the case of M.I.T., these offices engage in activities extending a bit beyond patent administration into business start-up and venture capital investments. The directors of in-house offices generally report to an administrative officer of the university (such as the vice president for research or the vice president for business affairs), but they may also take guidance from a patent committee.

PATENT COMMITTEES

Universities using any one or a combination of the administrative arrangements just described may also employ a patent committee as an element of patent administration. These committees can serve several functions, one of which is to act as scientific advisory panels to evaluate the commercial worth and patentability of university inventions. This function is important when no other such evaluation is provided or in situations similar to that of the University of Illinois, where the university wishes to identify potentially valuable inventions before they are turned over to an outside firm. Committees also may be used more directly in patent administration, perhaps exercising supervisory control over the director of an in-house patent office, hiring staff, or engaging outside patent counsel.

The most common use of patent committees is as policy-setting and oversight boards. In recent years, with so many changes taking place in patent policy, this has become an important function requiring considerable effort. Also, such committees often review cases of potential conflict of interest or university researchers' competing claims to ownership of patentable intellectual property. The patent committee is a convenient way to involve university faculty, researchers, and relevant administrators in the administration of intellectual property.

Functions of Patent Administration

Patent administration at major research universities is inherently complex. The patent application process is difficult in itself, often requiring both considerable knowledge of the science underlying an invention and extensive research into patents already granted. But the application process is only one element, and probably not the most important one in university patent administration today. Adding to this complexity is the fact that major research universities encompass a broad range of science and engineering fields, making it very difficult to focus expertise in a few areas. Furthermore, few university patent administration offices have a clearly defined mission. The "laundry list" preamble of most university patent policies (see chapter 5 for an example) gives little guidance to patent administrators, who are often pressured to maximize royalty income at the same time they are expected to render unremunerated service to the university. There are few agreed-upon standards for measuring the effectiveness of patent administration except perhaps for gross royalty income. Setting standards is difficult both because missions are unclear and because the elements of patent administration are so interrelated.

A review of the operations of patent offices indicates that they perform at least 11 separate but related functions:

1. Soliciting patentable ideas

2. Receiving and evaluating disclosures

3. Preparing and submitting patent applications and securing patents on inventions and discoveries

4. Maintaining patents

5. Marketing licenses

6. Receiving and distributing royalty income and administering the patent office budget

7. Auditing licensees

8. Negotiating rights with research sponsors

9. Administering, interpreting, and recommending changes in patent policy

10. Monitoring new developments

11. Keeping the university and university inventors out of trouble

In the remainder of this section, each of these functions is discussed and related to the others. There is also an evaluation of the effectiveness with which patent administration carries out each of these tasks.

SOLICITING PATENTABLE IDEAS

Most patent administrators consider it their responsibility to encourage university inventors to come forward with potentially patentable ideas. This is often an uphill battle because many university researchers are not interested in patents and do not want to spend the time involved in even the preliminary preparation of an invention disclosure. Effectiveness in soliciting ideas depends upon an internal marketing and education program that requires patent administration staff to invest a good deal of time-consuming effort to stay familiar with individual researchers and their work. Also affecting effectiveness are factors outside the control of the patent administration staff, including the general attitude of the university and its departments toward patent activity.

One measure of the effectiveness of this element of patent administration is the number of disclosures generated per year, compared, perhaps, to the number of researchers in the university, the total dollar volume of research, the number of disclosures in the previous year, or the number of disclosures per patent filing. But numbers are not everything. Soliciting the wrong kind of disclosures can be costly to the patent office and frustrating to the university researcher. Thus, the soliciting of new ideas must be joined with an education program that informs researchers about the nature of the patenting process and the kind and level of detail desired in invention disclosures.

RECEIVING AND EVALUATING DISCLOSURES

Evaluating invention disclosures for patentability and commercial value is possibly the most important and complex function performed by patent administration. Proper evaluation requires a thorough understanding of the science underlying an invention, of patent law, of similar patents that already exist, and of the potential commercial value of the invention. A decision to proceed with a patent application commits the institution to the costs not only of application and filing but also of patent maintenance and marketing.

Despite the study and effort that must go into the evaluation of invention disclosures, there is considerable pressure to perform such evaluations quickly. Delay in responding to invention disclosures was the most often cited reason for dissatisfaction with patent administration among the university researchers interviewed in this study. Publication deadlines also contribute to the pressure for rapid evaluations, since most researchers are eager to publish. But much of the value of an invention, particularly in foreign markets, is lost once the research that spawned the invention is published. Response time is thus a key indicator of patent administration competence.

Short response times create other problems, however. Often science proceeds in steps, with one breakthrough leading to others. Each step may involve findings worthy of publication. In such a case the patent administrator must decide at each step whether to protect the invention with a patent. Deciding to file for a patent risks not only rejection of the patent on the grounds that the invention is not mature enough but, more important, the opening of the technology to "improvement inventions" filed by others. Alternatively, deciding to wait for the next stage in development in the hope of securing a more defensible patent risks being "beaten to the punch" by competing patents from others in the same field.

Deciding not to patent an invention at all carries other costs. First, there are the opportunity costs and the possible loss of credibility if "a big one gets away." There are also the potential bad feelings and bruised egos of university inventors whose ideas are rejected. In universities where fewer than 10 percent of disclosures result in patent applications, patent administration is hard pressed to maintain its credibility. Administrators sometimes feel that it is easier to file a patent than to face the criticisms resulting from not filing. This attitude may spawn a tendency to mask an ineffective evaluation process with a large number of applications, since administrators figure that marginal inventions may not meet patentability requirements and thus may be rejected by patent counsel or the Patent Office itself. Of course, this can be an expensive strategy. A cheaper way to avoid or decrease criticism may be a speedy evaluation followed by a prompt release of the invention back to the inventor if the patent decision is negative.

Because quantity and quality are not synonymous, the number of patent applications filed is in itself a poor measure of the effectiveness of disclosure evaluation (although the ratio of applications filed to acceptances may give a part of the answer). The ultimate measure of evaluation effectiveness may be the number of licenses granted or the amount of royalty income received, but these measures also depend on other elements of patent administration, especially marketing.

SECURING PATENTS

Although effective administration calls for the patent office to be involved in the patent application process, much of the work of preparing and filing applications often is done by outside counsel. The patent administrator usually is involved in selecting and monitoring these outside attorneys. This function is important since good patent attorneys must also have a clear understanding of science and, indeed, often do so much work on an invention that they are almost coinventors. This is particularly true today in biotechnology, which has generated a new category of patent lawyers.[11]

Even more challenging than overseeing paperwork or selecting attorneys is the selling of an invention before it is patented. Here marketing and patenting intersect. If the patent administrative staff can find a potential licensee of a technology before a patent is obtained, licenses can be written to include the cost of patent application, thus freeing the university from having to make an initial investment for filing costs. Stanford, U.C., and M.I.T. all do a considerable amount of prepatent licensing.

Again, the number of patents obtained may not be a good measure of effectiveness in this category because it does not measure the value of the patents. The number of patents obtained without the use of university funds might be a better measure, although perhaps it measures the effectiveness more of marketing than of patenting. Another measure, the average cost of obtaining a patent, might also be appropriate if the patent administrative staff has some control over patent costs.

MAINTAINING PATENTS

Maintaining patents involves maintaining statutory filings, dealing with "interference" problems, suing infringers, and handling infringement suits brought by others against the university. Interference refers to the situation when two patents, owned by different individuals, are both necessary for the production of a product, or where two or more parties claim ownership of the same invention. The U.S. Patent Office uses an "interference proceeding" to determine the party, among two or more parties claiming the right to a patent, to whom it will award the patent. Normally a patent is issued to the first inventor unless that inventor failed to use "due diligence" in reducing the invention to practice or filing a patent application. Sometimes negotiations are necessary when the licensing of a valid patent is effectively blocked because of a competing patent or because the licensing of one technology depends upon the licensee's having a license to another technology owned by someone else. The patent administrator and/or outside counsel generally conduct such proceedings or negotiations.

Success breeds its own problems. The more successful a university's patent administration is, in terms of the number of patents filed and the value of patented inventions, the more likely it is that the university and university inventors will be involved in infringement suits, although the university may protect itself legally by placing the burden of such actions on licensees. The university may have a fiduciary duty to its licensees to protect university-owned patents vigorously, although it may pass the costs of such protection on to the licensees. Furthermore, as a holder of a patent, the university itself may be sued for infringement. As illustrated in the M.I.T.-RCA core memory conflict, described in chapter 4, the reputation of the university almost always suffers in these cases, regardless of the outcome. Again, universities usually employ outside counsel in these matters, but the patent administrator often plays an important role in prosecuting the case and in defending the interests of the university.

MARKETING LICENSES

From the technology transfer perspective of this study, the marketing of licenses to use university technology is of central importance in patent administration. In the four universities in this study, emphasis on this element of patent administration has increased significantly in the last ten years. The greater emphasis is the result of pressure on patent administration to produce more royalty income and on the related renewed interest in the technology transfer mission of the university. The marketing function is beginning to bring about the restructuring of the organization and staffs of university patent offices, and at U.C., Penn State, and perhaps M.I.T., that restructuring will continue for awhile.

Effective marketing of university technology requires a staff of people with a unique blend of talent and expertise. Above all, they must have the educational background or personal interest to become sufficiently grounded in the science underlying the types of inventions they will sell. In addition, they should be enterpreneurial, sales oriented, and able to develop and maintain networks of contacts with potential licensees. It is also desirable that they understand patent law and the patent application process and that they understand universities and be sensitive to the academic culture in which they work.

As one might expect, finding good people with these qualifications is not easy, and retaining their services is harder still. This is one of the basic problems faced by universities, especially since few of them are willing or able to compensate technology marketers competitively.

Another problem is related to the range of scientific fields represented by major research universities. Because effective marketing requires some knowledge of the science underlying an invention and because most universities produce inventions from a wide range of scientific and engineering disciplines, it is difficult to assemble and fund a staff large enough to cover a number of important fields or composed of individuals who are able quickly to understand almost any invention. Because of this dilemma, most universities must pick areas of specialization and hope that they can spot a "big one" in other areas or that they can farm out inventions in other areas to other technology licensing agencies. Also, scientists in areas not chosen for specialization are likely to form a negative impression of the patent administration process.

In the marketing area, conventional measures such as number of licenses issued and amount of royalty income are fairly reliable indicators of effectiveness. At Stanford and M.I.T., the marketing staff of the technology licensing office is organized into individual "responsibility centers," and annual royalty income per marketer ranges from $300,000 to over $1,000,000. To be sure, luck has something to do with such performances, and it is easier to make money in some subject areas than in others.

DISTRIBUTING ROYALTY INCOME AND ADMINISTERING THE PATENT OFFICE BUDGET

The patent office is usually charged with receiving and accounting for royalty income and, of course, administering and staying within its own budget. In large operations this is a complex task. When inventor shares are based on net royalty income, royalty accounting requires the maintenance of a cost accounting system to keep track of income and expenses associated with each invention. Inventor and, where applicable, research sponsor shares must be calculated and disbursed. To complicate matters, reserves for anticipated expenses often also must be calculated and maintained. Sometimes disputes between inventors must be adjudicated. Payments to outside counsel must be approved and made, and receivables from licensees must be maintained and collected. All this record keeping is costly and adds to the overhead required for an effective patent administration. However, failure of this function can have disastrous effects on the reputation of the patent office both inside and outside the university.

A number of conventional measures of budgetary effectiveness can be employed to judge this element of patent administration. Examples of such measures include adherence to budget standards, age of accounts receivable, turnaround time for accounts payable (including inventor shares), cost effectiveness of operations, and clarity and appropriateness of financial reports.

MAINTAINING RELATIONSHIPS WITH LICENSEES

Once a license has been issued, regular interaction with the licensee must be maintained. Periodic reports on sales or activity volumes must be collected and reviewed. Royalty income must be collected and must also be reviewed to determine whether it is being paid in accordance with the license agreement. Sometimes the books of the licensee must be audited. Receivables must be established and collected.

In addition to maintaining these financial relationships, the patent office may have to deal with other issues related to a license, including cooperating on infringement suits and providing technology updates to licensees. Occasionally, too, judgments of licensee effectiveness may need to be made. Most exclusive licenses provide for "march-in" rights by the university. If a licensee is not performing well, the march-in provision allows the university to void the license and regain control of the invention. This provision not only protects the financial position of the university but also prohibits the defensive use of patents—the withholding of an invention from the market in order to protect a market position based on other technology.

Perhaps most important, contact with licensees can lead to the sale of future licenses either to the current licensees or to other licensees related to them. Current licensees form an important part of the contact network that is so important in marketing.

There are few appropriate quantitative measurements of the effectiveness of this element of patent administration, but judgments can be made, particularly with regard to financial relationships and the completeness of files and reports from licensees. Maintenance of licensee relationships is an important follow-through element of patent administration.

NEGOTIATING WITH RESEARCH SPONSORS

Often research sponsors, especially those in industry, seek to acquire rights to inventions that might result from the research they are sponsoring. Patent administrators are frequently involved in the resulting negotiations, though in many cases the negotiation of research contracts, including rights to inventions clauses, is the responsibility of the sponsored projects (contract) office of the university. Although most universities have well-articulated guidelines governing the rights of sponsors, many sponsors seek exceptions to these policies. Sometimes, too, the circumstances of a contract do not fit neatly into the guidelines.

From a patent administration point of view, these negotiations freqently are disproportionately time consuming and unproductive. Few research contracts actually result in patentable inventions of value, so the whole exercise is most likely to be over an irrelevant point. Further, it is very difficult to construct sponsors' rights clauses in the absence of a clear notion of what the invention might be, how valuable it might be, and what fields of use it might cover. Finally, negotiations often involve sponsors who are unfamiliar with universities and thus need lengthy education in university requirements and practices.

For these reasons it is often preferable to agree to general sponsors' rights provisions and then work out specific licensing agreements after an invention is made. Most universities will grant research sponsors preferential treatment for licenses even in the absence of specific provisions. This sorting out of rights after creation of an invention also is usually left to the patent administrator.

Again, the effectiveness of this element is difficult to measure. Because in most instances sponsor negotiation is an "overhead" function not directly related to income generation or patent administration, keeping track of the amount of money spent on this function is useful. Failing to do so may make other cost- effectiveness measures inaccurate or misleading.

ADMINISTERING, INTERPRETING, AND RECOMMENDING CHANGES IN POLICY

The patent office is the place where patent policy failures, inadequacies, and difficulties will first be apparent. The patent administrator must make constant interpretations of policy and usually serves as a valuable resource to those who have questions about policy. Often the patent administrator deals not only with patent policy but also with policies governing other forms of intellectual property, such as copyrights, tangible research property, biological materials, and even trademarks.

Latitude in interpreting policy varies considerably from campus to campus. At Stanford, where policies are stated in very general terms, the patent administrator has considerable authority to make judgments that in other universities would be equivalent to making policy. At M.I.T., where a considerable body of policy exists and policy changes are frequent, the patent administrator frequently is involved in recommending amendments to policy, although in practice he operates with about the same administrative latitude that the Stanford administrator enjoys. At U.C., where the patent administrator serves a system of nine university campuses and three national laboratories, patent administration involves coordination of policy among the campuses and a much greater emphasis on education about policies.

Policy interpretation is another overhead function. Unlike negotiating with sponsors, however, this function is fully integrated as a natural part of the job of patent administration.

MONITORING NEW DEVELOPMENTS

In recent years the pace of change in the management of intellectual property has been rapid. New developments come from two directions—science and the regulatory environment surrounding intellectual property. Examples of scientific developments that have had significant impact are the advances in biotechnology and in computer software that have required new ways of marketing and pricing licenses. Regulatory changes include the 1980 patent law changes and the Harvard mouse and Chakrabarty patents.

Because universities are on the cutting edge of research and because of the broad scope of science within universities, they must make active responses to a wide range of new developments. The patent administrator usually must take the lead in making such responses. When this is done effectively, patent administration gains credibility. Failures in this area cost the university lost royalty income or place the university at legal risk.

KEEPING OUT OF TROUBLE

Keeping the university and university researchers out of trouble is less a requirement than a goal of patent administration. Although it is impossible to avoid all litigation surrounding patents, it is possible to minimize such litigation and its negative effects through careful patent administration practices.

Legal proceedings are not the only form that trouble can take, however. Because university involvement in technology transfer and economic development is coming under greater scrutiny, failures in other areas can draw criticism. Effective patent administration should protect university researchers and their intellectual property from criticism. Claims of prior discovery need to be taken seriously and sometimes should be pursued regardless of the economics of the case. Faculty researchers also need to be protected from charges of conflict of interest and often need advice on the appropriateness of proposed actions.

Conclusion

The relative importance of the various functions of patent administration is different from one university to another, but almost every university patent administration includes some aspect of each of these elements. The next chapter shows how the four selected universities have organized their patent administration and how they differ in their approaches to its various functions.

Notes

1. Research Corporation, *Research Corporation Report, 1984* (Tucson, Ariz.: Research Corporation, 1984), 11, 18.

2. Association of American Universities, *Trends in Technology Transfer at Universities* (Washington, D.C.: Association of American Universities, 1986), 18.

3. John Eckhouse, "Colleges Turn Research into Cash," *San Francisco Chronicle*, 29 May 1984, A17.

4. W. Stevenson Bacon, letter to author, 22 September 1987.

5. Philip Bloomer, "UI Faculty Get Details of New Patent Agent," *News-Gazette*, 26 October 1988, 1.

6. Ibid.

7. Association of American Universities, *Trends in Technology Transfer*, 27.

8. Ibid., 26–29.

9. Ibid., 29–33.

10. Ibid., 33–39.

11. Stephen Goode, "New Legal Species Born of Biotech," *Insight*, 19 September 1988, 54–55.

Bibliography

ASSOCIATION OF AMERICAN UNIVERSITIES. *Trends in Technology Transfer at Universities*. Washington, D.C.: Association of American Universities, 1986.

BLOOMER, PHILIP J. "UI Faculty Get Details of New Patent Agent." *News-Gazette*, 26 October 1988, 1.

ECKHOUSE, JOHN. "Colleges Turn Research into Cash." *San Francisco Chronicle*, 29 May 1984, A17.

GOODE, STEPHEN. "New Legal Species Born of Biotech." *Insight*, 19 September 1988, 54–55.

RESEARCH CORPORATION. *Research Corporation Report, 1984*. Tucson, Ariz.: Research Corporation, 1984.

7

Patent Administration at Four Universities

The descriptions in this chapter cover the period since 1980 and continue through current activities, organization, and issues related to patent administration at each of the four universities in this study. The chapter concludes by attempting to determine the meaning and impact of the changes that are taking place or are likely to take place soon at these universities. It is these changes that are most important to the story of the impact of technology transfer on universities. At U.C. and Penn State, changes are occurring so rapidly that the present discussion is likely to become outdated quickly. Nonetheless, certain conditions underlying patent administrative changes are common to all four universities. An explication of these conditions is the goal of this examination.

Comparison of Patent Administration at Four Universities

A brief sketch and comparison of patent administration at the four universities is useful to set the stage for this discussion. The practice at Stanford is closest to a model for effective patent administration in this new era of technology transfer. It has undergone few major organizational or policy changes since the early 1970s. Stanford's Office of Technology Licensing is well regarded internally and has developed a national reputation as well. In contrast, in 1985 M.I.T. undertook a major change in patent administration, making the transition from a relatively inactive, legally oriented patent office to an extremely aggressive, technology transfer oriented Technology Licensing Office (TLO). The TLO has quickly gained the confidence of

M.I.T. researchers and the business community. It is moving so rapidly that policy development and procedures are hard pressed to keep up.

Patent administration at U.C. is in a state of restructuring after a period of considerable turmoil and controversy. The difficulties of carrying out and coordinating the patent activities of nine campuses and three laboratories overwhelmed the Patent, Trademark, and Copyright Office (PTCO) which came under severe criticism both from the campuses and from the press. PTCO was placed under new leadership in April 1989, its staff was increased, and its functions were reorganized. These efforts appear to be correcting some of the difficulties, but many remain. One feature of this reorganization is the delegation of authority for the administration of patents to those campuses that wish to establish offices for that purpose. Other elements of policy, most notably inventor share provisions and the ownership of equity in start-up companies, are also under review.

At Penn State, where the rate of disclosure and the amount of patent activity are relatively low, patent administration appears not to have kept pace with the university's recent increase in emphasis on technology transfer and service to Pennsylvania's economic development. University inventors are dissatisfied with patent administration and with the services provided by RCT. This dissatisfaction centers on both the low rate of invention acceptance and the low rate of return to inventors and inventors' departments.

Levels of activity in certain areas are commonly used as measures of overall patent administrative effectiveness. Table 7-1 shows activity levels in a number of such categories for the four selected universities for the fiscal year 1986–1987. This table provides an overall basis for comparing the four institutions. Comparisons between the institutions on specific measures are difficult, however, and are likely to lead to erroneous conclusions because universities use different methods of accounting and because of timing differences. In the remainder of this section, each category listed in Table 7-1 is discussed and its validity as a measure of overall administrative effectiveness (alone or in combination with other measures) is considered.

DOLLAR VOLUME OF RESEARCH FUNDING

All four of the selected universities are among the top 20 recipients of research funding, with Stanford ranking second in total research funding dollars ($278 million in 1988), M.I.T. fifth ($271 million, not including affiliated laboratories), Penn State fifteenth ($188 million), and U.C. Berkely sixteenth ($186 million). The campuses of U.C. are usually taken separately for ranking purposes, but if the campuses are combined, the U.C. system is far and away the largest research institution in the country in terms of funding received (over $1.1 billion in 1988).

Annual research funding is often used as an "input" measure in evaluating the effectiveness of an institution in producing inventions. It is usually combined with an "output" measure, such as number of patents issued or

TABLE 7-1   Patent Administrative Activity at Four Research Universities, 1986–1987

	STANFORD	M.I.T.	U.C.	PENN STATE
Annual research funding (in millions of dollars)	$299[a]	$616[b]	$832[f]	$185[d]
No. of invention disclosures	172[a]	205[b]	307[g]	80[d]
No. of patent applications filed	40[a]	80[b]	110[g]	9[d]
No. of patents issued	44[a]	69[b]	69[g]	5[d]
No. of patents held	309[a]	1,000[b]	612[g]	66[d]
No. of licenses issued	21[a]	65[b]	68[g]	NA
Royalty income (in thousands of dollars)	$6,055[a]	$3,100[c]	$5,393[g]	$202[e]
No. of inventions producing income	116[a]	200[b]	161[h]	NA
No. of inventions producing x percent income	7[a]	NA	12[h]	NA
	75[a]	NA	66[h]	NA
No. of patent employees	14[a]	19	25	1
Patent administrative costs (in thousands of $)	$805[a]	$600[c]	$1,081[h]	$100 (est.)

[a]Sally Hines (Office of Technology Licensing, Stanford University), correspondence with author, August 1988.

[b]Massachusetts Institute of Technology, Report to the President, 1986–1987 (M.I.T. Archives).

[c]John Preston, interview with author, Cambridge, Mass., 7 July 1988.

[d]Pennsylvania State University, Invention Disclosure and Patenting Activity, 1986 Report to the Pennsylvania Research Corporation (University Park: Pennsylvania Research Corporation, 1987).

[e]Mark Righter, correspondence with author, 23 May 1988.

[f]Peat, Marwick, "University of California Patent, Trademark and Copyright Office, Review of Operations, Final Report, February, 1988" (Berkeley: University of California, 1988).

[g]Roger Ditzel, "Analysis of U.C. Patent Operations, 1986–87" (Berkeley: Office of the President, University of California, 1987).

[h]August Manza, "Patent Operations at Berkeley: An Analysis of Six Possible Strategies" (Berkeley: University of California, November, 1987).

royalty income, to arrive at an index that can be used to compare universities. For instance, using the data from Table 7-1, Stanford's "return on annual research funding" (to coin a measure) might be computed at 0.02025 ($6.055/299) and compared with the same measure for M.I.T., 0.00503 (3.1/616).

There are several problems with this kind of comparison, however. First, such measures do not take into account the lag between funding and invention output. It takes at least three years, and sometimes more than ten, for an invention to produce income. Where the amount of research funding is changing, such as at Penn State (which has enjoyed significant increases in funding in recent years), this kind of measure can be misleading.

Second, universities add up their research funding in different ways. They may include or exclude foundation grants, count funding awards rather than actual expenditures, and include or exclude affiliated research laboratories. Sometimes it is difficult to find two internal reports that list the same figure for a given year.

Third, and most important, total research dollars do not reflect the "marketable research base;" that is, the research funding that is most likely to produce inventions. For instance, M.I.T., with its concentration on engineering and the applied sciences, might be expected to have a higher marketable research base than Stanford. There are good arguments in favor of excluding classified research from the marketable base. The figures for Stanford and Penn State include classified research, unlike Berkeley and M.I.T. where classified research is performed in affiliated laboratories and is not included in the university totals. Also, Stanford and Penn State both have medical schools (and the U.C. total figures include funding for five medical schools). Universities in which high percentages of research funding go to the social sciences and humanities probably will produce fewer patentable inventions per research dollar than those in which most funding goes to the applied sciences.

Despite these problems, total research funding does show a rough correlation to patent output measures. For the four universities, the annual research expenditures measure provides a general description of relative size.

NUMBER OF INVENTION DISCLOSURES

The annual number of invention disclosures is both an input and an output measure. It is an output measure in that it is an indication of the effectiveness of the patent office and of the institutional culture in soliciting ideas. Researchers are less likely to come forward with their inventions at those universities in which patent administration is viewed as ineffective or where the institutional culture does not reward innovation and production of ideas with commercial potential.

Table 7-2 shows the number of invention disclosures at the four universities for the fiscal years 1979–1980 to 1986–1987. As the figures in this table show, except for a peak in 1982–1983, the height of excitement over advances in biotechnology, there was a steady increase in invention disclosures. This is one of the clearest indications that university researchers are becoming more aware of the potential commercial value of their research.

Invention disclosures are also inputs to the patent administration process, the basic material upon which much of the activity of the patent office is founded. The quality of these inputs is important to the patent office, since frivolous ideas or ideas that are too immature burden patent adminis-

TABLE 7-2 Number of Invention Disclosures at Four Research Universities, 1979–1988

	1987–1988	1986–1987	1985–1986	1984–1985	1983–1984	1982–1983	1981–1982	1980–1981	1979–1980
Stanford[a]	168	172	148	133	124	196	142	140	142
M.I.T.[b]	278	205	143	123	119	150	164	156	142
U.C.[c]	309	307	291	270	261	297	292	234	235
Penn State[d]	53	80	54	64	48	48	48	48	48
Total	808	764	636	590	552	691	646	578	567

Sources for the period from 1979–1980 to 1986–1987:

[a]Sally Hines (Office of Technology Licensing, Stanford University), correspondence with author, August 1988.

[b]Massachusetts Institute of Technology, Report to the President, 1986–1987 (M.I.T. Archives).

[c]Roger G. Ditzel, "Analysis of U.C. Patent Operations, 1986–87" (Berkeley: Office of the President, University of California, 1987).

[d]For 1986 and 1987, conversation with Richard Hansen. For 1979 to 1985, Robert F. Custard, "Report of Invention and Patent Activity for the Calendar Year 1985." Note that information for the years 1979 through 1984 presents an average of the four years.

Sources for 1987–1988:

[a]Sally Hines, letter to author, 13 September 1989.

[b]Lita Nelson (Technology Licensing Office, Massachusetts Institute of Technology), telephone conversation with author, June 1990.

[c]David Pierpont Gardner, "Annual Report Pertaining to University Patent and Other Intellectual Property Matters," May 1989.

[d]William Moir, (Pennsylvania State University), telephone conversation with author, June 1990.

tration with unproductive work and injure its credibility with university researchers. Thus the number of disclosures taken alone might be misleading; the quality in terms of degree of patentability or potential for commercialization is also an important dimension.

NUMBER OF PATENT APPLICATIONS FILED

Filing for a patent is relatively easy for the university patent administrator, especially since outside attorneys frequently handle such filings. Taken by itself, then, the number of filings is not a particularly good measure of effectiveness. A large number of filings can mask an ineffective evaluation process. On the other hand, where technology transfer is truly a goal of patent administration, it can be argued that a large number of filings is an appropriate measure of effectiveness since filing for a patent is a good way to disseminate an idea. Unless one accepts the technology transfer argument, the number of patent filings, taken by itself, is a poor measure of patent administration effectiveness.

Table 7-3 shows the number of patent filings for the four universities in this study from 1979–1980 to 1987–1988. Again, the table shows an increase in activity, although the increase is less even than in some of the other measures. To be most useful, the number of patent applications filed should be compared with the number of patents issued (to determine whether filings result in patents) or with the number of licenses granted (to determine how often patent applications result in something of commercial value).

NUMBER OF PATENTS ISSUED AND NUMBER OF PATENTS HELD

Patents issued can be viewed as additions to the "inventory" of the patent office. The size in numbers of the inventory may not be a good indicator of its value, however. M.I.T., for instance, has been awarded over 800 patents since 1969,[1] but they are earning $3.1 million annually, while U.C. earns about $5.4 million on 612 patents. Of course, neither university may be realizing the full potential value of its patent portfolio. In fact, most patents are not valuable. In a study conducted by the National Science Foundation in 1985, out of a sample of 248 patents obtained by universities and other research institutions, only 7 percent were found to have any value.[2]

Table 7-4 shows an increase in the number of patents issued to the four universities in this study between 1979 and 1988. Universities were issued about 800 patents in 1988, down from almost 900 in 1987.[3] In this area, too, there has been a clear increase in activity. Partly because of the passage of PL 96–517 (the Uniform Patent Act), the total annual number of U.S. patents issued to universities has increased from 379 in 1980 to 646 in 1986.[4] (Note that the four universities of this study accounted for about 30 percent of this total.)

TABLE 7-3 Number of Patent Applications Filed at Four Research Universities, 1979–1988

	1987–1988	1986–1987	1985–1986	1984–1985	1983–1984	1982–1983	1981–1982	1980–1981	1979–1980
Stanford[a]	20	44	28	38	35	14	57	45	28
M.I.T.[b]	NA	80	72	98	62	46	96	90	88
U.C.[c]	140	110	107	112	115	92	67	95	76
Penn State[d]	5	9	9	11	6	6	6	6	6
Total	NA	243	216	259	218	158	226	236	198

Sources for the period 1979–1986:

[a]Sally Kines (Office of Technology Licensing, Stanford University), correspondence with author, August 1988.

[b]Massachusetts Institute of Technology, Report to the President, 1986–1987 (M.I.T. Archives).

[c]Roger C. Ditzel, "Analysis of U.C. Patent Operations, 1986–87" (Berkeley: Office of the President, University of California, 1987).

[d]For 1986 and 1987, conversation with Richard Hansen. For 1979 to 1985, Robert F. Custard, "Report of Invention and Patent Activity for the Calendar Year 1985." Note that information for the years 1979 through 1984 presents an average of the four years.

Sources for 1987–1988:

[a]Sally Hines, letter to author, 13 September 1989.

[c]David Pierpont Gardner, "Annual Report Pertaining to University Patent and Other Intellectual Matters," May 1989.

[d]William Moir (Pennsylvania State University), telephone conversation with author, June 1990.

TABLE 7-4 Number of U.S. Patents Issued at Four Research Universities, 1979–1988

	1987–1988	1986–1987	1985–1986	1984–1985	1983–1984	1982–1983	1981–1980	1980–1981	1979–1980
Stanford[a]	46	44	28	38	35	14	7	12	13
M.I.T.[b]	66	69	NA	51	50	50	61	50	47
U.C.[c]	64	69	51	74	50	39	45	41	27
Penn State[d]	7	5	3	2	4	4	4	4	4
Total	183	187	NA	165	139	107	117	107	91

Sources for the period from 1979–1980 to 1986–1987:

[a]Sally Hines (Office of Technology Licensing, Stanford University), correcspondence with author, August 1988.

[b]Massachusetts Institute of Technology, Report to the President, 1986–1987 (M.I.T. Archives).

[c]Roger G. Ditzel, "Analysis of U.C. Patent Operations, 1986–87" (Berkeley: Office of the President, University of California, 1987).

[d]For 1986 and 1987, Robert F. Custard, "Report of Invention and Patent Activity for the Calendar Year 1985." Note that information for the years 1979 through 1984 presents an average of the four years.

Sources for 1987–1988:

[a]Sally Hines, letter to author, 13 September 1989.

[b]"Give and Take," *Chronicle of Higher Education*, 1 March 1989, A23.

[c]David Pierpont Gardner, "Annual Report Pertaining to University Patent and Other Intellectual Property Matters," May 1989.

[d]William Moir (Pennsylvania State University), telephone conversation with author, June 1990.

NUMBER OF LICENSES ISSUED

The number of licenses issued is a very good measure of the effectiveness of the marketing function of patent administration. However, it, too, can be deceptive. One invention may produce hundreds of licenses. For this reason, and because statistics are kept in different ways by the selected universities, it is difficult to draw meaningful conclusions from a license figure. Stanford, for instance, sometimes combines patents into related groups of technologies and issues licenses on the resulting technological "packages," counting each license only once, whereas some other universities do the same thing without the package concept. U.C. reported 21 licenses granted in 1977–1978, 100 in 1979–1980, and an average of 60 per year from 1982 to 1987.[5] Stanford averaged 40 licenses per year from 1982–1983 to 1986–1987, with a range of 21 to 57.[6]

M.I.T. shows the clearest pattern. This institution undertook a major reorganization of its patent administration in 1985. It increased the number of licenses issued from 12 in 1983–1984, 15 in 1984–1985, and 17 in 1985–1986 to 65 in 1986–1987 and 100 in 1987–1988. The new director of the Technology Licensing Office capitalized on a backlog of licensable inventions that previously had not been marketed effectively.[7]

If one were to accept technology transfer as the ultimate goal of patent administration, the number of licenses issued would be the most significant measure of patent administration effectiveness. It is a commentary on the importance given to technology transfer in universities, however, that the number of licenses issued is generally considered only in conjunction with the amount of income from royalties.

ROYALTY INCOME AND NUMBER OF INVENTIONS PRODUCING INCOME

The gross amount of royalty income is by far the best understood measure of patent adminstration effectiveness, and it often is considered the most important. By this measure, most university patent offices have improved markedly in the last five years. Table 7-5 gives a recent history of royalty income for the four universities in this study.

The combined royalty income of Stanford, M.I.T., and U.C. more than doubled from $5.932 million in 1982–1983 to $14.548 million in 1986–1987. At other universities the increase has been even more startling. The University of Washington, which established its Office of Technology Transfer in 1983, increased its royalty income from $35,000 in 1984 to $397,000 in 1987.[8]

Like other measures, however, royalty income can be deceptive, and its deceptiveness is all the more dangerous because of its widespread acceptance. First, as a way of comparing universities, it suffers from the same problems that attend gross research funding figures: universities include different elements in their definition of royalty income. For instance, at M.I.T.

TABLE 7-5 Royalty Income at Four Research Universities, 1979–1988 (in Thousands)

	1987–1988	1986–1987	1985–1986	1984–1985	1983–1984	1982–1983	1981–1982	1980–1981	1979–1980
Stanford[a]	$9,179	$6,055	$5,130	$3,935	$3,124	$2,055	NA	NA	NA
M.I.T.[b]	2,972	3,100	2,300	1,869	1,894	1,641	$1,815	$1,607	$1,441
U.C.[c]	6,822	5,393	3,407	3,074	2,689	2,236	1,731	1,253	737
Penn State[d]	329	202	457	NA	NA	NA	NA	NA	NA
Total	$19,302	$14,750	$11,294	NA	NA	NA	NA	NA	NA

Sources for the Period from 1979–1980 to 1986–1987:

[a]Sally Hines (Office of Technology Licensing, Stanford University), correspondence with author, August 1988.

[b]Massachusetts Institute of Technology, Report to the President, 1986–1987 (M.I.T. Archives).

[c]Roger G. Ditzel, "Analysis of U.C. Patent Operations, 1986–87" (Berkeley: Office of the President, University of California, 1987).

[d]For 1986 and 1987, conversation with Richard Hansen. For 1979 to 1985, Robert F. Custard, "Report of Invention and Patent Activity for the Calendar Year 1985." Note that information for the years 1979 through 1984 represent an average of the four years.

Sources for 1987–1988:

[a]Sally Hines, letter to author, 13 September 1989.

[b]Lita Nelson (Technology Licensing Office, Massachusetts Institute of Technology), telephone conversation with author, June 1990.

[c]David Pierpont Gardner, "Annual Report Pertaining to University Patent and Other Intellectual Property Matters," May 1989.

[d]William Moir (Pennsylvania State University), telephone conversation with author, June 1990.

and Stanford, royalty income includes income from copyrighted software, whereas at U.C., copyright royalties are dealt with at the campus level and are not included in the systemwide figures. U.C., however, includes patent cost reimbursement as income, whereas the other universities do not.

Second, an emphasis on gross income ignores the expenses associated directly with the production of the income, and these expenses can be considerable. For instance, again at M.I.T., gross dollar royalty income for 1987–1988 was about $2.9 million, down a bit from 1986–87. Part of the explanation for this decline was that an agreement had been reached with a company to handle fulfillment services and updating of a computer software package. This package had generated about $500,000 in revenue per year but had associated expenses of about $300,000. Assigning the package to a spin-off company decreased income but preserved the net return.[9]

Third, a strict use of royalty income as an evaluative measure ignores other important economic benefits that are derived from patent administration. For instance, licenses for university inventions frequently are traded for or associated with conventional research funding. That is, a sponsor may agree to fund research provided that it receives a license on inventions produced by that research, or the sponsor may trade current research funding for a license on a technology that already exists. At M.I.T., the only university in this study that keeps track of such funding, license-related research funding was approximately $1.4 million in 1987–1988[10] and ranged from $193,000 to $1,216,000 between 1980 and 1985.[11]

Equity taken in lieu of royalty income is another patent-related economic benefit. In 1987–1988, M.I.T. accepted stock in eight companies in lieu of or in addition to royalties on licenses.[12] A very speculative estimate of the value of this stock as of July 1988 is $3.5 million, more than the cash royalty income for the year. If even one of these eight firms is successful, the return to M.I.T. could be very high indeed.

There is an even more fundamental problem with using royalty income as an evaluative measure, however. The amount of royalty income generated in a university may be more a function of luck than of the effectiveness of patent administration. An illustration of this point also shows why an emphasis on royalty income, particularly as a source of funding for patent administration costs and for further research, carries other dangers.

In almost every instance, a very few inventions account for most of a university's royalty income. In WARF's first 50 years, that organization evaluated 1,702 invention disclosures, filed for 415 patents, obtained 270 of these, and licensed 62 inventions to 650 licensees (400 of which were associated with the Steenbock patents). Of these 62 licensed inventions, only 43 ever produced any income. Of these, 14 produced a total of between $10,000 and $100,000, nine produced between $100,000 and $1,000,000 and four produced over $1,000,000.[13] Thus, only one in 40 disclosures ever resulted in royalty income.

Table 7-6, which summarizes a 22–year history of royalty income and

TABLE 7-6   Income from Inventions at M.I.T., 1950–1972

ROYALTY RANGE (IN 000s)	NUMBER OF INVENTIONS	GROSS INCOME 1950–1972	INVENTORS' SHARE
$0–25	14	$ 97,000	$ 11,640
$25–50	4	140,000	16,800
$50–100	3	250,000	30,000
$100–4,000	7	7,615,000	913,800
Over $4,000	1	17,000,000	2,040,000
Total	29	$25,102,000	$3,012,240

Source: Massachusetts Institute of Technology, "Committee on Inventions and Copyrights, Proposed Change in Patent Policy for Consideration by the Executive Committee, June 9, 1972" (M.I.T. Archives), Office of the President and Chancellor 1971–1980 (Weisner/Gray) (81-27), Box 30.

inventions for M.I.T., makes the same point. It shows that most of the royalty income generated during those 22 years came from one invention (the Forrester core memory); only 29 inventions produced any income at all.

Table 7-1 indicated that this trend continues to the present. At U.C., 12 inventions accounted for 66 percent of the royalty income for 1986–1987. Even more striking, seven inventions accounted for 75 percent of Stanford's royalty income for that year, and two of these together brought in over $3 million, or almost half of the total income.[14] To quote Niels Reimers of Stanford: "With few exceptions, a ULO (University Licensing Office) is economically viable only if one or more 'big hit' inventions has come along."[15] Continued financial viability for the patent administration operation that must be funded entirely out of royalty income is quite tenuous, since "big hit" inventions can dry up in a short period of time, especially in this age of rapid technological change.

In spite of its drawbacks, especially in making comparisons between universities, royalty income is likely to remain the most commonly used measure of patent administration effectiveness, as long as such income must fund patent administration and is expected to fund other important university missions as well. Nevertheless, the difficulties outlined here support the argument that will be made in the conclusion of this chapter, namely, that patent administration should be placed in a broader context that incorporates other elements of technology transfer and interaction with industry.

ADMINISTRATIVE MEASURES

The remaining elements in Table 7-1, number of patent office employees and office operating costs, might be termed *administrative measures,* since they deal with traditional terms of economic comparison and accountability. Penn State, which contracts with RCT for most of its patent administration, obviously cannot be compared with the other three institutions on these measures. It is difficult enough to compare the others because the figures arise from such different institutional settings, although Stanford and M.I.T. are the most similar. Each institution also defines administrative costs differently. Furthermore, at M.I.T. and even more at U.C., the office staffs have been growing in recent years. The patent office staff at U.C. jumped from 20 in 1985–1986 to 25 in 1986–1987 and by early 1990 had climbed to 42. New staff members require time to become productive, so comparing current costs with current income, for instance, is invalid.

CONCLUSIONS

In this section the patent administration activities of the four universities were described and compared, and in the process a number of conventional measures of patent administration effectiveness were criticized. This discussion shows clearly that valid comparisons between universities using specific measures in isolation are exceedingly difficult. However, when these measures are combined with an understanding of institutional contexts and the perceptions of the university communities, judgments can be made.

The underlying problem with most of these measures is that the life cycle of an invention extends over many years, so year-by-year comparisons are bound to present a distorted picture of patent operations. It is virtually impossible to match income with related expenses in the same accounting period. The costs of patenting and of litigation are treated as current costs, even though they relate to future earning potential. Patents issued in a given year are usually associated with earlier patent applications. Royalty income is usually associated with patents issued at least five years previously. For instance, at U.C., about 96 percent of utility (as opposed to plant) patent income received in 1986 was derived from patents issued prior to 1981.[16] This discontinuity between the manner in which patent offices are evaluated (usually on a comparison of one fiscal year to the next) and their actual operating cycle will continue to lead to erroneous judgments and poor management decisions. Effective management of patent administration requires a more sophisticated accounting and management information system than presently exists at any university.

In spite of these caveats, this analysis makes it clear that the pace and volume of activities related to patent administration has increased markedly in the last eight years in at least three of the four universities in this study. This growth is reflected in other universities as well. Membership in the Society of University Patent Administrators has tripled in the last five years and by early 1989 included 400 members from more than 125 universities.[17]

The organization of patent adminstration offices in the four universities is described next. This description includes organizational and operational changes that have taken place at the universities.

Recent Changes and Present Patent Administration at Four Universities

Operating statistics such as those just examined are not sufficient to describe patent administration and its relationship to technology transfer. This section goes behind the numbers to see what is really occurring at the four universities: how patent administration is organized at each campus, what its operating context is, and what changes it has recently undergone.

STANFORD

Stanford is presented first because in many ways it provides a model for university patent administration. Since World War II or even earlier, Stanford has enjoyed extraordinarily close relations with industry. The Office of Technology Licensing (OTL) is a vital element in this pattern of relationships, although the initiating force for that office was its present director, Niels Reimers. From its beginning, the OTL was dedicated to technology transfer and the marketing of licenses rather than focusing on legally oriented and defensive issues. The OTL was able to capitalize on industrial contacts already in place, but it also served as a catalyst to further and support the technology transfer-oriented atmosphere of the university. In fairness, one should point out that, unlike U.C. and Penn State, Stanford is a private, single-campus university. This fact makes the operations of the OTL easier, but they are no less praiseworthy.

Perhaps the most distinctive feature of the context for the operations of the OTL is the Stanford policy of granting inventors the rights to inventions, instead of requiring that such rights be assigned to the university. This practice allows the OTL to assume a service rather than a policing role. This service role includes helping university inventors obtain rights from government sponsors and also evaluating and securing help in marketing inventions through agencies outside Stanford, such as professional patent management firms. The OTL's service posture also means that it must satisfy its inventor constituency in order to stay in business.

RECENT CHANGES. There have been no startling changes in the operations of the OTL, at least compared to the changes seen at the other universities in this study. One administrative change that did occur, however, was the assumption by the OTL in May 1987, of the negotiation of intellectual property rights (sponsors' rights) provisions in research contracts. This had previously been the responsibility of the Sponsored Projects Office. As that office came to rely increasingly upon the OTL for help with intellectual property issues, the consensus was that such negotiations should be handled by the OTL.

Although actual changes have been few, the OTL has responded to some new developments, demonstrating the leadership and competence that is important in increasing the credibility of the office. The first new development was software licensing, in response to which the OTL established a three-tier licensing structure that reflected the different levels of software support that customers might require, and also standardized and made intelligible the business arrangements that customers could expect to negotiate. Another new development, and a remarkable story, involved the Cohen-Boyer patent related to DNA splicing. Stanford's OTL took the lead in obtaining and marketing this extremely valuable and controversial technology. It faced many obstacles, including sorting out the rights of the many parties to the invention (the two inventors, U.C., NSF, National Institutes of Health (NIH), and the American Cancer Society), dealing with public concern over safety and ethics issues, and filing the patent application in a timely fashion. But the most complex issue, and the one most creatively solved by the OTL, was the structuring and marketing of the licenses for the patent. These activities involved both predicting the market for the new technology and the even more difficult task of setting a price on the licenses.[18]

Other new developments handled by OTL involved the ownership and licensing of tangible research property and the control of the transfer of biological materials. Although these functions are not technically a part of patent administration, it was natural that OTL should be involved in administering them.

ORGANIZATION. The director of the OTL, Niels Reimers, reports to the associate vice president for finance and oversees a staff of about 14 people. The office is organized around the marketing function. Each technology marketer is assigned a subject area of science and is responsible for managing the entire range of operations within that area, including evaluating disclosures, locating potential licensees, accounting for royalty income, and interacting with licensees. The OTL staff also includes an analyst, who makes market analyses for identified technology areas or specific technologies, as well as staff to administer a software distribution center and the biological materials registry.

The OTL was initially funded by an administrative grant of $125,000. For ten years it managed to finance its operations from this fund plus the income it generated from licenses.[19] Now it is funded entirely by license income.

FACULTY PERCEPTIONS. The OTL enjoys high visibility among Stanford faculty and generally receives high marks for effectiveness. Faculty members who expressed reservations seemed to doubt the effectiveness of the patent and licensing process in general rather than the specific operations of the OTL; they felt that technology was more effectively transmitted through publications or through personal contacts with companies than through the formal process of patenting. The reputation of the director and the OTL's quick attention to invention disclosures were cited as the most important elements in the good reputation of the OTL.

CONCLUSIONS. Supported by a favorable institutional context and academic culture and by an intellectual property policy that demonstrates a positive attitude toward the commercialization of research, Stanford's Office of Technology Licensing, under its first and only director, Niels Reimers, has established a well-deserved high reputation both inside the university and nationwide. Keys to its success have been its market and technology transfer orientation, its attention to serving rather than policing university researchers, and its ability to deal effectively with new developments related to intellectual property. Underlying all of this is a clear understanding, often articulated by Reimers, of the nature of a university and what it should and should not do. Reimers has stated, "From a pragmatic point of view, it would be fatal to the licensing program at this or any other university if an administrator delayed a scientist's publication in order to secure a patent position."[20] This sensitivity to the academic ethos and willingness to put aside business considerations when they conflict with university values elevates the OTL above the status of a mere logistical support organization and is the primary reason for its success.

M.I.T.

Like Stanford, M.I.T. is well known for its relations with industry. Its academic culture is possibly even more supportive of these relationships than Stanford's. Its Industrial Liaison Program is a model for and the envy of other universities. M.I.T.'s faculty are deeply imbued with a sense of the importance of industry interactions.

RECENT CHANGES. In contrast to its present active role, patent administration at M.I.T. languished through the 1970s and early 1980s. Perhaps influenced by the Forrester-RCA controversy and other problems that cropped up in the 1970s, the patent office adopted a narrowly legalistic view of its job, concentrating on the negotiation of favorable terms in license agreements. The office was staffed by patent attorneys, who tended to be reactive rather than proactive in marketing patents.

Faculty dissatisfaction with M.I.T.'s patent administration increased through the 1970s, as evidenced by at least one proposal in 1977 to disband the office and turn its work over to the M.I.T. Development Foundation. Nothing was done, though, until about 1984, when a new provost was appointed. The provost was beset with faculty complaints, including one from a productive inventor who threatened to go outside the institute rather than rely on its patent office. Sidestepping the standing faculty patent committee, the provost appointed an ad hoc committee, which quickly set about reviewing the situation and making recommendations.

The committee's most important recommendation was to hire Niels Reimers as a consultant. Reimers arranged a three-month leave from Stanford and began work at M.I.T. on September 1, 1985. The three months eventually turned into almost a year. During that time Reimers completely reorganized M.I.T.'s patent administration, helped to redraft policies and replaced all but one of the professional staff of the patent office with technology licensing and marketing specialists. To reflect the new emphasis on technology transfer and marketing, the name of the office was changed to the Technology Licensing Office (TLO).

Reimers also recruited a new director for the TLO, John Preston, who began to serve in March 1986. Preston has an MBA, had worked for the Industrial Liaison Program as associate director, and had been an enterpreneur, founding several start-up companies.

Preston moved quickly and aggressively to market M.I.T. technology, as attested by the increases in the number of licenses and the amount of royalty income shown in the last section. M.I.T. went beyond even Stanford in that, at Preston's urging, in 1988 it began to accept equity in companies that license M.I.T. technology. This change has broadened the horizons for patent administration, but it has also introduced a range of actual and potential problems that are still being sorted out.

ORGANIZATION. The director of the TLO reports to the provost, who is also advised by a faculty committee composed of some of the same people who served on the ad hoc committee that established the new office. The committee is actively involved in the review and modification of policy, and it also occasionally renders advice on operational issues or on actual or potential instances of conflict of interest related to intellectual property. The TLO's internal staff is organized, as one might expect, on the "Stanford model" of responsibility centers headed by technology professionals who manage a portfolio of licenses. Office operating costs are still higher than the stated goal of 15 percent of gross royalty income but this may be simply because the office staff has been increased recently and productivity has not caught up yet because of the unavoidable lag time.

FACULTY PERCEPTIONS. Word of the changes in M.I.T. patent administration and the new TLO spread fast, and interviews of faculty members and members of the venture capital community conducted in late 1987 and early 1988 made it clear that most people knew and approved of the new arrangements. However, those who understood the ramifications of the acceptance of equity in start-up companies expressed some reservations about the appropriateness of this step.

CONCLUSIONS. M.I.T.'s example shows just how rapidly and completely a negative perception can be turned around. As with Stanford, M.I.T.'s success is based on both an understanding of the academic culture and a clear conception of the appropriate orientation of patent administration in this age of technology transfer. The jury is still out, however, on the question of whether the new system's early triumphs can be sustained. The backlog of commercializable inventions that accounted for much of M.I.T.'s licensing in 1987 and 1988 may not be exhausted, so the "honeymoon" may not be over yet. A more important question is whether M.I.T.'s new aggressiveness, particularly in the acquisition of start-up equity, will cause the university trouble.

THE UNIVERSITY OF CALIFORNIA

As a large public university and as a system of campuses, U.C. presents a picture quite different from that of either Stanford or M.I.T. In contrast to the situation at the two universities just dicussed, patent administration at U.C. in the 1980s is a story of continuing turmoil and controversy. To begin with, until well into 1989 patent administration for the U.C. system was centralized. This meant that the difficulty of coordinating the patent activities of geographically separated campuses was added to the seemingly unavoidable conflict between a central office and field locations. Furthermore, U.C.'s public character renders it more visible and more subject to political scrutiny than either Stanford or M.I.T., particularly with regard to money-making activities. Its sheer size also intensifies certain problems. For instance, even a small percentage of increase in invention disclosures can overwhelm the patent administration staff.

RECENT CHANGES. U.C.'s systemwide patent administration function has been handled by the Patent, Trademark, and Copyright Office (PTCO). In 1979, PTCO, newly organized under its director, Roger Ditzel, had six employees. In 1980 the staff rose to ten and in 1981 to 18. Even this rapid increase in staff could not keep pace with the increase in work load, however. Apparently unable to develop administrative systems and procedures to carry out properly the difficult task of coordinating patent administration on nine campuses, PTCO began to come under criticism from the faculty and the campus administrations. One faculty member commented, for example, "The Patent Office is frustratingly slow and often nonresponsive to faculty submitting inventions for patenting."[21] In October 1982 a committee on university-industry relations reported: "Administration of patent policy appears to be too centralized, cumbersome and directed towards achieving uniformity. If the University is to maximize opportunities with industry, a better system is needed for negotiating and acting flexibly and expeditiously. Better general guidance and, perhaps, delegation of more authority to Chancellors should be considered."[22]

A "deficit" in PTCO's budget also caused concern, although this deficit was somewhat artificial. Through the 1970s all royalty income was deposited in a patent fund, out of which were paid PTCO's expenses, inventor and campus shares, the state of California's assessment for reimbursement of research costs, about $270,000 per year for graduate student support, and $500,000 per year to support university research. By 1981 all general funds had been removed from the budget of the office, and PTCO was required to be funded from royalty income. At the same time, the patent fund was expected to fund research and scholarships at the same rate as in the past. This change, not surprisingly, had a negative effect on the operations of the office that was characterized by one faculty member as "a domination almost to paralysis with the idea of getting back enough money to cover its expenses."[23] In 1985 and early 1986 the patent fund began running a deficit, which eventually amounted to almost $1 million.

In the summer and fall of 1985 the U.C. administration undertook a full-scale review of PTCO. Part of the review was conducted by the accounting firm of Peat, Marwick and Mitchell & Co. Their wide-ranging report recommended changes in policy, organization, and operating procedures. The recommendations included being more proactive in licensing, revising royalty sharing arrangements, and making "significant improvements . . . in the management and operations of the patent office."[24]

The PTCO review culminated in a proposal, adopted by the Council of Chancellors and the administration in October of 1986, that made three recommendations:

1. Let PTCO function as a relatively freestanding auxiliary enterprise, in control of income and expense of its own making.

2. Appoint campus and laboratory patent coordinators to coordinate local patent and intellectual property activities with the central PTCO.

3. Provide opportunity for campus- or lab-supported patenting and licensing activities.[25]

Reorganization based on these recommendations ran into trouble almost immediately. Strictly speaking, PTCO had no "income of its own making." Royalty income was derived from inventions made on particular campuses; over 80 percent of it, in fact, came from only two campuses, San Francisco and Davis. These two campuses, in effect, were paying most of the costs of PTCO. The reorganization plan stated that once the deficit in the patent fund had been paid off, any excess in the fund after PTCO costs and mandatory disbursements (state and inventor shares) would be distributed to the campuses in proportion to their inventor shares. Under this plan, the campuses, except possibly for San Francisco and Davis, could not expect to support their own patent and licensing activities from royalty income. It quickly became apparent that the administration was not willing to increase campus shares to fund these activities at the local level.

Furthermore, the reorganization proposals did not address the operational problems of PTCO, which began to capture public attention. "UC Losing Millions on Inventions," proclaimed a headline on the front page of the *San Francisco Examiner* on October 18, 1987.[26] The accompanying article cited lack of staff as the principal reason for the failure of PTCO "to develop the awesome array of inventions of its research labs," resulting in the loss of "millions of dollars in public revenue and countless jobs."[27]

Under continuing pressure from the faculty, the campuses, and the public, the administration commissioned another study from Peat, Marwick. The second report, completed in February 1988, concentrated on internal staffing and operational problems. Recognizing that the maintenance of a centralized patent administration was both desirable and in a great deal of trouble, the report proposed a long-range strategy as well as a short-term "turnaround" plan. Its primary recommendation was that PTCO staff should be increased.[28]

In May 1989 yet another Peat, Marwick report was issued. This report cited significant progress toward correcting some of the deficiencies reported earlier. Eight more positions had been added to the office, bringing the total number to 33, ten of these were professionals. A new associate director was hired and soon was appointed director, replacing Roger Detzel. Some of the operational issues that had plagued the office were reported to have been resolved.[29] Apparently unrelated to these changes, PTCO was relocated in September 1988 to a site in Alameda that is removed from both the Berkeley campus and the new location of the Office of the President. PTCO occupies 10,000 square feet of space donated rent free for ten years by a real estate developer in a growing industrial area.[30] Although this arrangement will save an estimated $1 million over the ten years, some observers, including faculty members at Berkeley, see the change in location as further evidence that PTCO is detached from campus concerns.

The Berkeley campus's response to these centralized changes has been a series of its own reviews. Given a partly opened door by the 1986 Council of Chancellors' action, Berkeley began studying the possibility of setting up its own patent office. A staff report completed in November 1987 outlined six possible alternatives for this.[31] In January 1988 a faculty committee proposed the establishment of a campus Office of Technology Transfer that would eventually include an urban extension coordinator, a software and copyrights coordinator, and a patent coordinator.

In early June of 1988, Chancellor Heyman of Berkeley announced his intention to press for an extensive decentralization of the activities of patent administration to the campuses. As he stated in a letter to the other chancellors, "The intellectual creativity of faculty, staff and students is most effectively stimulated and supported at the campus level. Campus management of patent operations could result in improved service to campus researchers in synergy with campus governance of campus-based research."[32]

After an earlier visit by Berkeley officials to Niels Reimers at Stanford, Heyman set about establishing the proposed campus Office of Technology Transfer. Reimers was hired as a consultant in the hope that he could duplicate his success at Stanford and M.I.T. In addition to performing some of the normal functions associated with patent administration, the new office, formally approved in early 1990, may assist faculty members in starting new businesses.

The Berkeley situation, obviously, is still in a state of flux. It is similar to earlier events at U.C. San Diego, which, after an extensive study, decided to hire a coordinator for biotechnology transfer.[33] U.C.L.A. has also established an office, and similar moves are being contemplated at U.C. Davis and U.C. San Francisco.

ORGANIZATION. The director of PTCO reports to the vice president for business affairs. Internally, PTCO is organized into five groups that combine both subject and function categories. Figure 7-1 shows PTCO's operational organization.

The separation of the disclosures and prosecution groups from the plant and utility licensing groups has created a number of serious problems in coordination for PTCO. No one person or group is responsible for coordinating the whole process, so steps crucial to the patenting and licensing process often are missed. The separation of administrative support in another unit has created other problems, although a recent reorganization has corrected some of these.[34]

A board of patents, appointed by the regents, was established in 1973 and functioned first as an overseer involved in patent evaluations and then as a general adviser on policy considerations. The patent board was abolished in 1985 and an Intellectual Advisory Council, composed of representatives from each campus, was formed to advise the president on a wide range of policy issues, including patent administration. This council in turn faded from the picture after the October 1986 Council of Chancellors' action. This action established a group, chaired by the director of PTCO, that was composed of the patent coordinators of the campuses and the labs. In its short history this group has served as a curious blend of advocate, adversary, and adviser to PTCO; its role and authority remain poorly defined. In 1989 this group was superceded by a PTCO steering committee composed of the vice chancellors for research from the nine campuses and their counterparts from the federal laboratories affiliated with the university. By early 1990 the role of this committee was still not well defined.

FACULTY PERCEPTIONS. As might be expected from the foregoing discussion, most faculty express dissatisfaction with patent administration at U.C., although recent changes may be altering these negative views. A 1987 faculty report to the dean of the School of Engineering at Berkeley concluded, "There is one point of strong consensus: that the current activities of the University's Patent, Trademark, and Copyright Office (PTCO) are disastrous from virtually everyone's viewpoint. That Office is seen as bureaucratic, inefficient and passive, impeding those who want to be unfettered in their commitment to the public domain, without bringing a compensating advantage of effective marketing or royalty production for those who want to see their inventions exploited."[35] Similar, though often less

FIGURE 7-1   PTCO Operational Functions

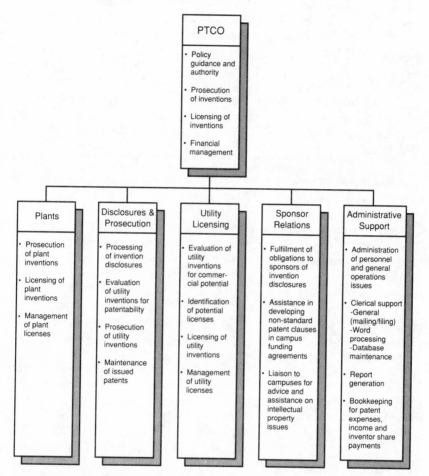

Source: Peat, Marwick, "University of California Patent, Trademark and Copyright Office, Review of Operations, Final Report, 1988" (University of California, 1988).

strongly stated, opinions were held by 11 of the 18 Berkeley faculty members responding to the question about PTCO. The overwhelming basis of their complaints was the failure of PTCO to respond in a timely fashion to requests for help. Also, most of the faculty members who had dealt with PTCO recounted some instance in which administrative errors or omissions had occurred.

CONCLUSIONS.   The U.C. experience, especially when contrasted with the situations at Stanford and M.I.T., offers several interesting lessons. At M.I.T, for instance, the operations of the patent office prior to 1985 clearly were not compatible with prevailing academic culture and administrative goals. The answer was a sweeping change that aligned not only the operations of the patent office but also university policy with the conceptions generally held by concerned faculty. No such sweeping changes seem possible at U.C. The prevailing academic culture is much less attuned to technology transfer. U.C.'s policies, many of which need to be changed if effective patent administration and technology transfer are to be encouraged, must be sifted through the sieve of public concern.

The conclusion that administrative responsiveness not only is a key element in the success of university patent administration but also is symbolic of university commitment to technology transfer is reinforced in the U.C. story. Faculty members, and increasingly members of the university's public constituencies as well, view the administration's inability to deal effectively with the long-standing and well-documented shortcomings of PTCO as evidence of a lack of commitment to technology transfer and economic development.

More directly germane to this study, developments at U.C. point to the importance of integrating patent administration into a broadly conceived framework of university involvement in technology transfer. Chancellor Heyman's suggestion regarding further decentralization to the campuses of responsibility and authority in patent administration, although prompted by frustration at the ineffectiveness of PTCO, instinctively recognizes that there is "synergy" in the conjunction of inventions, patents, licenses, relations with research sponsors, campus economic development and assistance efforts, and the generation of revenue.

Despite its problems, U.C.'s potential for developing marketable inventions is high. The potential number, range, and value of the university system's inventions suggest that a "critical mass" of in-house expertise feasibly could be assembled, organized, and dedicated to university purposes. The right combination of centralized expertise and a decentralized network of personal contacts could be a powerful engine, both financially and politically, for the university. Although U.C. seems to be moving in this direction, the prospects remain exceedingly uncertain.

PENN STATE

Penn State is quite different from the other three universities in its provision for patent administration, but many of the issues it faces are similar to those facing the others. In using RCT as its patenting and licensing organization, it avoids much of the day-to-day operational concerns that plague the other compuses. But, as we will see, it does not avoid criticism.

In recent years Penn State has moved to become an aggressive part of Pennsylvania's effort to revitalize its state economy. This revitalization has taken the form of a number of state and local initiatives, most notably the Ben Franklin Partnership, to save and create jobs and to encourage the establishment of new businesses.

RECENT CHANGES. For many years, Robert Custard served as Penn State's patent counsel. In addition to several other duties, including serving as the university's security officer for classified research, he acted as liaison between the university and Research Corporation. In late 1986 the university hired Richard Hansen, also a patent attorney, to replace Custard, who had announced that he was retiring in mid-1987. Hansen's job was more focused on patent administration, particularly on the solicitation of patentable ideas from faculty members, and he moved quickly to make himself known to the faculty and to familiarize himself with the research that was taking place on campus. He also began reviewing the backlog of ideas rejected by RCT to determine the economic feasibility of developing some of them.

In 1987, spurred by the continued interest in technology transfer, the university announced a reorganization in which those organizational units identified with technology transfer, including the patent office, would report to a new associate vice president. Disturbed by this reorganization, which he viewed as a lowering of status for patent administration, Hansen left his job at the university and returned to private practice. His work was turned over to outside counsel pending the hiring of the new associate vice president. In the meantime, the university began seeking a replacement for Hansen, this time emphasizing entrepreneurial and marketing skills rather than possession of a law degree.

ORGANIZATION. As indicated, the patent officer at Penn State reports to an associate vice president. Invention disclosures from faculty members are transmitted to RCT for review. Patents obtained through the efforts of RCT are held by the Pennsylvania Research Corporation (PRC), which receives the university's share of royalty income from Research Corporation and remits it to the university. Although the university is fully in control of PRC, PRC provides some legal shielding for the university. The patent counsel and associated clerical support are funded from general funds of the university.

FACULTY PERCEPTIONS. Although many of the faculty members interviewed were dissatisfied with patent administration at Penn State, citing most often the low percentage of inventions accepted by RCT and the lack of expertise by RCT in specific fields of science, a significant number were comfortable with the present arrangement. The surge of recent concern over technology transfer and economic development has not been accepted fully by some faculty members, who see the two goals of excellence in basic research and aid to the economic development of Pennsylvania as somewhat inconsistent. In particular, the administration's increased emphasis on state-wide economic concerns has been questioned by a group of faculty in the humanities, who feel that the university is developing two categories of researchers—applied and basic, and two categories of faculty members—

those who have a chance to be wealthy and those who do not. These faculty members were satisfied with the relatively passive stance taken by the university with regard to commercializing research and felt that the shielding performed by RCT was desirable.

CONCLUSIONS. Perhaps because it lacks a "critical mass" of inventions and therefore the financial base for supporting a full patent administration staff, Penn State has continued in its relationship with RCT, although this relationship is beginning to show the strains common to such situations throughout the country. RCT's relatively low acceptance rate and its distance from faculty members and from the concerns of the state are creating a gap between the realities of patent administration and the goals of the university. The lagging of patent administration behind other aspects of technology transfer at Penn State is puzzling because the university is so clearly controlled by a technology transfer-oriented administration. There are signs of changes in the wind, however, as Jack Yost, the new associate vice president in charge of technology tranfer, begins to outline a program. There has already been at least one plan to increase the size of the patent office staff, for example. Given the pressure on the university, significant changes can be expected soon.

Conclusion

Universities across the country, including those not considered to be major research universities, are facing a common set of issues related to the management of intellectual property, for which *patent administration* has been used here as a catchall term. The four case studies presented in this chapter have illustrated each of these issues and, in some cases, suggested action plans for dealing with them. The issues are financing patent administration, developing and retaining a critical mass of expertise for identifying and marketing university technology, establishing policy and procedures for participating in the equity of start-up companies, and fitting patent administration into the broader context of technology transfer.

This chapter concludes with a discussion of each of these issues.

FINANCING PATENT ADMINISTRATION

Successful patent administration obviously requires stable and secure funding, yet few universities have established such funding. Lack of financing is a particular problem in the establishment of new operations or in offices where growth of staff must precede growth in patent activity. Because most or all of the funding for patent administration usually is expected to come from royalties, patent administration in most universities is condemned to a tenuous existence for two reasons. The first is the long lag time—from three to ten years—before an invention begins to produce income. Consequently, for start-up patent administration operations, at least three years of full operation must be funded before any returns will be evident. The second is that a university usually has to wait for a "big hit" before its patent administration can achieve financial viability. Yet big hits are never predictable, and even the largest universities may have to wait ten years or more to get one.

Another funding issue has to do with inventor's share arrangements, which were discussed in chapter 5. Inventor share distributions reduce the amount of funding available for patent administration; thus, if the inventor's share is set too high, effective patent administration is impossible without subsidies from other sources. On the other hand, if inventor shares are set too low, patent administration may not get any material with which to work. Thus, striking a balance between these two extremes is a vital part of patent administration financing.

An appropriate definition of the mission and responsibilities of patent administration is also necessary. When the patent office is expected to perform functions unproductive of income and related only tangentially to patent administration, those functions are in effect being subsidized by royalty income. Examples are negotiating with sponsors of research, developing and administering policies, and resolving disputes.

Finally, the costs and benefits of supporting administrative systems are usually an issue in financing both start-up and continuing operations. Many such systems are indispensable to effective operations but require a significant one-time investment. Generating the funding to buy these systems is a challenge for many patent administration operations.

One plan for providing the fiscal base necessary for patent operations has been proposed by Niels Reimers. Under this plan an "investor" (usually a source within the university) advances the money to fund the start-up operation or the growth and is paid back over a period of years as returns from inventions begin coming in. The procedural accounting expression for this is "capitalizing" start-up costs. Most businesses do it, writing off initial costs against related income that comes in future years. Of course, this process presumes that the operation eventually does become profitable, something that is rarely assured.

DEVELOPING AND RETAINING MARKETING EXPERTISE

The case studies in this chapter clearly show the increasing importance of the marketing function in patent administration, as opposed to the legalistic emphasis of earlier times. Yet it is virtually impossible to find or assemble a single staff of experts large enough to understand effectively, let alone market, every one of the broad range of ideas likely to come from university researchers. People with the requisite combination of technical knowledge (in science or engineering) and marketing skills are hard to find. Compensating license marketers is another problem universities face, since few universities have compensation plans that allow sufficiently high or incentive-based pay.

The need for expert marketing staff is leading to a hybridization of the traditional marketing forms. For instance, the University of Illinois, under its new contract with RCT, may send its medically oriented inventions to RCT, which has a reputation in that field, but develop on its own inventions in engineering, where it might have its own contacts and expertise. Similarly, U.C. San Diego is concentrating on biotechnology and leaving development of inventions in other areas to PTCO.

PARTICIPATION IN EQUITY

Many universities are already participating in the equity of companies based on university technology, and many others are considering equity participation. The temptation to do this is very great because such participation both holds out the promise of large potential gains and seems to square with the "new" mission of the university to be involved with economic development. (Although this issue goes beyond patent administration and involves university technology and licensing, the patent office is usually a part of the process. At M.I.T., for example, the TLO is proactive in promoting licensing deals that include equity participation by the institute.) A broad range of problems and policy issues attend such equity participation, however. This subject is discussed in the next chapter.

FITTING INTO THE BROADER CONTEXT

The contrast between Stanford and M.I.T. on the one hand and U.C. and Penn State on the other illustrates the problems that can arise when patent administration does not fit the academic culture and institutional context of the university. Moreover, the occurrence of a great many problems associated with patent administration is usually symptomatic of an underlying confusion over the role of technology transfer in the university. Conversely, effective patent administration can symbolize a university's positive stance toward technology transfer; indeed, it can lead the university and its faculty to a more integrated view of the university's role. It might facilitate more technology development and commercializing activity through active solicitation of ideas and aggressive marketing, or it might take a more passive and protective stance, guarding the interests of those faculty who wish to engage in such activity and shielding those who do not want to become involved.

Whatever a university's situation, administrators should conceive of the role of patent administration in its broadest context as part of a coordinated effort composed of many elements. Patent administration and patent policy are parts of, and should be consistent with, the university's attitude toward intellectual property, its relations with industry and efforts to develop and serve research sponsors, and its efforts to be involved with its community's economic development.

Notes

1. Peter Dworkin and Andrea Gabor, "Academia Goes Commercial," *U.S. News and World Report*, 2 May 1988, 50–51.
2. SRI International, "NSF Engineering Program Patent Study, Final Report" (Washington, D.C.: National Science Foundation, 1985), 19.
3. "Give and Take," *Chronicle of Higher Education*, 1 March 1988, A23.
4. Dworkin and Gabor, "Academia Goes Commercial."
5. Roger Ditzel, "Analysis of U.C. Patent Operations, 1986–87" (Berkeley: Office of the President, University of California, 1987).
6. Sally Hines (Office of Technology Licensing, Stanford University), correspondence with author, August 1988.
7. John Preston, interview with author, Cambridge, Mass., 7 July 1988.
8. Gerald A. Erickson and Donald R. Baldwin, "The New Frontier of Technology Transfer," in *Research Administration and Technology Transfer*, ed. James T. Kenny (San Francisco: Jossey-Bass, 1988), 25.
9. Preston, interview.
10. Ibid.
11. Massachusetts Institute of Technology, *Report to the President, 1980–81 through 1985–86*, M.I.T. Archives.
12. Preston, interview.
13. Howard W. Bremer, "University Technology Transfer—Publish and Perish," in *Patent Policy*, ed. Willard Marcy (Washington, D.C.: American Chemical Society, 1978), 57.
14. August Manza, "Patent Operations at Berkeley: An Analysis of Six Possible Strategies" (Berkeley: University of California, November 1987).
15. Niels Reimers, "Survey of Directed Mechanisms for Innovation of University Research," *Les Nouvelles*, 15:2 (1980): 83.
16. Ditzel, "Analysis of U.C. Patent Operations," 12.
17. Gilbert Fuchsberg, "Universities Said to Go Too Fast in Quest of Profit from Research," *Chronicle of Higher Education*, 12 April 1989, A28.
18. Niels Reimers, "Tiger by the Tail," *Chemtech*, August 1987, 464–471.
19. Niels Reimers, interview with author, Palo Alto, Calif., February 1988.
20. Reimers, *Tiger by the Tail*, 467.
21. University of California, "Patent Operation to Be Reviewed by Administration," *Notice* 9 (July 1985): 1, 4.
22. University of California, "Report of the University-Industry Relations Project" (Berkeley: University of California, 1982), 23.
23. University of California, "Patent Operation."
24. Peat, Marwick, Mitchell & Co., "Review of University Patent Policy and Function of the Patent, Trademark, and Copyright Office" (Berkeley: University of California, 1985).
25. University of California, "Proposal for Restructuring Patent Operations" (Council of Chancellors item, 29 October 1986).

26. Eric Best, "U.C. Losing Millions on Inventions," *San Francisco Examiner*, 18 October 1987, A1.
27. Ibid.
28. Peat, Marwick. "University of California Patent, Trademark and Copyright Office, Review of Operations, Final Report, February 1988" (Berkeley: University of California, 1988).
29. Peat, Marwick, Main & Company, *University of California Patent Program: Models for Campus Involvement* (Oakland: University of California, Office of the President, 21 April 1989).
30. University of California, "Patent Office to Move to Harbor Bay Business Park," *Intercom* 1 (January 1988): 1.
31. Manza, "Patent Operations at Berkeley."
32. Ira Michael Heyman, letter to Chancellors Atkinson, Hullar, Krevans, and Young, 3 June 1988.
33. University of California, *A Study of the Biotechnology Transfer Process* (San Diego: School of Medicine, University of California, 1987).
34. Peat, Marwick, "Final Report," 8–16.
35. Ad Hoc Committee to Advise the Dean on Patent/Copyright Policy, "Report" (Berkeley: College of Engineering, University of California, Berkeley, 1987).

Bibliography

AD HOC COMMITTEE TO ADVISE THE DEAN ON PATENT/COPYRIGHT POLICY. "Report." Berkeley: College of Engineering, University of California, 1987.

BEST, ERIC. "U.C. Losing Millions on Inventions." *San Francisco Examiner*, 18 October 1987, A1.

CUSTARD, ROBERT. "Report of Invention and Patent Activity for the Calendar Year 1985." University Park: Pennsylvania State University, 1986.

BREMER, HOWARD W. "University Technology Transfer—Publish and Perish." In *Patent Policy*, edited by Willard Marcy. Washington, D.C.: American Chemical Society, 1978.

DITZEL, ROGER. "Analysis of U.C. Patent Operations, 1986–87." Berkeley: Office of the President, University of California, 1987.

DWORKIN, PETER, and ANDREA GABOR. "Academia Goes Commercial." *U.S. News and World Report*, 2 May 1988, 50–51.

ERICKSON, GERALD A., and DONALD R. BALDWIN. "The New Frontier of Technology Transfer." In *Research Administration and Technology Transfer*, edited by James T. Kenny. San Francisco: Jossey-Bass, 1988.

FUSCHBERG, GILBERT. "Universities Said to Go Too Fast in Quest of Profit from Research." *Chronicle of Higher Education*, 12 April 1989, A28.

GARDNER, DAVID PIERPONT. "Annual Report Pertaining to University Patent and Other Intellectual Property Matters." Oakland: University of California, Office of the President, May 1989.

"Give and Take." *Chronicle of Higher Education*, 1 March 1989, A23.

LACHS, PHYLLIS S. "University Patent Policy." *Journal of College and University Law*, 10 (1983–1984): 263–292.

MANZA, AUGUST G. "Patent Operations at Berkeley: An Analysis of Six Possible Strategies." Berkeley: University of California, November 1987.

MASSACHUSETTS INSTITUTE OF TECHNOLOGY. "Committee on Inventions and Copyrights, Proposed Change in Patent Policy for Consideration by the Executive Committee, June 9, 1972." M.I.T. Archives, Office of the President and Chancellor, 1971–1980 (Weisner/Gray) (80–27), Box 30.

————. *Report to the President, 1980–81 through 1985–86.* M.I.T. Archives.

————. *Report to the President, 1986–87.* M.I.T. Archives.

PEAT, MARWICK, MITCHELL & Co. "Review of University Patent Policy and Function of the Patent, Trademark and Copyright Office." Berkeley: University of California, 1985.

PEAT, MARWICK. "University of California Patent, Trademark Copyright Office, Review of Operations, Final Report, February, 1988." Berkeley: University of California, 1988.

PEAT, MARWICK, MAIN & Co. *University of California Patent Program: Models for Campus Involvement.* Oakland: University of California, Office of the President, 21 April 1989.

PENNSYLVANIA STATE UNIVERSITY. "Invention Disclosure and Patenting Activity, 1986 Report to the Pennsylvania Research Corporation." University Park: Pennsylvania State University, 1987.

REIMERS, NIELS. "Survey of Directed Mechanisms for Innovation of University Research." *Les Nouvelles* 15 (1980): 79–88.

————. "Tiger by the Tail." *Chemtech,* August 1987, 464–471.

SRI INTERNATIONAL. "NSF Engineering Program Patent Study, Final Report." Washington, D.C.: National Science Foundation, 1985.

UNIVERSITY OF CALIFORNIA. "Patent Office to Move to Harbor Bay Business Park." *Intercom* 1 (1988): 1.

————. "Patent Operation to Be Reviewed by Administration." *Notice* 9 (1985): 1, 4.

————. "Proposal for Restructuring Patent Operations." Council of Chancellors item, 29 October 1986.

————. "Report of the University-Industry Relations Project." Berkeley: University of California, 1982.

————. *A Study of the Biotechnology Transfer Process.* San Diego: School of Medicine, University of California, 1987.

8

The University as Venture Capitalist: Owning Equity in Research-Based Companies

On a "crisp October morning" in 1980, Derek Bok, president of Harvard, woke up (as he put it) "to find my own likeness staring forth from the front page of the *New York Times* under the caption 'Harvard Considers Commercial Role in DNA Research.' . . . Once again, the *Times* had managed to transform Harvard's quiet intramural discussions into a public issue of national proportions."[1] The headline that startled Bok referred to Harvard's proposed investment in a company that was to be formed to exploit Professor Mark Ptashne's discoveries in biochemistry, which, it was believed, would enable scientists to produce drugs and other products by manipulating the genetic makeup of bacteria. In November 1980, after over a month of public debate and widespread publicity about Harvard's plans, Bok announced, again on the front page of the *Times:*

> The preservation of academic values is a matter of paramount importance to the university, and owning shares in such a company would create a number of potential conflicts with these values. . . . I have concluded that Harvard should not take such a step, even on a limited experimental basis, unless we

are assured that we can proceed without the risk of compromising the quality of our education and research.[2]

On September 15, 1988, just about eight years later, this time on page 37 of the *Times*, another headline appeared: "Harvard, in Reversal, Will Seek Research Profits."[3] The article announced that Harvard was establishing a fund that would invest in companies formed to bring Harvard faculty members' research to market.

What happened in the eight years intervening between these two decisions by Harvard's president is not Harvard's story alone. Rather, it is symbolic of the changes that many universities have been forced to deal with in the last few years. This chapter describes those forces, the issues in the debate concerning university ownership of research-based companies, and the way in which a few universities are coming to terms with such ownership. After considering some of the factors that are pushing universities into ownership of research-based companies and some of the forms such ownership can take, this chapter recounts a number of case histories that illustrate pitfalls and policy issues created by university ownership. Finally, the problems of university ownership are analyzed and recommendations for minimizing them are made.

Reasons for University Ownership of Research-Based Companies

There are many good reasons why Harvard, Johns Hopkins, M.I.T., the University of Chicago (in a partnership with Argonne Laboratory), and almost every other major university have recently taken or are considering taking ownership in companies based on university research. These can be placed in three categories: financial reasons, reasons related to technology transfer, and faculty-related reasons.

FINANCIAL REASONS

Once embarked upon attempts to exploit their research commercially through the patent and licensing activities described in the three previous chapters, universities logically are drawn into considering the ownership of stock in companies that either are buying the licenses or are being formed to develop the technology. The logic of the progression from patent licensing to equity ownership is based on several factors. First, patent licensing is in many ways an unsatisfactory means of capturing the value of the university's intellectual property. Patents lose value by being "invented around," improved upon, infringed, or rendered obsolete. These risks can be avoided to some extent, however, if universities accept stock in licensee companies instead of or in addition to "running royalties" (royalties based on future sales of a product that utilizes the license), because the value of the stock is based on the company's operations, not the performance of a particular product.

Second, it is often easy for universities to negotiate for stock in licensee companies, particularly start-up companies. Such companies are sometimes unwilling to spend scarce capital on initial license fees or to tax future income with high running royalties, but they are willing to give up some equity, especially to a high-prestige investor such as a university. As one licensing officer put it, "Universities can often get the equity for free." In other words, once the university has negotiated the best royalty agreement, the only thing the company has left to bargain with is equity.

However, the most compelling reason for universities to consider owning stock in research-based companies is the hope of enormous financial return—the dream that the next equity stake will turn into another Xerox or Polaroid Corporation, both of which were based on university-born inventions. Such hopes may be quite reasonable for the best research universities. For example, a recent study found that 404 companies in Massachusetts had been founded by M.I.T. graduates. These companies employed 160,000 people and had gross revenues of $27 billion—about one-fifth of Massachusetts' gross income.[4] Another study found that 156 of the 216 high-technology companies in the Boston area were created at M.I.T. facilities.[5] Small wonder, then, that M.I.T. has embarked on an aggressive program of equity ownership of start-up companies.

M.I.T.'s aggressiveness pales, however, in comparison to its neighbor across the Charles, Boston University. Since 1980 Boston U. has committed $50 million of its $166 million endowment to Seragen, Inc., a company that is developing treatments to fight leukemia and other diseases. The university owns 73.4 percent of Seragen's stock and is reportedly ready to spend up to $100 million to keep the company viable until its first products reach the market. Boston U. is gambling a large portion of its endowment on its estimate that the company will be worth $1 billion by 1999.[6]

REASONS RELATED TO TECHNOLOGY TRANSFER

Ownership of start-up companies can be an effective way of associating a university with the economic development efforts of a region and of demonstrating involvement in technology transfer. Because much university research requires extensive development even to bring it to the point where it can be evaluated for commercial potential, the university is often the only source of early development funding. Funding of such early development and ownership of stock in subsequently formed commercial enterprises (usually through development companies or other entities designed to buffer the university as much as posssible from the negative aspects of ownership) both show significant involvement in technology transfer and economic development processes. Through such ownership the university also can influence the location of companies, tying them to local regions. As shown in the case of Penn State, this visible involvement can have favorable political consequences.

FACULTY-RELATED REASONS

One of the advantages of the 1980 proposal for a Harvard-owned company cited by Daniel Steiner, general counsel to Harvard University, was that the university, through its ownership of such a company, could be of assistance to the faculty members involved and "to the functioning of the academic community," serving to "help obtain fair treatment for the faculty members, help assure that the business arrangements are sound, and assist in making certain that commercial motives do not lead management of the company to divert faculty attention from their primary research and teaching obligations at the University."[7] Steiner went on to say, "Harvard can be both an effective advisor and negotiator for the faculty and a buffer between the faculty and the company when operations begin."[8] This may be stretching things a bit; certainly there are many dangers when universities serve at the same time as advocates of and monitors for faculty, especially through membership on the board of directors of a company or through stock voting rights. However, active knowledge of actual or proposed business arrangements and active counseling of faculty members is certainly to be desired, and, insofar as university ownership promotes these activities, it does have advantages.

The most significant faculty-related advantage associated with university ownership of companies, however, is the role that the university might play in attracting and retaining faculty members. Such ownership might play the same role that medical practice plans play in medical schools. These plans are based upon individual negotiations between the school and the faculty member that specify the extent of clinical practice the faculty member will undertake and the shares of the resulting income that will go to the faculty member and the university. These plans are crucial in faculty recruitment and retention in medical schools, providing a supplement to faculty salaries that is necessary for universities to compete for qualified faculty as well as an opportunity for medical school faculty to remain in touch with patients. They also clarify the rights and duties of faculty members, making explicit the extent to which activities other than teaching and research are permitted and providing the university with a share of the value of these activities. Such plans are not easy to work out or administer, but their existence is evidence that such arrangements are feasible in and useful to universities.

University stock ownership in start-up companies has some interesting parallels to medical practice plans, although there are some significant differences. Most important of these is that in medical practice plans the faculty member/physician is selling his or her own skill and services; this is not the case with faculty stock ownership. The similarities include that both stock ownership and practice plans focus attention on the extent of faculty activity and financial involvement outside the university and force agreement between the faculty and the university about such activities. Also stock ownership might provide the same kind of financial incentive as practice plans to induce successful faculty to remain with the university. Studies of faculty in fields related to biotechnology indicate that the most common reason such faculty leave the university is to found or be involved in the founding of start-up companies.[9] Yet many of these faculty members leave the university reluctantly and remain connected to the institution in some way after they have left. A closer examination of individual circumstances and active involvement of the university in commercial projects might "save" some of these researchers who would otherwise leave.

Another similarity is that joint university-faculty involvement in commercial ventures related to faculty research can help faculty members stay in touch with practical applications in their fields in the same way that medical practice plans provide doctors with the opportunity to stay in touch with clinical practice. The interviews of faculty members reported in chapter 12 indicate that many researchers are anxious to see their discoveries developed and that some left the university, not because they wanted to get rich, but because they saw their involvement in the development process as the only way their discoveries could be made useful.

Increased opportunity for sharing research facilities and instruments provides another faculty-related reason for university ownership of research-based companies. So does the potentially positive influence of the university on the activities and policies of a company.

A university can find many potential advantages in owning companies based on the research it has sponsored. The case histories examined later, however, make clear that there are also many dangers in such ownership.

Forms of Ownership

Many other universities experimented with ownership of research-based companies long before Harvard's much-publicized rejection of such ownership. University ownership has taken many forms, the most common of which are ownership through direct investment and through buffer organizations.

OWNERSHIP THROUGH INVESTMENT

University treasurers, in managing an endowment portfolio, often have invested in highly speculative start-up companies, either directly or, as is more common, through venture capital funds or partnerships. M.I.T. invests about 10 percent of its portfolio, or about $150 million, in venture capital projects.[10] Stanford invests a like percentage, and the University of

California invests about 2 percent of its total portfolio in venture capital partnerships.[11] Often these investments, by accident or design, include companies that are based on research conducted by the university. Decisions related to such investments usually rest with the treasurer, who is required to exercise only investment criteria in making those decisions and to ignore other factors, such as the presence of faculty members among company shareholders. Theoretically, the treasurer is shielded from the rest of the university and, in turn, keeps the rest of the university out of the financial decision-making process. Such independent investments normally fall outside the subject matter of this chapter.

OWNERSHIP THROUGH BUFFER ORGANIZATIONS

Universities often have entered into ownership of research-based companies through "buffer" organizations formed for this purpose. These organizations have several advantages. First, they shield the university from commercial concerns, much as the treasurer does in handling endowment investments. This shielding can be important both legally and in terms of public relations. Second, buffer organizations can provide both outside expertise in investing in and developing new companies and a compensation plan for this expertise that is unfettered by university policy. Finally, they can provide an effective structure within which the players in new business development can communicate and negotiate.

These development or buffer organizations exist in infinite variety. They may be for-profit or nonprofit, partnerships or corporations, autonomous or university controlled. Their goals may be to produce revenue for the university, to render service to the university and the community, or both.

The University of Rochester may have been the first university to establish a wholly owned, tax-exempt venture capital arm. University Ventures, Inc., a subsidiary of the university with a board of directors composed mostly of university officials, was founded in November 1981 with $67 million of endowment funds.[12]

A more recent example is Johns Hopkins University, which has just formed a for-profit enterprise called Triad Investors Corporation to commercialize university research. Triad will be owned by Dome Corporation, which in turn is owned jointly by Johns Hopkins and the Johns Hopkins Health System. Triad will try to attract outside investors willing to contribute $2 million each. With this funding, estimated at between $10 million and $30 million, Triad will evaluate the results of Johns Hopkins research, obtain the rights to patents on promising inventions, and then help to develop business plans and provide first-stage financing for new ventures.[13]

In contrast to Triad, which has been formed to make money from university research, the Pittsburgh Foundation for Applied Science and Technology, although it performs many of the same functions that Triad does, was founded primarily to support state economic development programs and to foster high-technology industry in Pennsylvania. It receives about one-third of its $12 million budget from the state of Pennsylvania.[14]

Case Histories

In the histories that follow there are more failures than successes. This is because failures, especially big failures, are more likely to come to light and be reported than are successes. Universities are not likely to brag about making a lot of money on a speculative venture; rather, this kind of success is likely to be hidden in the back pages of a treasurer's report. However, it should be emphasized that equity ownership has paid off for universities in a number of cases, even though the payoff was less than the dreams of a Polaroid or Xerox. For instance, Lawrence Klein, professor of economics at the University of Pennsylvania and a Nobel Prize winner, set up the Wharton Economic Forecasting Association (WEFA). In return for the use of the Wharton name and a $250,000 line of credit, the university received equity in the venture. The university ultimately received about $150,000 per year in dividends from the the WEFA, and in 1980 it sold 80 percent of its stake in the venture for an estimated $7 million.[15] In spite of such successes, the following discussion concentrates mainly on failures because they reveal most clearly the challenges that universities face when they decide to follow the path of equity ownership.

HARVARD

The Harvard example that began this chapter is important because it prompted the first full-scale public debate over the issues involved in university ownership of equity in commercial ventures and the outcome, at least at present, represents an interesting turnabout.

The late 1970s will certainly be known as the time of the biotechnology "gold rush." At the center of this rush was Walter Gilbert, a Nobel Prize-winning biologist at Harvard. His work in gene cloning, particularly on what came to be known as the lambda repressor, which was potentially significant in the fight against cancer, propelled him to the forefront of the new field. He was instrumental in founding Biogen, one of the first companies formed to exploit the commercial potential of DNA research. Eventually Gilbert's preoccupation with this company forced him to resign his American Cancer Society chair at Harvard, and he became Biogen's chief operating officer. In January 1980, after a press conference conducted by Gilbert and other scientists from Biogen at which a breakthrough in gene cloning was announced, Biogen's stock climbed rapidly, adding $50 million to the paper value of the company and considerably to Gilbert's personal net worth.[16]

Mark Ptashne, whose Harvard laboratory was located in the same

building as Gilbert's on Divinity Street in Cambridge, took a different approach, both scientifically and commercially, to the development of the
lambda repressor. When his work led to the development of a method of
inducing bacteria to produce particularly large amounts of an inserted
gene's protein, Harvard officials had to convince him to cooperate in patenting the process.[17]

In April 1979 General Electric Corporation's venture capital subsidiary
offered Harvard $500,000 for the rights to the Ptashne patent.[18] By this
time, however, Harvard, with the Biogen experience clearly in mind, was
developing other plans. Harvard officials approached Ptashne with the idea
of forming a biotechnology company that would be owned jointly by Harvard and by Ptashne and several other Harvard professors. Harvard would
license Ptashne's process to the new company in return for a 10 percent
equity stake. In the eight months during which negotiations concerning the
new company were in progress (from February to October 1980), Ptashne
rejected all outside offers: "People asked me to join existing corporations
and to set up new ones. I turned down every one because I was not sure it
was in the interests of the university."[19]

While negotiations continued, Harvard officials held private meetings
with members of the faculty to gauge their opinions of the proposed venture. Word began to leak out. On October 9, 1980, Daniel Steiner, general
counsel to Harvard, circulated a memorandum to faculty entitled "Technology Transfer at Harvard." In this memorandum he listed both the advantages and the disadvantages of Harvard's becoming a minority shareholder
in companies based on Harvard research. The memorandum did not mention Ptashne or disclose the details of the proposed arrangement with him.
Rather, it stated the issues in general terms and invited comment. There was
some urgency in the call, however: "Harvard has made no decision about
participation as a minority shareholder in a company working in the recom-
binant DNA area. However, because participation in such a company with
a faculty member has been under active consideration over the summer and
Harvard may have to make a decision before the end of October, a description of the possible structure of such an enterprise and a discussion of the
issues seem appropriate before the president makes a decision."[20]

The sense of urgency of the memorandum, its lack of details, and the
impression it conveyed that the president had already decided to go ahead
with the scheme prompted considerable opposition among the faculty. The
strongest criticism came not from Ptashne's colleagues but from biologists
working in other fields. They warned of a decline in academic freedom, of
secrecy invading laboratories, and of the prospect that commercial concerns
would distort the university's research agenda and injure its reputation for
objectivity. For instance, they warned, Harvard scientists might not argue
as strongly for adequate laboratory and regulatory safeguards if they became part of the commercial world. Then, too, those faculty members felt
there was a certain contradiction, even insolence, in Steiner's assertion that

by owning a minority interest in a company the university could help to "make certain that commercial motives do not lead management of the company to divert faculty attention from their primary research and teaching obligations" and that Harvard's involvement would prevent undue secrecy.[21]

Added to these specific arguments against the proposed arrangement there was a general feeling that the arrangement somehow violated the concept of the university. In the opinion of Harvard biologist Woodland Hastings,

> The whole matter violated the role of the university in our society so extensively and so terribly that I don't see how anything can come of it. The university would no longer be a nonprofit organization. It would mean that in everything we do, in our laboratories, in our scholarship, we are joining with the university to make a profit.[22]

In a comment the irony of which was not lost on close observers of the situation, Walter Gilbert also complained:

> I think the idea is extraordinarily unwise. . . . I have my own company and I would resent being put in the position of having to compete against Harvard for the best people and the best work. I might have to push Harvard to the wall in some cases of competition. That shouldn't have to happen.[23]

Faced with both considerable internal opposition and significant public attention, President Bok backed away from the venture. He left the door open for similar possibilities in the future but said that he would not again consider such an idea "unless we are assured that we can proceed without the risk of compromising the quality of our education and research."[24]

Apparently Bok felt he had received those assurances by September 1988, when Harvard announced that it was seeking $30 million from outside investors to form a partnership to finance scientific research and spin-off companies for selling marketable products resulting from it.[25] This proposal appears to be very similar to the Johns Hopkins arrangement described earlier. Harvard will participate in the venture through a subsidiary and will receive 10 percent of all profits. It will not own stock in spin-off companies directly. All business decisions will be left to the managers of the partnership, who will also receive 10 percent of net proceeds. A special faculty panel will review all proposals to assure that university interests are upheld. No faculty member will be forced to cooperate with the new venture.

There are some significant differences between this arrangement and the one that Harvard considered early in the decade. First, the new arrangement apparently has been subjected to considerable scrutiny by the faculty and has gone through a slow and orderly process of evaluation and examination. Second, because the managers of the partnership are expected to act independently, there is apparently distance between the venture and the university, and between the subsequent potential ownership interests of faculty members and the university, to protect academic freedom and the "quality of teaching and research." Third, the partnership is established to serve the whole faculty, with benefits ostensibly available to all based on the scientific and commercial merits of their discoveries as determined by an independent invention review committee, rather than involving one specific technology and named professors.

The end of the Harvard story is still to be told, but it has some interesting footnotes. After several years as chief operating officer of Biogen, Walter Gilbert resigned. He was accepted back at Harvard as a tenured professor. Mark Ptashne, by contrast, remained a Harvard professor and went on to form a company called Genetics Institute, which eventually raised between $200 and $300 million in capital and expects to bring its first product to market in 1990. Harvard has realized considerable income from license fees paid by Genetics Institute.[26]

M.I.T.: THE M.I.T. DEVELOPMENT FOUNDATION, INC.

The Harvard venture capital partnership has some interesting antecedents at other universities, although the legal form of most of these earlier developments was quite different. In the late 1940s, for example, M.I.T. was a shareholder in the American Research and Development Corporation, perhaps the first venture capital firm and certainly a predecessor to the modern venture capital partnership. It was founded by M.I.T.'s President Karl Compton, the institute's treasurer Horace Ford, M.I.T. professor Edwin Gilliland, and Vermont senator Ralph Flanders, who was an M.I.T. Corporation member.[27]

A more current example is the M.I.T. Development Foundation, Inc. (MITDF) which presents some interesting parallels to recent university venture capital experiments. MITDF was founded in April 1972 and went out of existence in December 1977. It was the brainchild of Richard S. Morse, the inventor of frozen orange juice and one of the founders of National Research Corporation. Morse had a distinguished career in business and government. In 1972, at the age of about 61, he turned his attention to M.I.T. and to the problem he saw in commercializing M.I.T.'s research. He convinced the members of the board of the M.I.T. Corporation to establish a nonprofit subsidiary to help technology transfer at the institute.

The original concept of MITDF was broad and high-minded. Its purpose was to "assist in the generation of new enterprises and serve as a communication link between Government, Industry and Venture Capital sources as they may be interested in M.I.T. technology and other activities."[28] It would do this by sponsoring courses and symposia on the technology transfer process and by being a clearinghouse for M.I.T. technology. It would try to "match" this technology with a combination of resources needed to bring the technology to commercialization. It would also directly involve itself in technology transfer projects by "advising technical entrepreneurs on business planning and market forecasting and providing limited amounts of

'seed funding' to assist these functions" and by participating "in the formation of new, profit-making enterprises where such are determined to be the
most effective mechanisms for technology transfer."[29] The founders "intended that the Foundation will keep an equity position" in these start-up
companies "for the benefit of M.I.T."[30]

The foundation was started with a $50,000 grant from the M.I.T. Cor-
poration and eventually attracted a total of about $470,000 from corpora-
tions, which pledged $25,000 each. These were strictly contributions; the
donating companies gained no equity in the new venture, and MITDF was
controlled entirely by M.I.T., which appointed its directors. Richard Morse
was the president of MITDF, and John Flender, who had extensive small
company experience, was hired as treasurer.

The MITDF soon ran into problems. The high-minded principles behind
its founding were not well understood either by people inside M.I.T. or by
the general public. MITDF began to look to some like a money-making
venture. Further, the mid-1970s were not particularly favorable to new ventures. It is remarkable that MITDF was able to fund the three enterprises
that it did. These three were probably the only successful businesses started
in the Boston area during this time.

Then there was the Sala Magnetics affair. Magnetic Engineering Associates, later known as Sala Magnetics, one of the companies with which
MITDF became involved, received a license for a certain technology from
M.I.T. Unfortunately, another firm in the Midwest, Magnetics International, was extremely interested in licensing the same technology. Angry at
being turned down by M.I.T., an officer of the company began questioning
the relationship between M.I.T. and Sala Magnetics. Upon learning of
M.I.T.'s ownership of Sala, the officer complained to Congress and to the
National Science Foundation, which had funded the research that resulted
in the discovery. His charge was that M.I.T. was attempting to line its pockets with profits from publicly owned intellectual property. He insisted that
M.I.T. owed it to the public to license the discovery to the company best
able to bring it to market. NSF eventually called for an investigation. Although Morse blamed M.I.T.'s patent office for the bulk of the trouble, the
embarrassment was widely attributed to MITDF.[31]

This publicity and the "money-making" reputation of MITDF may have
led to an IRS investigation that in turn led to a ruling that the foundation
would lose its tax-exempt status if it received compensation for rendering
services to start-up companies. This effectively meant that unless MITDF
could cash in on some of its start-ups, its ability to fund its operations
would stop when it ran out of its original donated stake.

The MITDF had other problems as well. Many of the faculty interviewed in this study associate the name of the MITDF both with M.I.T.'s
failed attempt at developing start-up companies and with an instance of
conflict of interest involving a faculty member who had developed a technology of interest to MITDF. M.I.T. professors also soon made clear that

they were not anxious to cooperate with MITDF. In December 1976, Morse complained in a letter to Chancellor Paul Gray:

> We are very discouraged with reluctance of the faculty to place the long term financial interest of the Institute on a par with, much less ahead of, their own personal financial goals. There is often little interest in providing an equity position for M.I.T. if by so doing a faculty or staff member reduces its [sic] own participation. Consulting activities are also often in direct conflict with the financial interests of the Development Foundation at M.I.T. . . . The nature of the academic system and often the naivete and material aggressiveness of the faculty members create an environment in which it is difficult for the Foundation to function in maximum effectiveness.[32]

Faced with these problems, Morse sought a fuller commitment on the part of the M.I.T. Corporation, including an operating budget of $100,000 per year, to sustain MITDF's operations. He was turned down. Following a letter dated November 16, 1976, from Weisner to Morse informing him of this decision, the board of directors of MITDF voted to terminate the activities of the foundation on January 14, 1977. MITDF returned $133,000 in cash to M.I.T., plus its equity stake in the three corporations it had founded: Rheocast, Sala Magnetics, and Surftech Corporation.

M.I.T.: THE TECHNOLOGY LICENSING OFFICE

Like Harvard, M.I.T.'s early retreat from commercial ownership was reversed in 1988. M.I.T.'s Technology Licensing Office (TLO), described in chapter 7, issued about 100 licenses in the fiscal year that ended on June 30, 1988. In eight of these license agreements M.I.T. received equity in the licensee. The value of this equity stake, based on first-round financing (and therefore highly questionable), was about $3.5 million, an amount in excess of total cash license revenue for the year. All these start-up investments are capitalized in excess of $1 million, and two of the companies created by the TLO are the largest start-ups for the year in the Boston area—American Superconductor Corporation ($4.5 million) and Immulogic Pharmaceutical Corporation ($3.25 million).[33]

Unlike the situation at Harvard, M.I.T.'s retreat from the MITDF and later institution of new ventures under the TLO was not accompanied by extensive publicity or faculty protest. The director of the TLO, John Preston, and a committee of knowledgeable faculty members have been trying to "manage" quietly any conflicts and problems that might be anticipated from M.I.T.'s ownership of start-up companies. Because this practice is so new, its full story cannot be told, and all possible problems cannot be anticipated. However, a description of the steps M.I.T. has taken so far and the way it has attempted to avoid potential problems illustrates many of the issues at work.

To begin with, the TLO determined to found start-up companies with care and according to established criteria.

> Our criteria for deciding to set up a new company are based on the upside

potential for the technology. It must be an invention that can create a business that is greater than $100 million. Also it is preferable to avoid single product based company. . . . In our opinion, it is the same amount of work to create a "high flyer" million dollar plus startup as a $50,000 startup. . . . We seek out venture capitalists with a proven track record of taking very early stage deals into successful companies.[34]

It also took steps to handle such potentially controversial internal matters as conflict of interest, royalty sharing, and management control.

CONFLICT OF INTEREST. The most serious potential problems facing the TLO have to do with conflict of interest with faculty members. The problems in this area experienced by the MITDF arose primarily because MITDF was putting its own money into companies that had obtained M.I.T. licenses and were founded and owned by the professors who had made the underlying discoveries or inventions. M.I.T. thus appeared to be entering into collusion with faculty members to divert institute resources into commercial ventures and to enrich the faculty members. An undercurrent of favoritism was also implicit in such arrangements.

The institute and faculty stockholders also may come into conflict as start-up companies face the difficult early stages of development. The TLO tries to avoid this problem by obtaining equity only in return for the value of intellectual property (and not for an actual dollar investment) and by discouraging faculty involvement in the management of the company. Usually the amount of equity distributed to faculty members through royalty-sharing arrangements is too small to arouse conflict-of-interest issues of this kind.

There had been a tacit policy at M.I.T. that the institute would not accept research funding from licensees in which either it or the inventor had an equity interest. Given the expected increase in such licensees because of the new practices of the TLO, however, this policy is under review and probably will be modified.

Another area of concern, made very evident in the MITDF-Sala Magnetics dispute, has to do with "pipelines" for M.I.T. technology. Pipelines are formal or informal arrangements whereby a particular licensee gains a favored position for future licenses, either by virtue of its association with a faculty member, financial ties to the institute, or the ownership of related licenses. Such pipelines foster the fear that the institute will not license the "best" licensee and thus will hurt the public interest. M.I.T. has tried to calm such fears by establishing policies that explicitly do not commit future inventions to particular licensees. The TLO also has established practices for publicizing inventions and evaluating licensees.

ROYALTY SHARING. When cash royalties are involved, royalty sharing between the university and the inventor is easy. When shares of stock in the licensee are involved, however, especially when there is no market for the stock (as is usually the case with start-up companies), royalty sharing be-

comes more difficult. The TLO attempts to solve this problem by distributing shares to inventors upon receipt of them. A related problem arises when
there are many inventors or when a department is to receive a royalty share.
This problem has not been fully resolved.

MANAGEMENT CONTROL. Start-up companies often require a great deal
of attention, but universities are not experienced in the management of such
investments. M.I.T. deals with this potential problem in two ways. First, in
most such situations, it relies heavily on a reputable venture capital partner
to handle early management decisions and assemble the start-up package.
M.I.T. will not go into a start-up venture without being completely confi-
dent in its partners. The institute relies on the venture capital partner to
protect the name of M.I.T. and not use it improperly in attracting other
investors or in other financial transactions.

Second, M.I.T. relies on its own venture capital portfolio manager to
make decisions necessary to manage the investment. When the TLO receives
stock, it gives the stock to the M.I.T. treasurer, who then handles it just as
he would any other venture capital investment—voting the shares, deciding
to buy more shares to maintain an undiluted stake in the company, and so
on. The treasurer continues this management function on behalf of the TLO
until the company goes public or there is a clear market for the shares, at
which time the treasurer either sells the stock or retains it and adds it to the
established venture capital pool. In either case, the proceeds go back to the
TLO to be used in accordance with the guidelines established for funding
the TLO.

The accommodations M.I.T. has made to manage its ownership share
in companies based on its research are possible because of the institute's
unique culture and administrative structure. Its trust in venture capital
firms, its access to an experienced in- house venture capital portfolio manager, and its ability to alter its policies quickly are advantages that few other
universities have. Nevertheless, what M.I.T. has learned and what it has
done so far provide significant examples for other universities.

PENN STATE

In September 1986, Dennis Costello, a principal in the seed venture capital firm Zero Stage Capital of Cambridge, Massachusetts, which had offices
just across the street from M.I.T., saw an article in *The New York Times*
describing the success of a team at Penn State's Materials Research Laboratory in producing diamond coatings. Costello called the lab and set in mo-
tion the development of a business venture that was to become the first
start-up company in which Penn State owned stock—Diamond Materials
Institute, Inc. (DMI)—and to launch Penn State's efforts to develop new
businesses in Pennsylvania.

The events leading to the founding of DMI began in the early 1980s and
came from two directions. The first events were related to the science and

technology involved. In 1984, Rustum Roy, professor of materials science at Penn State, returned from a trip to Japan where he had witnessed the demonstration of a new technique for producing diamond films. He established a research team at the MRL, and by February 1986 this team had succeeded in producing thin diamond coatings by using a new process that could be performed at relatively low temperatures and pressures. In September 1986, Roy and his colleagues formed the Diamond and Related Materials Consortium with a number of industrial firms to attempt to transfer the technology to the companies.[35]

The second group of events related to DMI's founding were economic and political. The early 1980s saw the state of Pennsylvania engaged in efforts to revitalize its economy. An element crucial to that development, venture capital, did not seem available in the state. In 1983, for example, firms in Pennsylvania raised only $32 million in venture capital, less than 1 percent of the nation's total.[36] Officials turned first to the state's pension funds, passing legislation that enabled these funds to invest up to 1 percent of their assets in venture capital projects. However, pension fund investments normally did not go to very early stage (seed) ventures, which are the most speculative and problematic of such investments but also are exceedingly important for developing technology from university research. To fill this gap, the state allocated $3 million of a $190 million bond issue to "seed" venture capital efforts. Each of the state's four advanced technology centers, which administered programs under the Ben Franklin Partnership Board, was to receive one-fourth of this money to establish a venture capital fund, provided it could find competent fund managers and could attract at least three times the state's capital contribution.[37] Penn State, as administrator of one of the advanced technology centers, thus was pushed into the venture capital arena. It first pledged $250,000 of its own "endowment-like" funds to the new entity that was to be established. This pledge, the university's first allocation of its investment portfolio to venture capital, represented less than 1 percent of its total assets. Penn State then secured the services of the venture capital managers at Zero Stage Capital in Massachusetts. The newly formed Pennsylvania-based Zero Stage Capital II soon became the most successful of the seed venture capital endeavors sponsored by Pennsylvania, eventually attracting almost $10 million in funds.

Costello and his colleagues at Zero Stage Capital II faced a number of problems after the initial formation of the new diamond technology company. First, the intellectual property rights underlying the new technology were not protected by patents or other means since they were products of Japanese and Soviet technology. The expertise resided in the heads of two of Professor Roy's colleagues who were doing the diamond film development work. They had not been consulted about starting the company. Worse, because of some loosely worded agreements between the university and the corporate members of the Diamond and Related Materials Consortium, it appeared that these corporate members had some rights to the technology.

Because the researchers refused to be involved with DMI due to the Material Research Laboratory's commitment to the consortium members, a long and acrimonious negotiation ensued, including meetings with the highest levels of university administration. The faculty members finally agreed that DMI could join the consortium and that the faculty would not consult with other companies on the subject technology.

The next steps were to secure additional capital for the new company and to find someone to run it. Zero Stage used its contacts to perform these two steps successfully, and DMI was incorporated in February 1987.[38]

Penn State received "founders' stock" in DMI in return for its rights to the intellectual property acquired by DMI. The university decided to invest additional funds in the new company to protect its founders' preferred shares against dilution and also, apparently, to test the waters of public and faculty opinion regarding such an action. Penn State announced that it had invested $15,000 in the new venture, but, significantly, it did not announce that this investment was but the first of four "gates" of financing and that the university expected to meet at least three subsequent cash calls, adding up to a considerably larger investment. The university also did not make public the role of its affiliated Ben Franklin Partnership program in the development of DMI.

"PSU Research, Funds, Spawn New Firm" was the headline on the December 23, 1987, front page of the *Centre Daily Times,* the main newspaper in State College.[39] An editorial in the same paper in February began, "Penn State's $15,000 investment in a new State College company, Diamond Materials Institute Inc., marks what should be just the beginning of a more aggressive approach to technology transfer that will have many benefits."[40] It continued, "The University's increasing efforts to put its research to work in the private sector are welcome and commendable."[41]

Not all public comment was so favorable. Under the headline "Unprecedented Investment Raises Questions of Ethics," a state legislator asked: "Where did Penn State get the money that it spun off? . . . We give them $200 million in state appropriations and we give it to them primarily to keep tuition down, not to use it as venture capital."[42] An article in the same edition quoted Rustum Roy, the originator of Penn State's involvment in diamond technology, as speaking against the investment in a faculty senate meeting: "Are we in the manufacturing business? . . . Why is it that money-making becomes a value for the university?" The article added that "Roy's remarks were greeted with applause by members of the Senate."[43]

These protests apparently will have little effect. Bolstered by an "economic development initiative" developed by the university administration and endorsed by the board of trustees, which includes the development of programs designed to "assess potential venture capital and equity investments by the University to promote economic development,"[44] Penn State plans to continue its venture capital investments. Partly as a result of the DMI experience, Penn State has adopted a low-profile approach to subsequent ventures. By September 1989, Zero Stage Capital II had raised $10 million in funds and had started 11 companies, three of which (including DMI) were based on technology developed by Penn State.[45]

Prospects for DMI's success are still questionable, however. By September 1989, DMI had completed product development and was seeking further financing which, considering its backlog of orders, it is believed will push it to a break-even position.[46] Professor Roy and his colleagues still feel that the whole experience was destructive of collegial relations and take every opportunity to point out the extent of diamond technology conveyed by the MRL to Pennsylvania and other companies.

U.C. BERKELEY

In the late 1960s, two U.C. Berkeley computer scientists developed software in the emerging area of time-sharing systems that they believed would revolutionize the computer world. They formed a company, Berkeley Computer Corporation, to exploit this new technology. The technology seemed so promising that the ultraconservative treasurer of the Regents of the university invested about $1 million in endowment funds in the company. However, after a series of management problems and difficulties between the two principals, into which the university as a principal stakeholder was inevitably drawn, the company failed, and the treasurer had to write off the university's investment. The loss of time and effort, the demonstration of the university's obvious lack of expertise in dealing with start-up companies, and the bad public relations caused by this debacle resulted in the adoption of a new policy: Henceforth, venture capital investments would be made only through established funds managed by experienced venture capitalists.[47]

In the late 1970s, Harvey Blanch, professor of chemical engineering at U.C. Berkeley, made some findings of considerable potential value in the field of fermentation. Because the products of many commercial fermentation processes, such as organic acids, alcohols, and solvents, are toxic to the organisms that produce them, the processes are considerably slowed and the recovery of pure products is difficult. Blanch, his colleague Channing Robertson at Stanford, and several other researchers at M.I.T. had been working on ways to remove the products quickly from the fermentation process, thus speeding the process and improving the degree of product concentration and purification.[48]

Capturing the attention of a number of large chemical manufacturing companies and capitalizing on the wave of interest in biotechnology, Blanch, his colleagues, and university officials developed a widely publicized plan to support further research and, at the same time, secure for the researchers and the universities involved a share of the anticipated commercial value of the developed processes. Under the plan a nonprofit Center for Biotechnology Research was formed to fund research at the three universities. A for-profit corporation named Engenics, Inc. was also formed to carry the research to commercial application. Thirty percent of Engenics stock was owned by the Center, 35 percent by Blanch and his colleagues, and the remaining 35 percent by six corporate sponsors, which each contributed $1.7 million to fund the two ventures. Of the $10 million contributed by these sponsors, $2.4 went to the Center for the immediate support of research and $7.6 went to Engenics to fund start-up costs and subsequent research and development.[49] In return for five years of research funding from the Center, the university, in a contract signed on September 1, 1981, agreed to grant Engenics an exclusive, royalty-bearing license for the life of any patent that resulted from the research supported by the Center. Significantly, this agreement was signed on the Berkeley campus; the treasurer of the Regents was not involved in the transaction. Blanch became a member of the board of directors of Engenics and a consultant to it.[50]

This arrangement was thought to avoid most of the pitfalls of university start-up ownership. The university's interests were protected by its ownership of patents, its licensing arrangements and income under the agreements, and the fact that the Center, which was dedicated to funding university research, would be the recipient of dividends and capital gains appreciation if the venture was commercially successful. The university researchers gained an equity stake in the new company while at the same time maintaining their independence and autonomy (at least theoretically) by receiving their research funding from the Center. The corporate sponsors, in addition to their equity stake in the new company, got a "window" on the new research and, presumably, the right to benefit from the technologies developed by Engenics.

The outcome was that Blanch and his colleagues did receive a considerable quantity of research funds, although after the first round of funding, money came in smaller amounts and with more conditions and restrictions attached—just as was happening in other parts of the biotechnology field. The board of directors of Engenics, apparently dissatisfied with the pace of new product development, replaced the first president and chief executive officer. Then, in 1987, after having trouble raising enough money to finance the production of its first products, Engenics was purchased by a larger corporation for the value of its physical facilities and the potential of the products it had rights to, even though none of them had achieved commercial success.[52]

MCGILL[53]

In November 1980, while commuting to work, Irving DeVoe, chairman of McGill University's department of microbiology, and his young associate, Bruce Holbein, began discussing how their work in removing iron from growth medium to control the growth of bacteria might be applied to producing pharmaceuticals and stopping the spoilage of foods. Rushing to their laboratory, they succeeded in removing iron from a liquid with their process. Later they concentrated on using the process to remove toxic metals from industrial waste and to remove gold from mine tailings. In 1982 they received a patent on the process and began trying to interest venture capitalists in their work. All the deals the two were offered, DeVoe felt, left them with too little equity and too little say in the management of the resulting company. As he later put it, "They offered to skin us alive."

Then, through their lawyer, DeVoe and Holbein were introduced to another potential investor, Irving Knott. Knott had considerable experience in organizing companies, but he did not have a high reputation. In 1976 he had pleaded guilty to a stock fraud scheme, paying a $500,000 fine, and at the time of the meeting with DeVoe and Holbein he was under indictment for defrauding a Montreal factoring concern of $4.4 million. DeVoe apparently was aware that Knott had had "some problems in the past," but turning to Knott and his "financial consulting firm" was "the only way we could raise capital at the time." When Knott visited the university laboratory, he was introduced only as "Mr. K." Selling the deal involved extensive visits by businesspeople to the laboratory and disrupted laboratory work. "The guys in three-piece suits were in and out of the office all the time," one graduate remembered. "The whole department was basically taken over."

Under the plan, McGill made an exception to its policy of not investing in faculty-owned companies and agreed to take 20 percent of any profits resulting from university inventions and accepted 8 percent of the stock of the newly formed DeVoe-Holbein, Inc. The two scientists acquired a 31 percent stake in the new company, and McGill agreed to rent a part of its lab to the company. When plans for a public offering failed to materialize, the company's stock was sold internationally through a Burmuda investment house controlled by Knott. The stock was promoted in part through an investment newsletter, also controlled by Knott, and in part through a prospectus that noted the presence of the company on the McGill campus and the university's holding of 100,000 shares of DeVoe-Holbein stock.

The initial offering, at 50 cents per share, raised $1.6 million. Subsequent sales of the stock by Knott-controlled companies and by the two scientists were at $5, and later $9, per share. The story of the subsequent sale and acquisition of DeVoe-Holbein stock is complex and involved a holding company called Belgium Standard. A series of investigations by the Quebec Securities Commission later resulted in a suspension of trading in Belgium Standard's stock. Meanwhile, DeVoe-Holbein cut its ties to Knott and, running out of its original $1.6 million stake and with no product yet on the market, began looking for additional capital. At this point it had 21 million shares outstanding.

All this was not secret. Articles began to appear in McGill's student paper and in the local newspapers alleging conflicts of interest and the misuse of government research funds. McGill hired a Montreal lawyer to investigate the charges; he concluded that there were indeed conflicts of interest caused by the professors' simultaneous associations with their company and with the university. He also found that government grant money had been used to purchase equipment for the company and to pay DeVoe's wife under her maiden name. DeVoe and Holbein went on unpaid leave, DeVoe resigned as department chair, and 11 members of the department of microbiology wrote a letter to the university administration stating, "It is highly unlikely that either professor could return to our department and regain any degree of trust, confidence, or collegiality."

Problems Associated with University Ownership of Research-Based Companies

The preceding case histories illustrate many of the problems associated with university ownership of companies based on university research. They also support Stanford's policy of not accepting stock in a company in which one of its professors has a stake. This section summarizes and comments on these problems.

CONFLICTS WITH FACULTY MEMBERS

Financial involvement of universities in companies based on the research of their faculty members inevitably confuses the relationship between the institution and its faculty. Of critical concern is the possible influence that such ownership might have on promotion and tenure decisions. When a faculty member is crucial to the success of a university investment, can the university possibly exercise, let alone preserve the appearance of exercising, only academic criteria in granting promotions or tenure to that faculty member? It is impossible to hide university-faculty financial involvement from those making promotion and tenure decisions, and in any case it seems unwise to deny knowledge affecting the institution to concerned individuals, for example, the university president. Because such financial arrangements are sure to become general knowledge, jealousy and charges of favoritism from members of the faculty not willing or able to produce commercially valuable research will inevitably result.

The action of a university joining with a faculty member on a commercial venture seems to contradict many aspects of the relationship that is supposed to exist between a faculty member and the university. The institution seems to be aiding and abetting the very type of behavior that it has developed policies and procedures to prevent or control—that is, it appears to be encouraging the faculty member to neglect the duties of teaching and research in favor of promoting financial interest. It also may appear to be sanctioning the use of its facilities for direct commercial benefit, diverting them from teaching (as in the McGill case); exerting an influence on the research agenda by encouraging research that would be helpful to commercial ventures; fostering secrecy in university laboratories as it tries to keep discoveries to itself until it can get them into the hands of a university-owned company; and perhaps fostering conditions for the exploitation of graduate students as well.

These same arguments might be used against university exploitation of patents, but there are important differences between selling patent licenses and owning significant stakes in the kinds of companies discussed above. In selling patent rights, the university is benefiting directly from the research performed by its faculty member in the normal course of his or her duties. In owning a company dedicated to exploiting such research, however, the university is removing itself from the academic work of the faculty member and relying for profit on the faculty member's nonacademic skills.

Finally, when a university and faculty members share ownership of a speculative, start-up company, there is ample opportunity for disputes between the university and the faculty members over the management of the company. These disputes can easily spill over into and damage other aspects of the relationship.

FACULTY OPPOSITION

Many faculty members remain opposed to university ownership of research-based companies. In the interviews of faculty members reported in chapter 10, the idea of direct investment in university-related companies rated the lowest on faculty attitudes toward the itemized transfer mechanisms. Surprisingly, faculty members at Berkeley, although opposed to this kind of equity holding overall, showed a more favorable attitude toward it than faculty members of the other three universities of the study. An analysis of responses by field indicated a generally positive response only among faculty in the "other engineering" category. There did not seem to be any logical reason for these two anomalies in the responses, except that the concept of university ownership of start-up companies was so novel to Berkeley faculty that perhaps they did not comprehend the possible negative consequences of such ownership.

As shown in the Harvard case, faculty opposition and misunderstanding of the issues is a major obstacle to policies favorable to university ownership of research-based start-up companies.

FINANCIAL AND MANAGEMENT CONFLICTS

Universities rarely have the skills and knowledge needed to recognize good start-up investments or to make the kinds of decisions necessary to guide a new company successfully through its crucial early stages. This was clearly demonstrated in several of the case histories presented earlier in this chapter; certainly the Berkeley Computer Corporation was an example, as was, perhaps, Engenics.

Another potential problem is that because universities usually have large pools of money to invest, they are expected to participate in subsequent rounds of financing—they must meet "cash calls." This means that initial investments often obligate a university to an unknown amount of future investment, that the university must closely monitor the activity of and prospects for a company in which it has invested in anticipation of requests for more financing, and that it can be accused of torpedoing the company if it refuses to participate in later rounds of financing.

THREATS TO REPUTATION

A crucial element in attracting investors to venture capital projects is the stature of the other investors and the extent of their involvement in the company. When university research forms the basis for a company, university and faculty financial involvement in the venture sends a powerful signal to potential investors. This gives such a company a competitive advantage, but it also makes the university vulnerable if the investment fails. Not only is the university's name associated with the failure, but investors who relied on the university's association with the investment also are angered, and any irregularities associated with the failure, as in the McGill case, are identified with the university.

In fact, the university's reputation is at risk even when such investments are successful. Universities are not supposed to speculate in risky ventures, as people pointed out in regard to Penn State's DMI investment. Furthermore, the university's image may be harmed if it supports a company that markets a controversial product, such as a pesticide, or that makes controversial managerial decisions concerning, for example, the pricing of a life-saving drug, the hiring of nonunion employees, the transaction of business in South Africa, or the disposal of toxic wastes. In order to preserve its public image, the university may have to take a stand on internal management decisions that is opposed to the views of the rest of the stockholders, perhaps coming into conflict with its own faculty members.

Another related problem, somewhat apparent in the MITDF-Sala Magnetics case and clearly evident in the Penn State-DMI controversy with the research consortium members, is that universities that go into start-up business investments may appear to be refusing to cooperate with the very business community to which they should be transferring technology. In fact, as Walter Gilbert observed in relation to the proposed Harvard genetics company, the university may come to be viewed as a competitor to established firms in the field.

THREATS TO INDEPENDENCE

Perhaps the most serious and philosophical concern about university ownership of stock in research-based companies is that such ownership seems to contradict the traditional privileged role that universities occupy in our society. Universities are supposed to exist for the pursuit of truth;

they are supposed to be bastions of objectivity and to serve the public by offering critical analyses uncolored by commercial concerns. When universities own stock in companies, especially when faculty members also have a stake in the companies, the institutions step down from their lofty perches and their ability to continue to render objective judgments comes into question.

Certain recent controversies in the pharmaceutical industry, even though most involve only individual university researchers, shed light on this problem. The January 26, 1989 *Wall Street Journal* reported a class action suit by the stockholders of Spectra Pharmaceutical Services, Inc. against Sheffer Tseng, formerly of the Harvard Medical School, who was accused of withholding evidence that Spectra's eye treatment did not work very well. According to the allegations, which also implicated another Harvard researcher, Tseng sold his rights to the product and his extensive holdings of Spectra stock after conducting studies that revealed the product's shortcomings but before the results of the studies were released. Because of this controversy and because of evidence that some of the studies were inappropriately altered, Tseng and his Harvard supervisor lost their National Institutes of Health funding.[54]

The apparently widespread practice by pharmaceutical companies of providing researchers with stock or stock options has raised questions about the independence of a number of other university researchers, including Burton Sobel of Washington University, who was given stock options in Genentech Corporation while he was conducting clinical trials comparing Genentech's heart drug, TPA, with a rival drug, streptokinase.[55] Although these controversies involve only individual researchers, the potential for damage to the reputation of their universities is clear, and such potential obviously would be intensified if the university itself were a stakeholder in the companies involved.

Another threat to the university's independence and reputation that may occur when the university owns stock in research companies was discussed briefly in relation to the M.I.T.-Sala Magnetics dispute and was also present in the Penn State-DMI controversy and the McGill case. This is the matter of "pipelining"—the appearance that the university is giving unfair advantage to a particular company during transfer of technology. University ownership of equity in a company provides an obvious motive for such favoritism. The public and, even more, competitors in the field are likely to complain that funneling promising technology to a university-owned company is blatantly self-serving and inappropriate for a public or quasi-public institution. This is especially true when the research that spawned the technology was supported by public funds. As was charged in the Sala Magnetics affair, such a situation provides the appearance that the university has used public money to make a profit for itself.

Recognition of these problems, pitted against the allure of considerable financial return that this type of investment might bring, has led or is likely to lead to a number of policy changes, adaptations, and philosophical shifts in universities. This chapter concludes with an examination of some of these changes.

Mitigation of Problems

Despite the potential problems of university investment in research-based companies and the many well-publicized cases in which universities have been criticized for such investment, almost every major university in the country and many smaller institutions as well have undertaken some form of company ownership. Given what seems to be an inevitable increase in this phenomenon, it seems important to consider the possible ways of avoiding or mitigating its negative aspects that can be drawn from the preceding examples and discussion. Many of these methods are being considered or are already being applied at various universities.

USE OF BUFFER ORGANIZATIONS

It is clear that in most cases where a long-range commitment to institutional involvement in start-up companies has been made, universities employ buffer organizations such as those described for Harvard and Johns Hopkins. Even M.I.T., which does not presently use a formal buffer organization, provides itself with some measure of buffering through the Technology Licensing Office, the treasurer's office, and selected venture capital firms. Although complete shielding of a university from controversy and trouble is probably impossible, buffer organizations do provide an important layer of insulation between the institution and the commercial activity surrounding the development of research. The degree of insulation is measured by the degree of independence from the university enjoyed by the managers of the buffer organization.

Buffer organizations also have other advantages that make them the option of choice for universities desiring to become active in the commercial arena. First, they can attract professional managers who are equipped by background, experience, and temperament to make the technology evaluations, investment choices, and hard management decisions that are necessary to make a start-up company successful. Buffer organizations also can provide liberal enough compensation plans to attract the best managers.

Buffer organizations provide a convenient separate entity for attracting outside investors as well, thereby preserving the conservative nature of university endowment management. (Neither the Harvard nor the Johns Hopkins plan involves any dollar investment by the university; the university's stake in the venture capital firm is based on its contribution of patents or other intellectual property, and the compensation for the venture capital managers is based on the success of the ventures, sometimes involving an equity stake in the companies started by the fund.) Finally, buffer organizations provide legal protection to the university, and, perhaps more important, they also provide some protection from negative public relations.

In forming buffer organizations, universities must balance between providing the organizations with enough autonomy to supply the required insulation from university activities, on the one hand, and retaining enough control so that the organizations do not engage in activities inimical to the university's interest or cause the university embarrassment, on the other. Buffer organizations must anticipate "ethical investment" challenges. For instance, investing in a company whose business is the production of poison gas clearly would be considered inappropriate. Control is usually exercised by hiring managers and filling seats on the board of directors of the organization with people whose judgment the university administration trusts and who are sensitive to university concerns.

PROACTIVE ESTABLISHMENT OF PROGRAMS AND POLICIES

In most of the negative examples discussed earlier, the university involved was prompted to change established policies or to enter into ownership of a research company because of a single, very promising discovery or technology. The lure of significant financial return was overwhelming. Derek Bok, in describing his initial reaction to the proposed Ptashne company, said, "The prospects seemed all but irresistible."[56] Irresistible prospects also were obviously involved in the cases of Berkeley Computer, Engenics, McGill, and perhaps DMI.

The fact that most of these investments did not actually result in significant returns to the universities and that they led the universities to reconsider relevant policies is instructive. Reaction to a single target of opportunity, no matter how promising, causes problems. Universities need to learn from Harvard's experience: After a long and careful consideration of the matter, the university established a comprehensive program, a process for evaluation and development of opportunities as they arose, and a supporting set of policies consistent with these practices. Instead of being reactive and subject to the charge that it is being shaped to favor a particular researcher, such a program is proactive, well defined in advance, and applicable to every university researcher on the same terms.

DEVELOPMENT OF POLICY MODIFICATION AND "MANAGING" MECHANISMS

When universities become active investors in start-up companies based on faculty research, and particularly when faculty members also are owners of such companies, considerable policy development or reconsideration must take place. University ownership opens a Pandora's box of possible conflicts and difficulties that must be dealt with. The wide variety of possible situations requires that the policy be flexible.

Policies that allow possible conflicts of interest to be "managed" on a case-by-case basis are being developed at some universities, including M.I.T. The essence of these policies is disclosure of all proposed financial relationships between the faculty member and the company and between the company and the university. These disclosures are made to a panel of academics who rule on the propriety of the arrangements. As decisions are made by these panels, a body of "case law" is developed to provide continuity and consistency as membership on the panels changes. Such flexible provisions are the only effective way of achieving a balance between the financial goals of university ownership of start-up companies and the maintenance of the university's traditional values and place in society.

Conclusion

Universities cannot ignore the potential financial returns that might be realized through the ownership of research-based start-up companies. Other factors supporting such ownership include the wide acceptance of technology transfer as a legitimate goal of the university and the recognition that the university can play an effective part in bringing embryonic technology to commercial fruition. The next five years will see universities maturing in their attitude toward equity ownership in university-related companies. They will become more realistic about the potential for financial return and, as the kind of policies just described are proven effective, less concerned about the potentially damaging effects of company ownership on their internal relations and public image. The increasing acceptance of university involvement in the ownership, and even the operation, of for-profit companies is perhaps the most important piece of the mosaic we are assembling to make a picture of the effects on the university of technology transfer. When universities engage in high-risk venture capital activities they are signaling an important change from their traditional role. These signals are clearly recognized both inside and outside the academy.

Notes

1. Derek C. Bok, *Beyond the Ivory Tower* (Cambridge, Mass.: Harvard University Press, 1982), 136.
2. Michael Knight, "Harvard Rules out Role for Now in Genetic Engineering Company," *The New York Times,* 18 November 1980, A1.
3. Allen R. Gold, "Harvard, in Reversal, Will Seek Research Profits," *The New York Times,* 15 September. 1988, A37.
4. John Preston, "Comments at the European Venture Capital Association Meeting, Brussels, 26 February 1988" (Cambridge: Technology Licensing Office, Massachusetts Institute of Technology, 1988), 1.
5. Marshall I. Goldman, "Building a Mecca for High Technology," *Technology Review* (May-June, 1984), quoted in Barbara J. Bird and David N. Allen, "Faculty Entrepreneurship in Research University Environments" (Cleveland, Ohio:

Department of Organizational Behavior, Case Western Reserve University, 1986), 8.

6. Gilbert Fuchsberg, "Boston U. Gambles on Its Big Investment in Biotechnology Company," *Chronicle of Higher Education,* 12 April 1989, A29.

7. Daniel Steiner, "Technology Transfer at Harvard University," *Bioethics Quarterly* 2 (Winter 1980): 208.

8. Ibid.

9. Martin Kenney, *Biotechnology: The University-Industrial Complex* (New Haven, Conn.: Yale University Press, 1986), 96–100, 133.

10. Preston, "Comments at the European Venture Capital Association Meeting," 1.

11. Information obtained by author from University of California Regents, Office of the Treasurer, Berkeley, Calif., 17 February 1989.

12. John Hillkirk, "Univ. of Rochester First School to Set Up Venture Fund," *Venture,* March 1982, 6.

13. George Malloan, "How Johns Hopkins Bridges the Lab-to-Market Gap," *Wall Street Journal,* 24 January 1989, A19.

14. Association of American Universities, *Trends in Technology Transfer at Universities* (Washington, D.C.: Association of American Universities, 1986), 31.

15. Herbert B. Chermside, "Some Ethical Conflicts Affecting University Patent Administration," *SRA Journal* 16 (Winter 1985): 15.

16. Nicholas Wade, *The Science Business* (New York: Priority Press, 1984), 42.

17. Ibid.

18. Kenney, *Biotechnology,* 78.

19. Nicholas Wade, "Harvard Marches up Hill and down Again," *Science* 210 (5 December 1980): 1104.

20. Steiner, "Technology Transfer," 207.

21. Ibid., 208.

22. Nicholas Wade, "Gene Goldrush Splits Harvard, Worries Brokers," *Science* 210 (21 November 1980): 878.

23. Peter Hilts, "Ivy-Covered Capitalism," *Washington Post* (10 November 1980), quoted in Wade, *The Science Business,* 44.

24. Knight, "Harvard Rules Out Role."

25. Gilbert Fuchsberg, "Harvard Sets up $30 Million Fund to Back Profitable Discoveries Made by Professors," *Chronicle of Higher Education,* 21 September 1988, A35.

26. Peter Dworkin, "Academia Goes Commercial," *U.S. News and World Report* 2 May 1988, 51.

27. Philip A. Trussell, letter to author, August 1989.

28. Massachusetts Institute of Technology, Office of the President and Chancellor, 1971–1980 (Weisner/Gray) (80-33), M.I.T. Development Foundation, Inc., "Exhibit D," 15 September 1972, M.I.T. Archives, Box 10, File: M.I.T. Development Foundation, Inc.

29. Massachusetts Institute of Technology, Office of the President and Chancellor, 1971–1980 (Weisner/Gray) (80-33), M.I.T. Development Foundation, Inc., description dated 19 July 1972, M.I.T. Archives, Box 10, File: M.I.T. Development Foundation, Inc.

30. Richard S. Morse, "New Enterprise Generation," transcript of talk delivered at the Conference on the Public Need and Role of the Inventor, sponsored by the

Office of Invention and Innovation, National Bureau of Standards, U.S Department of Commerce, Monterey, Calif., 16 June 1963. Available in M.I.T. Archives, Massachusetts Institute of Technology, Office of the President and Chancellor, 1971–1980 (Weisner/Gray) (80-33), Box 10, File: M.I.T. Development Foundation, Inc.

31. The material in this paragraph has been obtained from a variety of sources, including interviews with M.I.T. faculty members, parts of documents located in the M.I.T. Archives, and interviews with people involved in the M.I.T. Development Foundation.

32. Richard S. Morse, letter to Paul Gray, Cambridge, Mass., 13 December 1976, M.I.T. Archives, Massachusetts Institute of Technology, Office of the President and Chancellor, 1971–1980 (Weisner/Gray) (80-33), Box 10, File: M.I.T. Development Foundation, Inc.

33. Preston, "Comments," 2.

34. Ibid.

35. Lisa B. Hawkins, "PSU Research, Funds Spawn New Firm," *Centre Daily Times*, 23 December 1987, A1.

36. Walter H. Plosila and David N. Allen, "State Sponsored Seed Venture Capital Programs: The Pennsylvania Experience," *Policy Studies Review* 6 (February 1987): 531.

37. Ibid., 533.

38. Dennis Costello, interview with author, State College, Pa., 18 May 1988.

39. Hawkins, "PSU Research."

40. "PSU Initiatives Boost Economy," *Centre Daily Times*, 4 February 1988, A4.

41. Ibid.

42. Phil McDade, "Unprecedented Investment Raises Questions of Ethics," *Centre Daily Times*, 31 January 1988, A4.

43. Phil McDade, "Faculty Criticize PSU Initiative," *Centre Daily Times*, 30 March 1988, 1.

44. Pennsylvania State University, "Board of Trustees Hears Economic Development Report," *Intercom* 16 (21 May 1987): 1.

45. Information obtained by author from the Office of the Treasurer of the Regents of the University of California, 15 February 1989.

46. Dennis Costello, telephone interview with author, 28 August 1989.

47. Ibid.

48. University of California, "University of California Research Agreement" (Berkeley, California: 1 September 1981).

49. "This Biotech Alliance Focuses on Processing," *Chemical Week*, 19 May 1982, 23.

50. University of California, "Agreement."

51. Harvey Blanch, interview with author, Berkeley, Ca., 22 November 1988.

52. Ibid.

53. All the material in this section is from Alan Freeman, "McGill Uproar: How a Top University Unwittingly Became Stock-Promotion Bait," *Wall Street Journal*, 27 April 1984, 1, 22.

54. Marilyn Chase, "Bad Chemistry: Mixing Science, Stocks Raises Question of Bias in the Testing of Drugs," *Wall Street Journal*, 26 January 1989, A1, A6.

55. Ibid.

56. Derek C. Bok, "Business and the Academy," *Harvard Magazine* 83 (1981): 26.

Bibliography

ASSOCIATION OF AMERICAN UNIVERSITIES. *Trends in Technology Transfer at Universities.* Washington, D.C.: Association of American Universities, 1986.

"Backlash Against DNA Ventures? More Thought at Harvard on Business Link." *Nature* 288 (1980): 203–204.

BIRD, BARBARA J., and DAVID N. ALLEN. "Faculty Entrepreneurship in Research University Environments." Cleveland, Ohio: Department of Organizational Behavior, Case Western Reserve University, 1986.

BOK, DEREK C. *Beyond the Ivory Tower.* Cambridge, Mass.: Harvard University Press, 1982.

———. "Business and the Academy." *Harvard Magazine* 83 (1981): 23–35.

CHASE, MARILYN. "Bad Chemistry: Mixing Science, Stocks Raises Questions of Bias in the Testing of Drugs." *Wall Street Journal,* 26 January 1989, A1, A6.

CHERMSIDE, HERBERT B. "Some Ethical Conflicts Affecting University Patent Administration." *SRA Journal* 16 (1985): 11–17.

DAVIS, BERNARD. "Sounding Board: Profit Sharing between Professors and the University?" *New England Journal of Medicine* (14 May 1981): 1232–1235.

DWORKIN, PETER. "Academia Goes Commercial." *U.S. News and World Report,* 2 May 1988, 50–51.

FREEMAN, ALAN. "McGill Uproar: How a Top University Unwittingly Became Stock-Promotion Bait." *Wall Street Journal,* 27 April 1984, 1, 22.

FUCHSBERG, GILBERT. "Boston U. Gambles on Its Big Investment in Biotechnology Company." *Chronicle of Higher Education,* 12 April 1989, A29.

———. "Harvard Sets up $30 Million Fund to Back Profitable Discoveries Made by Professors." *Chronicle of Higher Education,* 21 September 1988, A35–36.

GOLD, ALLEN R. "Harvard, in Reversal, Will Seek Research Profits." *The New York Times,* 15 September 1988, A37.

"Harvard Finally backs Off Gene Venture," *Nature,* 288 (1980): 311.

"Harvard Keeps its Hands Clean," *New Scientist* (20 November 1980): 497.

HAWKINS, LISA B. "PSU Research, Funds Spawn New Firm." *Centre Daily Times,* 23 December 1987, A1.

HILLKIRK, JOHN. "Univ. of Rochester First School to set up Venture Fund. *Venture,* March 1982, 6.

KENNEY, MARTIN. *Biotechnology: The University-Industrial Complex.* New Haven, Conn.: Yale University Press, 1986.

KNIGHT, MICHAEL. "Harvard Rules out Role for Now in Genetic Engineering Company." *The New York Times,* 18 November 1980, A1.

MALLOAN, GEORGE "How Johns Hopkins Bridges the Lab-to-Market Gap." *Wall Street Journal,* 24 January 1989, A19.

MASSACHUSETTS INSTITUTE OF TECHNOLOGY. Office of the President and Chancellor, 1971–1980 (Weisner/Gray) (80–33), M.I.T. Development Foundation, Inc. Description dated 19 July 1972. Cambridge, Mass.: M.I.T. Archives, Box 10, File: M.I.T. Development Foundation Inc.

———. Office of the President and Chancellor, 1971–1980 (Weisner/Gray) (80-

33), M.I.T. Development Foundation, Inc. "Exhibit D." Cambridge, Mass.: 15 September 1972. M.I.T. Archives, Box 10, File: M.I.T. Development Foundation, Inc.

McDADE, PHIL. "Faculty Criticize PSU Initiative," *Centre Daily Times,* 30 March 1988, 1.

———. "Unprecedented Investment Raises Questions of Ethics." *Centre Daily Times,* 31 January 1988, A4.

MORSE, RICHARD S. Letter to Paul Gray. Cambridge, Massachusetts, 13 December 1976. M.I.T. Archives, Office of the President and Chancellor, 1971–1980 (Weisner/Gray) (80–33), Box 10, File: M.I.T. Development Foundation, Inc.

———. "New Enterprise Generation." Talk delivered at the Conference on the Public Need and Role of the Inventor, sponsored by the Office of Invention and Innovation, National Bureau of Standards, U.S. Department of Commerce, June, 1963 at Monterey, Ca. Available in the M.I.T. Archives, Office of the President and Chancellor, 1971–1980 (Weisner/Gray) (80–33), Box 10, File: M.I.T. Development Foundation, Inc.

PENNSYLVANIA STATE UNIVERSITY. "Board of Trustees Hears Economic Development Report." *Intercom* (21 May 1987): 1.

PLOSILA, WALTER H., and David N. Allen. "State Sponsored Seed Venture Capital Programs: The Pennsylvania Experience." *Policy Studies Review* (February 1987): 529–537.

PRESTON, JOHN. "Comments at the European Venture Capital Association Meeting, Brussels, 26 February 1988." Cambridge: Technology Licensing Office, Massachusetts Institute of Technology, 1988.

"PSU Initiatives Boost Economy." *Centre Daily Times,* 4 February 1988, A4.

STEINER, DANIEL. "Technology Transfer at Harvard University." *Bioethics Quarterly* 2 (Winter, 1980): 203–211.

"This Biotech Alliance Focuses on Processing." *Chemical Week,* 19 May 1982, 23.

UNIVERSITY OF CALIFORNIA. "University of California Research Agreement." Berkeley: 1 September 1981.

WADE, NICHOLAS. "Gene Goldrush Splits Harvard, Worries Brokers." *Science* 210 (1980): 878–879.

———. "Harvard Marches Up Hill and Down Again." *Science* 210 (1980): 1104.

———. *The Science Business.* New York: Priority Press, 1984.

9

Industrial Liaison Programs at Four Universities

University-industry liaison programs, sometimes known as industrial affili-ate programs, are the most recent, obvious, and important means by which research universities, or units within universities such as departments or schools, have organized relations with industry. In a typical liaison pro-gram, corporate members, in return for an annual fee, are provided with "facilitated access" to the university and the research being done by the university or department. Annual fees typically range from $10,000 to $40,000. "Facilitated access" usually means the right to send one or two corporate representatives to an annual conference summarizing the research being performed by faculty and students in the university; provision of re-search publications (including the privilege of reviewing materials before official publication dates); the right to attend conferences, workshops, and lectures on topics of special interest; and special arrangements that encour-age exchange of information with faculty and students. There are many variations on this basic model—so many that defining liaison programs is difficult.

Liaison programs are a common but relatively recent development in research universities. In both of the only two studies of liaison programs, by Tamaribuchi (1982) and Peters and Fusfeld (1983), about half of the universities were found to have liaison programs (19 of 39 for Peters[1] and 23 of 40 for Tamaribuchi),[2] and both studies indicated that universi-

178

TABLE 9-1   Liaison Programs at Four Research Universities by Year of Initiation

Year of Initiation	M.I.T.	Stanford	U.C. Berkeley	Penn State	Total
Before 1950	1				1
1950–1959		3			3
1960–1969		8		1	9
1970–1975	1			1	2
1976–1978	1	2		1	4
1979	1	3	2	1	7
1980	2	1	2		5
1981		4			4
1982		3			3
1983	2	3		2	7
1984		1	2		3
1985	1	2	1	2	6
1986		1	1	1	3
1987	1			1	2
1988	1				1
Total	11	31	8	10	60

ties that did not have such programs were considering them. Although some liaison programs date back to the 1950s or earlier, most presently existing programs were started after 1970. There has been an especially rapid proliferation of liaison programs during the 1980s. In the Peters and Fusfeld study, 52 of the 92 liaison programs identified were found to have been started after 1978 (in other words, they were less than three years old), and only 19 were in existence prior to 1972.[3] In the study described in this chapter, 68 percent (41 out of 60) of the liaison programs examined were begun after 1978. (Table 9-1 shows the distribution of initiation years for these programs.) This recent proliferation is yet another clear indication of increased university interest in relating to industry, in obtaining industry support for research, and, by implication, in technology transfer.

This chapter extends the findings of Tamaribuchi and of Peters and Fusfeld by examining in detail the liaison programs at the four research universities in this study. First, the liaison programs at each university were identified (see Table 9-2 for a list of the programs identified at each campus). At M.I.T., U.C. Berkeley, and Penn State, where no systematic information about these programs existed, several programs were selected for review, and detailed information was collected on them. The directors of selected programs were interviewed using a summary questionnaire (see Exhibit 9-1). At Stanford extensive information about university liaison programs was available in a report prepared by the development office.[4] In addition,

TABLE 9-2   Liaison and Related Programs at Four Research Universities

Name of Program	Year Started	Approximate Number of Members
M.I.T.		
General-purpose program		
Industrial Liaison Program	1948	325
Focused programs		
Collegium of the Materials Processing Center*	1980	60
Marine Industry Collegium*	1975	100
Biotechnology Process Engineering Center*	1985	45
M.I.T. X Consortium*	1988	24
Industry Power Electronics Collegium*	1983	14
Collegium of the Laboratory for Manufacturing and Productivity*	1980	30
Energy Laboratory Electric Utility Program*	1976	28
Research Laboratory in Electronics Industrial Collegium*	1987	6
Transportation Studies Affiliate Program*	1979	21
Chemical Sciences Industry Forum*	1983	36
Construction Industry Affiliate Program		
Economics Forum		
Communications Forum		
Physics Industry Forum		
Media Technology Forum		
Related programs		
Polymer Processing Center		
Microsystems Industrial Group		
Center for Information Systems Research		
Laboratory for Computer Sciences		
Earth Resources Founding Members		
Center for Energy Policy Research		
Multi-electric Utility Program		
Stanford		
Focused programs		
Graduate School of Business*	1959	154
Aeronautics and Astronautics*	1955	11
Applied Mechanics*	1979	2
Biochemistry*	1979	13
Biological Sciences*	1981	2
Center for Computer Research in Music and Acoustics*	1985	6
Chemistry and Chemical Engineering*	1969	14
Computer Forum*	1969	72

*Programs examined in this study.

TABLE 9-2 *continued*

Name of Program	Year Started	Approximate Number of Members
Construction Institute*	1960	22
Design Division*	1969	15
Earth Sciences*	1950	9
Earthquake Engineering*	1976	18
Energy Modeling Forum*	1980	14
Engineering-Economic Systems*	1982	4
Geothermal*	1981	7
Industrial Engineering/Engineering Management*	1976	7
Information Systems*	1969	15
Materials Science and Engineering*	1969	5
Northeast Asia-U.S. Forum on International Policy*	1979	13
Operations Research*	1983	6
Petroleum Geology*	1981	13
Rock Physics and Borehole Geophysics*	1983	15
Solid State*	1960	48
Space Telecommunications and Radio Science*	1981	5
Stanford-China Geosciences*	1986	4
Stanford Exploration Project*	1983	12
SUPRI-A*	1982	12
SUPRI-B*	1982	13
SUPRI-C*	1985	10
Thermosciences*	1968	14
X-Ray*	1984	1
U.C. Berkeley		
General-purpose program (for College of Engineering)		
Industrial Liaison Program*	1979	325
Focused Programs		
Berkeley Computer Sciences Affiliates	1984	3
Berkeley Microelectronics Affiliates	1984	10
Berkeley Sensor and Actuator Center	1986	12
CAD/CAM Consortium	1980	10
Department of Electrical Engineering and Computer Sciences Program	1979	142
Department of Civil Engineering Program	1980	47
Department of Mechanical Engineering Program	1985	50

*Programs examined in this study.

TABLE 9-2 *continued*

NAME OF PROGRAM	YEAR STARTED	APPROXIMATE NUMBER OF MEMBERS
Penn State		
Focused programs		
Industrial Coupling Program of the Materials Research Laboratory*	1962	40
Center for Dielectric Studies*	1978	30
Consortium on Chemically Bonded Ceramics*	1985	20
Diamond Consortium*	1987	27
Center for Advanced Materials*	1986	10
Biotechnology Institute*	1983	3
Cooperative Program in Coal Research*	1979	8
Cooperative Program in Enhanced Oil Recovery*	1974	4
Institute for the Study of Business Markets*	1983	20
Project Fermi*	1985	10
Related programs		
Center for the Management of Technological and Organizational Change*	1985	8
Center for Automation and Robotics*	1981	8
Corporate Associates Program*	1977	5
Cooperative Program in Metallurgy*	1935	11

*Programs examined in this study.

faculty were asked about their views of liaison programs (see chapter 12). Sixty-five of the 81 faculty members interviewed responded to the question about liaison programs.

The purpose of this chapter is to examine the role and functions of liaison programs at the four selected universities, to describe the programs' operations, and to discuss the implications of this information for changes that are taking place in the role of universities in the technology transfer process. This examination begins by defining liaison programs as well as the many variations and related programs that are sometimes described as liaison programs. It outlines and compares the history of and the institutional setting for liaison programs at the four universites in this study. This is followed by a description of certain other aspects of the liaison programs studied and by a summary of faculty perceptions and comments about such programs. The chapter concludes with a discussion of the important issues and problems facing universities with liaison programs and the role of these programs in the technology transfer process.

Liaison Program Types and Variations

Liaison programs are difficult to examine because of their seemingly infinite variety of organization and operational emphasis and also because they are often confused with, subsumed under, or made to encompass other forms of interaction with industry. This section describes some of the common forms that liaison programs take and distinguishes them from other organizational forms of relations between universities and industry.

Although the terms *liaison* and *affiliate* are often used interchangeably, some see a subtle distinction between them. *Liaison* seems to describe best the standard form of these programs, but in practice this term is used in the title of only a few general-purpose programs. In the four universities in this study, it is used only in the title of M.I.T.'s general program and in Berkeley's College of Engineering umbrella program. *Affiliate* is a much more common term and is used by Stanford and Penn State as a generic title for their focused programs. At M.I.T., where the greatest variety of university-industry interactions exists, the terms *collegium* and *forum* are used to name the focused liaison programs that have developed recently. These terms perhaps reflect M.I.T.'s closer ties to industry. A university with such ties is more likely to recognize industry scientists as colleagues (interacting in a collegium) and believe that the university can serve effectively as a convenor of meetings (forums) dedicated to the joint examination of research issues. Other terms used in titles of liaison programs include *center, consortium, institute, program, cooperative program in——*, and *associates*. Use of these terms is confusing because they commonly describe other kinds of university-industry interactions, some of which have elements of liaison programs within them.

GENERAL-PURPOSE LIAISON PROGRAMS

In the earliest study of liaison programs, Tamaribuchi identified two types: universitywide or general-purpose programs and "focused" programs. General-purpose programs are designed to provide corporate members with access to a wide segment of the university and its research and to provide the university with the funds to operate the programs as well as unrestricted money for the furtherance of research or the support of graduate students. They appeal to companies that seek overviews and updates on new developments in a broad range of science and engineering areas and are most successful in universities that are recognized in many fields.

General-purpose programs usually have "liaison officers" who arrange events, conferences, and meetings to bring corporate members into contact with university research. These officers typically are responsible for keeping up to date with the research in one or several university departments and for maintaining relationships with a caseload of member firms.

M.I.T. has the oldest, largest, and best known general-purpose liaison program. It was started in 1948 and by 1987–1988 had over 325 members, a budget of nearly $8 million, and a staff of about 57, including some 20 liaison officers. Roughly 40 per cent of its income is spent on administrative costs.[5] Despite the obvious success of this program and similar ones at the California Institute of Technology and the Georgia Institute of Technology,

EXHIBIT 9-1   Liaison Program Summary Questionnaire

University: _____

Record locator: _____

Name of program: _____

Name of director: _____

Address of director: _____

Telephone number: _____

When established: _____

Purpose of program: _____

_____

How and why was the program started? _____

_____

Present number of participating companies: _____

_____

Comments about the history of membership: _____

_____

Is the program interdisciplinary? _____

_____

Was the program started with "seed funding"? _____

_____

What type of company belongs (large, small, domestic, foreign)? _____

_____

Annual membership fee: _____

Annual budget: _____

How many employees staff the program? _____

How many students are involved? _____

How many faculty members are involved? _____

What services are provided to members? _____

_____

How is the program directed? _____

_____

Does the university subject the program to review? _____

EXHIBIT 9-1 *continued*

What problems are associated with this program? _____

Remarks, distinguishing features: _____

Peters and Fusfeld reported a significant number of failed attempts at such programs.[6] The three other universities in this study do not have general-purpose programs (although U.C. Berkeley has an umbrella program covering the College of Engineering).

Certainly part of the reason that general-purpose programs are not more common is that heavy, long-term commitment is required to put such a program together and sustain it. Another part is that companies, particularly medium-sized and small companies or divisions of larger companies, view general-purpose programs as being too superficial to justify the annual fee.[7] As a result of these problems, special or focused liaison programs have become far more numerous than general-purpose programs.

FOCUSED LIAISON PROGRAMS

Most focused liaison programs operate in the same way as general-purpose programs but within a more narrowly defined subject area or technology grouping. Like general-purpose programs, they provide companies with information about research and developments in the defined research field for an annual fee. Unlike general-purpose programs, however, they often also provide close association between research faculty and technical people from industry. Their list of benefits is likely to include exchanges of faculty and industry scientists, agreements for licensing the results of sponsored research, and access to students for in-house projects, summer employment, and recruitment. Focused liaison program staffs are generally clerical support; faculty members handle the higher-level administrative and liaison duties themselves. Such programs are more likely than general-purpose programs to be viewed by both sides as two-way collaborations, and faculty members involved in focused programs are more likely to become involved in research directed at solving industry or company problems.

Because of the low (or rather, hidden) overhead involved in administering focused liaison programs and the low out-of-pocket costs of providing member benefits, most of the money received from the programs' membership fees can be spent at the discretion of the sponsoring department. Usually these funds are used to support research and graduate students. This funding feature gives focused liaison programs certain aspects of research consortia, which are discussed in the next section.

One example of a focused liaison program is the Diamond Consortium at Penn State. This program is one of several associated with the university's Materials Research Laboratory. It was founded by Rustum Roy, who saw that the United States was considerably behind Russia and Japan in diamond technology. Roy and a number of his colleagues first attracted "seed" funding from the Office of Naval Research by demonstrating their expertise in diamond technology. The consortium now has about 30 corporate members, each of whom pays an annual fee of $25,000. Members attend two research workshops per year, during which faculty and graduate students make presentations on the latest research and developments in the field. Members also receive a newsletter four times per year and periodically attend informal "information transfer sessions" in which various technological developments are demonstrated. Some 17 faculty members are available to "help our members in any way we can." About seven graduate students per year are funded directly by the consortium.

Focused programs are easier to start and require less institutional commitment than general-purpose programs. However, they do require significant commitment from enough faculty members in a particular area to form a recognizable group. They are more likely to have limited life spans, since both the internal and the external commitments involved in them are less stable and enduring. Because they stress direct interchange between university and industry researchers, focused liaison programs are closer to technology transfer than are general-purpose programs.

RELATED ORGANIZATIONAL FORMS

Certain other organizational forms of university-industry interaction are related to liaison programs. A brief description of some of these forms is necessary to clarify the nature and functions of liaison programs.

RESEARCH CONSORTIA. Research consortia in this context are programs of university research funded by a number of corporate sponsors and often by the federal and state governments as well. Typically, corporations involved in such consortia provide annual funding for ongoing investigation in a specific field of research and are involved to some degree in setting the research agenda. Universities stay in close touch with the corporate sponsors and make periodic reports on research progress. The difference between a liaison program and a research consortium is that in a liaison program the industry member is paying primarily for "membership" and access to research being performed, whereas in a consortium the sponsor is paying primarily for research. However, because liaison program membership fees often fund research and because a consortium's established annual funding level and set of "deliverables" from the university look very much like liaison program membership fees and member benefits, the two forms are often hard to distinguish in practice. In fact, most consortia with more than four or five members are virtually indistinguishable from liaison programs except in the proportion of corporate contributions that is used to fund research. Peters and Fusfeld found that "a key to the development of many successful research consortia is an industrial affiliate program"[8] through which member relations can be maintained.

RESEARCH CENTERS. University research centers are even harder to define than consortia or liaison programs. The term *center* is frequently used to group university researchers in a defined field. Often the center exists in name only and functions mainly as a marketing device to attract funding by calling attention to the combined strength of colleagues who are doing related research. It is particularly useful for describing interdisciplinary groups. Centers can be very simple administrative structures, requiring only that faculty members allow their names to be listed as affiliated with them on grant proposals or fund solicitations, or quite elaborate, with separate and extensive staffs and research facilities. Where centers are well defined and have a distinct administrative existence, their activities in organizing and managing relationships with corporations can appear very similar to those of liaison programs. As with consortia, centers may encompass liaison programs.

Contrasting examples of the relationship between liaison programs, consortia, and centers can be found at M.I.T. Merton Flemmings, a professor at M.I.T., conceived and initiated the Materials Processing Center by first securing seed money from NASA ($300,000) and then soliciting the affiliation of research colleagues with the center. These researchers in turn brought research contracts and grants to the center. Flemmings then established the Materials Processing Center Industry Collegium (a liaison program) to organize the center's relations with industry. Several research consortia (multisponsored research programs) were formed from subsets of collegium members. In M.I.T.'s Biotechnology Processing Center, on the other hand, membership in the center's collegium is conditional on membership in the center's single consortium—the collegium exists to serve consortium members.

"MEGA-AFFILIATE PROGRAMS." Another organizational variation related to liaison programs might be called the "mega-affiliate program." It combines elements of liaison programs, research consortia, and research centers but is distinguished from all of these by the size of corporate commitment, by the combination of state and federal funding with corporate funding, and by the fact that such a program usually is associated with the development of a research facility of some complexity. Stanford's Center for Integrated Systems (CIS), Berkeley's Microelectronics Innovation and Computer Research Opportunities (MICRO) Program and its Computer Aided Design/Computer Aided Manufacturing (CAD/CAM) Consortium, and M.I.T.'s Microsystem Industrial Group (MIG) are examples of this organizational form. Planning for Stanford's CIS began in 1977 when several engineering faculty members saw the need for a facility devoted to continuing research in the areas of integrated circuit and systems engineering in a research facility dedicated to the design, fabrication, and testing of very large scale integrated (VLSI) chips. By 1984, 20 corporate sponsors had pledged $750,000 each for the construction of a building and a research facility and $100,000 per year for its operation and maintenance. This funding, combined with federal support, resulted in the construction of a $15 million facility, research projects of about $12 million annually, and support for the work of 71 faculty members from seven departments as well as 30 Ph.D. and 100 master's candidates per year.[9] Naturally, relations between the contributing companies and the university in such a program are exceedingly close and go well beyond the "member benefits" usually associated with liaison programs. Programs of this size have a significant impact on the operation of smaller programs and on the atmosphere surrounding research on a campus.

However difficult it is to make clear distinctions among all these forms of university-industry interactions, it is important to understand that there are significant differences in the level of commitment (financial and otherwise) by industrial sponsors, the degree to which such sponsors are involved in and informed about the research work of the university, and the level of commitment required from the university and its faculty. In what follows, *liaison program* is used as a general descriptor for the programs under examination, but care will be taken to describe variations in these elements.

Description and Institutional Setting of Liaison Programs

Information was collected on two general-purpose and 58 focused programs at the four universities of this study. Five more programs were identified, and a brief examination of an additional 11 "liaison-like" programs also was made. All these programs are listed in Table 9-2. A summary of the information collected for the focused programs appears in the following section (the two general-purpose programs at M.I.T. and Berkeley and the liaison-like programs are not included).

As Table 9-3 shows, the number of members in the examined programs ranged from one to 154, with an average membership of 23. Annual fees ranged from $1,000 to $120,000.

Table 9-4 shows the distribution of the focused programs by field. Of the 58 programs examined, most were in earth sciences (11), with a significant number in electrical engineering (10) and materials science (8). The category "Other engineering" included 17 programs in widely dispersed fields of engineering.

Table 9-5 shows the types of member benefits in the focused programs examined. The benefits most frequently reported were an annual review-of-research conference and free research reports and publications (including, for most programs, access to reports before publication). Direct faculty liaison, including free consulting time in some cases, was also a common benefit. The opportunity for a member to send a "visiting scholar" to the univer-

TABLE 9-3 Focused Liaison Programs: Number of Members and Annual Fees

	M.I.T.	STANFORD	U.C. BERKELEY	PENN STATE
Number of focused programs examined	10	31	7	10
Range of number of members	6–100	1–154	3–142	3–40
Average number of members	36	18	37	17
Range of annual fees[a]	$1,000–100,000	$1,000–50,000	$20,000 and up	$1,000–25,000
Average annual fee[b]	$22,000	$14,000	N/A[c]	$15,300

[a]Annual fee data based on only those programs with established fee schedules.

[b]Computed by adding the annual fee (or, in those cases in which a range of fees was offered, the average fee) for each program and dividing by the number of programs.

[c]Berkeley programs all have variable schedules of annual research support levels.

189

TABLE 9-4 Focused Liaison Programs: Distribution by Field

FIELD	M.I.T.	STANFORD	U.C. BERKELEY	PENN STATE	TOTAL
Electrical engineering/ computer science	2	3	5		10
Biology and related fields	1	2		1	4
Materials science	1	1	1	5	8
Earth sciences		9		2	11
Other science	2	1			3
Other engineering	3	12	1	1	17
Other	1	3		1	5
Total	10	31	7	10	58

TABLE 9-5 Benefits Offered through Liaison and Liaison-Like Programs

	M.I.T.	STANFORD	U.C BERKELEY	PENN STATE	TOTAL
Number of programs examined	10	34	7	24	65
Number of programs offering:					
Annual conference for review of research	5	29	7	10	51
Publications	10	34	7	7	53
Special conferences	7	8	3	6	24
Visiting scholars	2	10	3	2	17
Faculty consulting, visits	3	25	1	2	31

sity was generally restricted to those programs with relatively high annual fees or was made available for an additional, negotiated fee.

Most programs provided benefits in addition to those shown in the table, including access to students and help for the companies in recruiting good students. In some cases access to university equipment, data bases, or special facilities was also provided.

Member benefits in most programs involved relatively small out-of-pocket costs for the university. In some cases, however, indirect costs and the hidden costs of faculty time and effort were substantial.

For the focused programs examined, the number of administrative staff ranged from 0 to 7, with an average of 2.2. Focused programs typically served from 4 to 20 graduate students directly and involved from 3 to 20 faculty members.

This summary tells only part of the story, however. Following is an examination of the institutional setting and history of liaison programs at each of the four universities, information necessary to understand the nature and purpose of the universities' present programs.

PROGRAMS AT M.I.T.

M.I.T. is examined first because it has the oldest liaison program in existence and because developments at this university in recent years represent an early indication of trends that are now being seen on other campuses. Furthermore, M.I.T.'s Industrial Liaison Program (ILP) is a prototype of general-purpose, universitywide liaison programs.

HISTORY. M.I.T. was established on the principle of close cooperation with industry. William Barton Rogers, M.I.T.'s founder, drawing on his experience at the College of William and Mary, sought the guidance of the industrial community in teaching students how to apply scientific principles to industrial problems. The institute went further than other universities of its time in aiding "the advancement, development, and practical application of science in connection with the arts, agriculture, manufactures and commerce" and in fulfilling its land-grant charter.[10] In 1918, in response to a challenge from a then-anonymous donor (George Eastman) to match a $10.5 million gift for the construction of the first buildings at the newly acquired Cambridge site for the Institute, M.I.T. established the "Technology Plan." This plan, a forerunner of later liaison programs, was the first direct link of a university with industry. As part of the plan, in return for donations, M.I.T. agreed to help companies identify and recruit engineering talent by maintaining a file of resumés, to allow companies access to the institute's libraries and files, and to arrange for faculty to provide advice on planning research programs and solving industrial problems. The Division of Industrial Cooperation was established as the administrative structure to carry out the plan. Over 180 companies pledged over $1.2 million as a result of this effort.[11] The division existed into the 1940s, after which it evolved into the Office of Sponsored Projects. In its later years the division was concerned primarily with contract and policy administration, although it also continued to perform other tasks including maintenance of a *"Tell You Where Directory* of staff activities" designed to "assist in the prompt and accurate handling of telephone calls and letters."[12] The Division of Industrial Cooperation seems to have been an early, scaled-down version of the present ILP.

THE INDUSTRIAL LIAISON PROGRAM. In the late 1940s, E. V. Murphee of Esso Development Corporation suggested to M.I.T. President Killian the idea of offering certain services to companies in return for financial support of the institute. Based on Murphee's suggestions, in August 1948 Killian established a program with seven founding members. The purpose of the Industrial Liaison Program, as it was called, was to foster an exchange of information between M.I.T. faculty and researchers and the industrial community, helping industry to stay abreast of rapid developments in science and the institute to stay informed of the needs of industry. A year after the program's founding, its membership had grown to 13, and by 1953 it had 60 members.[13] Today the program has about 325 members, having seen significant growth during the 1980s. The expansion of the program is shown graphically in Figure 9-1.

The ILP has not been uniformly successful throughout its life. By the late 1960s the program was in obvious difficulty, having become a bureaucracy dissociated from faculty interests and concerns. The institute apparently considered changing the program radically or even discontinuing it. Instead, in about 1970, Samuel Goldblith, a respected member of the faculty, was made director of the program. He took a number of steps to restore its usefulness and credibility in the eyes of the faculty, the most important of these being the institution of a point system by which 10 percent of the annual budget of the program was returned to faculty members or their administrative units in proportion to the time and effort each devoted to the program.[14] These steps were quite successful in restoring confidence in the program.

The ILP faced another challenge in the late 1980s as a number of "mini-ILPs" began to form around the institute. The circumstances surrounding the development of these smaller, more narrowly focused programs and their eventual relationship to the general-purpose ILP is described in the next section. As a result of the emergence of these focused programs, the mission and organizational position of the ILP have shifted a bit. The ILP is now viewed as more closely associated with M.I.T.'s corporate development efforts than as the separate service and liaison program of the 1970s and 1980s. The ILP director now reports to the institutional officer responsible for directing M.I.T. development operations. This is a natural evolution: as the focused programs began attracting the attention and money of specific corporate scientific and product development staffs, the ILP assumed the broader and more comprehensive task of relating to top corporate management, providing a wide range of institutional (as opposed to departmental) resources.

Because the ILP has always been associated with corporate "friend building," the shift is not particularly dramatic. The ILP is an organizational form that many other institutions are just now in the process of developing in recognition of the fact that corporate relations involve a complex of interactions that require coordination and the services of competent staff. M.I.T.'s ILP program is ahead of the efforts of most universities.

However, the most distinctive advantage of the ILP in promoting effective corporate relations is the manner in and degree to which service to corporate sponsors is integrated into faculty and researcher roles. The main reason for the continuous and active support of M.I.T. faculty and researchers for the ILP program is the "point system" previously mentioned—per-

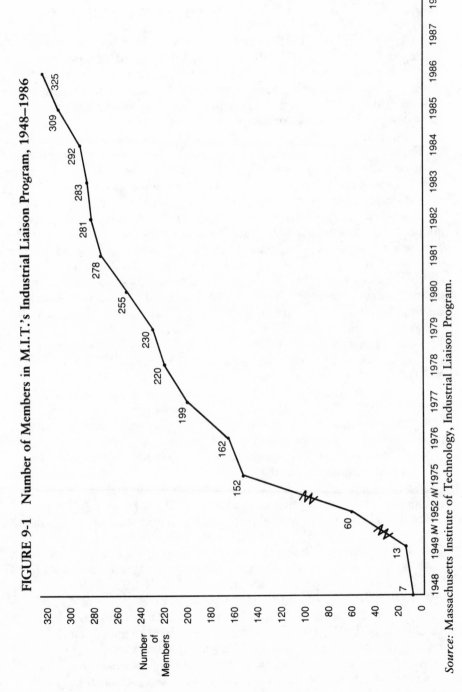

FIGURE 9-1 Number of Members in M.I.T.'s Industrial Liaison Program, 1948–1986

Source: Massachusetts Institute of Technology, Industrial Liaison Program.

haps the most elaborate existing institutional expression of support for faculty involvement with industry and, indirectly, with technology transfer. Through this system, the ILP "pays" institute personnel for the time and effort spent with ILP member corporations.

In fiscal year 1985–1986, over $800,000 was distributed as revenue sharing to faculty and staff, with each point being valued at about $34.[15] The funds are not paid directly to faculty members, but rather are placed in separate accounts for the faculty members' professional development or support of the work or research of the members' contributing department, laboratory, or other administrative unit. The point system extends also to lecturers, technical instructors, research scientists, engineers, and associates.

The activities for which points are awarded and the relative number of points awarded for each activity give a clear indication of the industrial relations activities desired by the institute and their relative importance. The rules governing the point system are spelled out in an ILP publication, "Working with the Industrial Liaison Program: A Guide for Faculty and Staff of the Massachusetts Institute of Technology."

The kinds of activities compensated by the ILP fall into three broad categories: conference participation, visits, and other activities. Conferences are clearly the most visible and important ILP activity. Qualified members of the institute can earn up to 20 points for chairing a symposium or a research briefing, 8 points for speaking in such a briefing, and up to 20 points for conducting a seminar for ILP member institutions.

Visits by representatives of ILP corporate members earn institute researchers 2 points for a 1-hour visit and up to 8 points for a full day. If an M.I.T. researcher visits a corporate member's location, 12 points are awarded. Other activities that earn points include submissions to the *M.I.T. Report* (an ILP publication), telephone discussions with ILP members (1 point), and travel time. In addition, the ILP director may approve points for other activities that are viewed as ILP services.

The history of the ILP illustrates some necessary adjustments to the external environment as well as the internal situation. The ILP was originally, "designed for large corporations in the United States, Europe, and Japan,"[16] and therefore was not very helpful to small companies. To remedy this deficiency, an ILP spin-off group called the M.I.T. Associates Program, dedicated to serving "new growth companies and well-established firms with a limited commitment to basic research technology,"[17] was established in 1961. By 1979 half the members of the Associates Program were based in New England, and the program was fulfilling M.I.T.'s "special responsibility to contribute to the scientific and technological development of the community in which it resides."[18] The Associates Program was administered by the director of the ILP and was similar to the earlier program in almost all respects. In 1981 the Associates and Liaison Programs were merged.

Along with the ILP's membership, both the size of the program's annual budget and the number of its staff members have seen considerable growth in recent years. Income to the program also increased, from about $4.7 million in 1981 to about $8.2 million in 1986.[19] In recent years about 40 percent of the ILP's budget has been spent on administrative costs, 10 percent has been used as faculty incentives, and the remainder has become unrestricted funds to "help further the basic goals of the Institute."[20] The number of "liaison officers" increased from 11 in 1980[21] to 15 in 1981[22] and over 20 in 1987, supported by an additional 35 clerical and administrative staff members.[23]

The ILP staff of about 60 people is headed by a director (now usually a member of the faculty) who reports to the vice president for resource development, and an associate director supported by a clerical and administrative staff of five. The liaison officers are divided into four groups, each headed by either an assistant director or a senior liaison officer and supported by clerical staff. There is also a four-person office in Japan and a staff of 11 dedicated to the production of the *M.I.T. Report* and other publications and to the presentation of ILP-sponsored symposia and conferences.

The liaison officers are crucial to the success of the program. Each serves as the representative of a number of member companies and is charged with facilitating the relationship of these companies with M.I.T. and encouraging the use of ILP services by the member companies. An officer typically has an advanced degree in a technical area (usually from M.I.T.), is extremely knowledgeable about what is happening at M.I.T., and is expected to become thoroughly familiar with the needs of each member company to which he or she is assigned.

Member firms pay an annual fee that is based on a sliding scale determined by annual sales. The exact fee is subject to negotiation, however. Figure 9-2 shows that, as of 1985–1986, most members had annual sales of between $1 and $5 million, were in the manufacturing, chemical, or electronics industries, and were in the United States or Canada, although there was also a significant number of members from Japan and Europe. The most important benefit that member firms receive is numerous opportunities for direct interaction with M.I.T. faculty and students. These interchanges occur through specially arranged meetings on the M.I.T. campus, visits by M.I.T. people to member firms, and symposia on topics of special interest arranged under the auspices of the ILP. Members also receive an annual directory of current research, which typically contains abstracts of over 3,000 research projects, and the monthly *M.I.T. Report,* which also contains research abstracts and bibliographic listings.

M.I.T.'s Industrial Liaison Program is the most centralized, well-organized, and well-operated university organization for managing and facilitating relationships between industry and the university of any of the formally established "transfer mechanisms" in this study. Its size and the extent of the resources provided to sustain it are clear indications of M.I.T.'s commitment to serving and maintaining relations with industry. The program is a

FIGURE 9-2 M.I.T. Industrial Liaison Program Members, 1985–1986

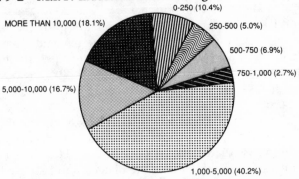

MIT INDUSTRIAL LIAISON PROGRAM

MEMBERS BY INDUSTRY CATEGORY

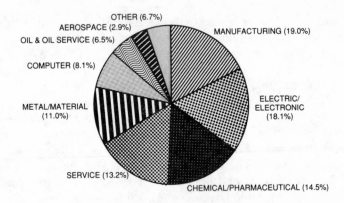

MIT INDUSTRIAL LIAISON PROGRAM

MEMBERSHIP BY REGION - JUNE 30, 1986

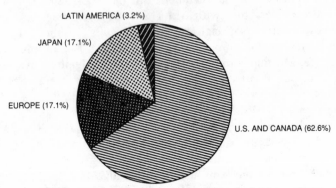

Source: Massachusetts Institute of Technology, "Industrial Liaison Program," undated report of activity, 1983–1986.

significant factor in sustaining faculty orientation toward and interest in relations with industry and applied research.

Nonetheless, the staff of the ILP must struggle continuously to overcome the stigma associated with bureaucracies and to maintain the personal relationships with company representatives and faculty members that are the key to the success of the program. The most frequent criticism of the ILP voiced by M.I.T. faculty was that the program was too bureaucratic and monolithic and tended to foster only superficial interactions with industry. These faculty members felt that significant relationships with industry depended on faculty contacts that already existed.

"MINI-ILPs." M.I.T.'s ILP program is well known and universally admired. Less well known or understood are the "mini-ILPs" or focused liaison programs that the institute has developed in recent years. An early program of this type was the M.I.T.-Industry Polymer Processing Program (IPPP), which was started in 1973 with seed funding from NSF under the Experimental R&D Incentives Program, a forerunner of the now-familiar NSF "centers" programs. These programs make continuation of NSF funding conditional on the generation of research support from industry. The IPPP quickly acquired 12 industrial members, each of whom pledged between $46,000 and $146,000 per year in research support.[24] As with other research consortia, the IPPP had to develop administrative practices to serve its members. Many of its services were similar to those provided by the ILP except that they were limited to the field of polymer processing. Significantly, at least until 1986, the IPPP had very little contact with the ILP.[25] The success of the IPPP caught the attention of many M.I.T. faculty and administrative staff members, and a number of similar research consortia soon developed.

Another early development occurred in the Department of Chemistry, which for a time appeared to have lost its industrial base of support. The dean of the college sought closer and more direct ties with the chemical industry than those provided through the ILP. He felt that the ILP's organization and fee structure inhibited departmental interaction with the research components of chemical companies, and he therefore established a "collegium" with a lower fee structure. He was able to retain its membership fees for departmental use.[26]

The real start of the mini-ILP movement, however, came with the establishment of the Materials Processing Center (MPC) in February 1980 as an independent, interdisciplinary research center within the School of Engineering. It had strong ties to the Department of Materials Science and Engineering. Professor Merton Flemmings perceived that American industry in his field did not have a sufficient scientific base and was in danger of falling behind that of other countries. He developed the center as described earlier in this chapter and then, in late 1981, formed the center's Industry Collegium, with a membership fee lower than that charged by the ILP. Flemmings had served as director of the ILP and was well aware of the administration and resources that the successful operation of this collegium required. The logic of his move, his stature in the university and the industrial community, his understanding of the politics of the situation, and his familiarity with the staff of the ILP and their concerns allowed him to found the collegium successfully. However, the relationship of his collegium to the ILP was a serious issue, and the program's establishment created concern among the ILP staff and others about the future role of the ILP and the possibility of the gates opening to a flood of similar programs.[27]

In fact the floodgates were opened, not by Flemmings' program, but by the proliferation of multisponsor research programs and the increase in industrial sponsorship of research. Both M.I.T. and industrial firms were beginning to see the utility of leveraging research funding by directing funding from a number of research sponsors toward a common goal. The idea that "$25,000 will buy you $250,000 of research" had great appeal for industrial sponsors. The institute, for its part, noted that such consortium arrangements, in addition to providing more research funding, also usually meant that the institute had more control over the research agenda than was possible with single-sponsor research contracts. Industrial sponsors of research needed and expected closer involvement with research and more extensive interaction with university researchers than did, say, the federal government, and much of the research funding from larger corporations was initiated through informal contacts between university researchers and their industrial counterparts and colleagues (often former graduate students from the M.I.T. departments) that bypassed the official, formal channels established by the ILP. The collegium (focused liaison program) was an obvious way for departments, laboratories, or centers of the institute to organize and manage research relationships with a number of corporate sponsors. The springing into existence of M.I.T.'s mini-ILP programs and the working out of relationships between these programs and the "grandfather" ILP is an interesting indication of a trend in university-industry relations.

At first, relations between the ILP and the collegia were awkward and sometimes strained. Early fee-splitting arrangements proved cumbersome. Furthermore, it was clear that the collegia were "harvesting" ILP members, that is, using existing ILP membership as a base for marketing collegium membership. In order to impose some control and limit what was then considered a possible hemorrhage of ILP membership, it was decided in 1985 that the new collegia had to have the approval of the president of the institute. There is some evidence that this led to a title ploy whereby what might have been named a collegium was instead called a consortium, a form that did not require presidential approval. Not surprisingly, corporate sponsors and even institute faculty and administrators were initially confused by these different organizational forms, and the institute had some difficulty in explaining itself. For example, ILP members expected to receive some of the benefits that collegium members thought were theirs by exclusive right.

Although some confusion still exists, most of these problems now have been resolved, and a mutually supportive relationship appears to exist between the ILP and the collegia and consortia. Fee problems have been solved by usually allowing collegium fees to be lower for ILP members. For instance, for the MPC Industry Collegium, fees are $7,500 per year for ILP members and $10,000 per year for non-ILP members.

M.I.T. has recognized the value of collegia and consortia and has incorporated the fostering and support of these groups into the mission of the ILP. This action has had a number of advantages. First, the ILP can offer the smaller collegia help and support. The ILP now advances seed funding to initiate programs where a demand appears to exist. It also helps in marketing the programs and lends administrative and clerical support to collegia in, for instance, the presentation of symposia and conferences. Second, the collegia often bring to the university new corporate sponsors who can be enticed into ILP membership and/or solicited for broader research or other financial support. Third, the ILP can exercise a monitoring and coordinating role, serving as a clearinghouse for information.

Early confusion among corporate members also has been dispelled for the most part. It has gradually become clear that the ILP and the collegia appeal to two different markets. The ILP appeals to companies that want an overview of current developments in a broad range of scientific fields. The collegia, on the other hand, appeal to companies that are intensely interested in developments in a particular field. In companies that are members of both the ILP and one or more collegia, the company contact for the ILP is usually in the headquarters division, while the collegium contacts are in particular divisions or research laboratories. It also may be true that companies see the opportunity offered by most collegia to suggest research projects and thus be involved in setting the research agenda as a distinct and valuable benefit that cannot be expected from membership in the ILP.

M.I.T.'s liaison programs are firmly embedded in a tradition and culture that is supportive of relations with industry, and they are clearly an important part of that culture. With a high level of industrial research sponsorship as compared to other universities ($36.6 million in 1987,[28] representing 14.3 percent of total research funding), a relatively high percentage of industrial sponsorship in the form of multisponsored arrangements ($11 million in 1987,[29] almost one-third of the industrial total), and two-thirds of its research being conducted in extradepartmental units (centers, laboratories, groups, and the like),[30] M.I.T. has had good reason to develop both a general-purpose liaison program and a number of focused programs to manage and facilitate the many relationships it needs to maintain with its industrial sponsors. In the development of these programs, M.I.T. is the most advanced of the four universities studied and probably of any university in the country. It represents a model to which other universities can be compared.

PROGRAMS AT STANFORD[31]

In contrast to M.I.T., Stanford has never had a general-purpose liaison program. Instead, it presents what might be called the "decentralized" model of liaison program organization. Stanford's 33 focused liaison programs were initiated at the department level; the central administration contributed to the growth of these programs primarily by allowing them to be formed.

The first focused liaison program at Stanford was established in 1950 in mineral sciences. It was followed by a program in aeronautics and astronautics in 1955. Frederick Terman conceived of the program in solid state electrical engineering in 1958. Each year, under Terman's leadership, the electrical engineering department held a two-day research review for representatives of its government sponsors of research. Terman suspected that duplicating this review for industry would allow the university to make contact with potential industrial sponsors and, for little additional effort, might generate "free" money for the department. It was Terman who decided to call the program an "affiliate" program.[32] The annual review of research has become the foundation benefit of most of Stanford's focused liaison programs.

At about the same time (1959), the affiliate program of the Graduate School of Business (GSB) was started. It began with 19 members, each of whom contributed $1,000. Today it has 150 members, who together contributed over $1.6 million to the school in 1987–1988. The GSB program differs from Stanford's other focused liaison programs in that it is viewed by both its corporate members and the faculty as a mechanism for providing regular financial support for the school; that is, it represents the school's development office. Whereas the faculty play substantial roles in other programs, the GSB program is staffed by a development officer and the faculty have little to do in its administration.

Since 1950, 41 affiliate programs have been established at Stanford, and eight have been discontinued. Three programs were established in each decade from the 1950s to the 1970s; since the academic year 1979–1980, 27 additional programs have been started and eight discontinued. Figure 9-3 shows the year of initiation of presently existing programs.

Figure 9-4 shows that the number of members of Stanford industrial affiliate programs and the total amount of member contributions have increased significantly since 1979. After a dip in both the number of members and the total contributions in 1986–1987, both measures increased in 1987–1988. The rather dramatic increase in contributions came primarily from eight newly established programs, four of which were in earth sciences. Member contributions represented about 22 percent of corporate giving in 1987–1988, in increase from around 17 percent during most of the 1980s.

Each Stanford affiliate program is independent of the others. Except for the GSB program, the programs are conducted by the faculty. The director of each program is an ex-officio member of the Council of Affiliate Directors, which is chaired by the vice provost and dean of research. This council is convened once each year for a discussion of topics of common interest.

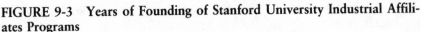

FIGURE 9-3   Years of Founding of Stanford University Industrial Affiliates Programs

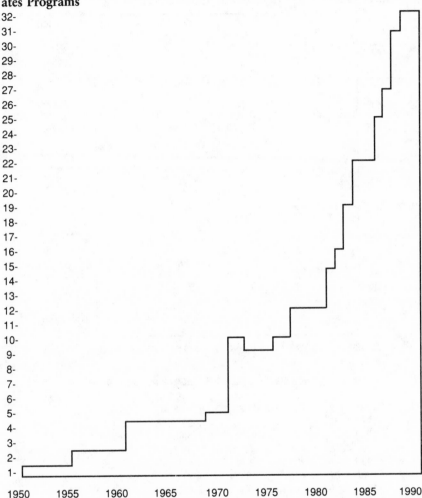

Source: Nancy L. Hay, "Stanford University Industrial Affiliates Programs, 1987/ 1988" (Stanford, Calif.: Stanford University, Office of Foundations and Corporate Relations, March 1989), 10.

New programs must be approved by both the vice provost and the administrative head of the appropriate academic department. Although the Development Office has no role in administering the programs, it does serve as a clearinghouse for information on them.

The situation of liaison programs at Stanford raises some interesting issues. First, the declines in membership and membership income shown in 1986–1987 in Figure 9-3 suggest that perhaps the market for liaison programs is becoming saturated. The recent increase in the number of such

FIGURE 9-4   Growth of Stanford University Industrial Affiliates Programs, 1979–1980 to 1987–1988

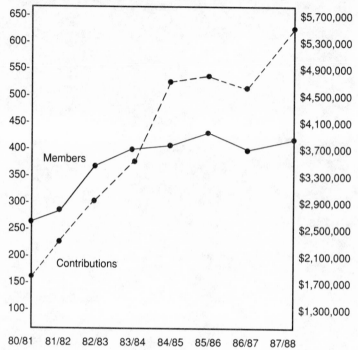

Year	Members	% Increase	Contributions	% Increase
1980-1981	263		1,746,870	
1981-1982	284	8.0%	2,334,397	33.6%
1982-1983	357	25.7%	2,997,988	28.4%
1983-1984	393	10.1%	3,684,556	22.9%
1984-1985	416	5.9%	4,842,833	31.4%
1985-1986	439	5.5%	4,898,919	1.2%
1986-1987	402	(8.4%)	4,717,406	(3.7%)
1987-1988	420	4.4%	5,615,717	19.0%

Source: Nancy L. Hay, "Stanford University Industrial Affiliate Programs, 1987/1988" (Stanford, Calif.: Stanford University, Office of Foundation and Corporate Relations, March 1989), 11.

programs at the four universities in this study undoubtedly is echoed in other universities, but there is some evidence that corporations are beginning to control more tightly their membership in these programs and to be more selective in the programs they join. The recent large number of mergers and acquisitions in industry also has contributed to the decline in corporate memberships. Where programs are viewed primarily as mechanisms for corporate giving, corporations are beginning to favor local institutions, a trend that works to the disadvantage of nationally recognized universities such as M.I.T. and Stanford. The tendency toward membership and contribution decline is even more clearly illustrated in the experience of the GSB program, which has declined from a membership of 220, contributing over $1.7 million in 1979–1980 to its present 150. After falling steadily for almost six years, the total contributions for 1987–1988 came close to the total for 1980–1981.

With so many programs Stanford may be experiencing internal as well as external competition. Some broad-based corporations may bridle at the prospect of having to join a number of different programs in order to gain access to all the research that is relevant to their needs. It is interesting to note that the Center for Integrated Systems (CIS), described earlier in this chapter, is not included in Stanford's listing of affiliate programs. This omission is primarily because the financial commitment required by the program pulls it clearly into the category of research consortia; contributions to it are not membership fees in the normal sense of the term. Still, the initiation of this mega-program caused some concern among directors of other programs, who felt that the size of the commitment demanded of companies belonging to CIS would use up the funds that the corporations might otherwise be willing to apply to membership in the smaller programs.

Despite these problems, Stanford expects membership in its liaison programs and member contributions to continue to increase. This contribution increase will have to come from one or more of three sources: increase in the size of membership fees, initiation of new programs, or increase in the membership of existing programs. Many of Stanford's programs have increased fees recently, apparently with no adverse effects on membership. Although the pace of new program initiation seems to be being maintained, the number of programs may have reached its limits. In the last two years, five programs were initiated and three were discontinued. The programs initiated since 1979–1980 have accounted for about 112 percent of the growth of total membership and 63 percent of the growth of member contributions. It is unlikely that this pace can be continued, however.

Increasing membership in present programs is viewed primarily as a marketing challenge. Again, the experience of the GSB program is instructive. In one year, 1979, GSB program membership increased from 184 to 236, primarily due to efforts to attract small, local, high-growth corporations. This market segmentation approach holds out hope for other programs.

The liaison programs at Stanford are important elements in the university's efforts to relate to industry and to solicit funds for its programs. The Development Office monitors membership in these programs and aids in the programs' efforts to increase membership, viewing such membership as perhaps the first step in what is hoped to be a long-term and more extensive relationship with a corporation. Stanford's liaison efforts are not the result of an overall administrative commitment to liaison programs, however. Rather, they arise from the faculty as natural elements in a culture that is supportive of relations with industry. The success of the programs rests to a large degree on the commitment of faculty and on the continued attractiveness of the liaison concept to corporations.

PROGRAMS AT U.C. BERKELEY

Liaison programs at Berkeley are almost all in the College of Engineering, although there are a few scattered programs in other colleges. For example, the College of Chemistry organizes its development efforts in a liaison-like program called the Industrial Friends Program. The Center for Real Estate and Urban Economics also has a program for real estate and development firms. This program has over 60 members, each of whom contributes between $5,000 and $12,000 per year to belong. (The real estate program is not included in this study because its activities are not related to technology transfer.)

The College of Engineering programs represent an interesting organizational hybrid between the M.I.T. and Stanford patterns and, in fact, were begun partly in response to the success of the programs at those two institutions. The engineering programs began in 1979 with the establishment of the first annual review-of-research conference. A collegewide general-purpose liaison program evolved from this conference.[33] At about the same time, then Governor Jerry Brown vetoed funds for the expansion of research facilities for the Department of Electrical Engineering. Rather than abandon their plans for the facilities, the highly regarded electrical engineering faculty went to industry for support, forming the CAD/CAM Consortium mentioned earlier. This appeal was very successful: by 1983 the $5 million fabrication facility had been built, and 15 industrial sponsors had contributed more than $18 million in cash and equipment to support the work of the consortium.[34] Membership in the consortium was closed after the initial financial requirement was met. Demand for membership was so great, however, that the department began the Berkeley Microelectronics Affiliates (BMA) in 1984. The success of the electrical engineering programs led to the formation of other liaison programs in computer science and in other departments of the College of Engineering.

The feature that distinguishes the Berkeley engineering liaison programs from those at Stanford and M.I.T. is the relationship between the college's general-purpose Industrial Liaison Program (ILP) and its focused programs. Membership in the focused programs automatically brings membership in the ILP. Corporations can join a focused program and thus become members of the ILP, or they can join the ILP without becoming members of the smaller and generally more expensive focused programs. The general-purpose ILP thus provides a floor of benefits that the focused programs can use as part of their benefits packages. Membership fees for Berkeley's programs are in the form of either unrestricted gifts or direct research support. The requirement for ILP membership is an annual contribution of either a minimum of $20,000 in unrestricted support or $100,000 in research contracts or grants. Members may direct their contributions to any one or more of seven academic departments, five organized research units, or nine interdisciplinary programs and centers, although only seven of these units have active programs. Thus virtually all research patrons automatically become members of the ILP and simultaneously can become members of one or more of the focused programs in fields they are supporting with research dollars. The 1986–1987 roster listed 28 college-level (ILP only) members and 281 departmental members. The fact that a number of departmental programs have members but are not really active makes a comparison of Berkeley's focused programs with those of the other universities in this study difficult in some respects, however.

The ILP and the departmental programs relate to one another under an operating policy in which the ILP defers to the more "grass-roots" programs.[35] That is, the ILP exists to support departmental programs and to provide companies with a means of obtaining broad access to the college without having to join a number of focused programs. The ILP has a small staff consisting of an administrator, who reports directly to the dean of the college, and clerical support, whose primary functions are to coordinate activities between programs, present the annual research review conference, serve the needs of the college-level members, act as a clearinghouse for information about the programs, foster the development of new programs, and help in the marketing of departmental programs.

This arrangement has a great deal of flexibility, providing potential members with a variety of options that allow companies or, more important, parts of companies to relate to the college at the most natural level. It also aids in the establishment of programs in departments, organized research units, and interdisciplinary centers by providing seed money and the previously mentioned base level of member benefits.

However, this organizational scheme is not without its problems. First, the college sometimes appears to be offering programs that really do not exist and for which faculty support may not be sufficient to fulfill member expectations. Second, the membership fee (or, more correctly, contribution) structure presents some difficulties. Some of the programs (CAD/CAM, BMA, Computer Sciences Affiliates) require unrestricted contributions of $120,000 annually and a pledge of a continuation of such contributions for five years, while others require a much lower commitment. The high-priced programs do provide the added opportunity for each member to send a visiting scientist to Berkeley, but most of the other benefits are similar in all the programs regardless of price, which irritates companies that make comparisons. Underlying these problems is the fact that the programs are not really "membership" programs at all but rather are simply a means for the college to organize its development efforts and set development targets by discipline, based in some measure on departmental reputation. There is nothing wrong with this unless the benefits of membership in these programs are oversold.

U.C. Berkeley's College of Engineering started its liaison program late, but with both M.I.T. and Stanford available as models, it developed a flexible and effective organizational arrangement for its programs that combines both the general-purpose and focused forms. Significantly, the push for the formation of Berkeley's programs came in the field of electrical engineering, which has had a long history of effective interaction with industry and participation in technology transfer. Eschewing a large bureaucracy for a centralized ILP, the college, through its focused programs and its visiting scientist benefits, has emphasized one-on-one contacts between faculty and their industry counterparts.

PROGRAMS AT PENN STATE

Liaison programs at Penn State present a less coherent and successful picture than the programs at the other three universities in this study. Of the 14 liaison and liaison-like programs identified at Penn State, only about six can be considered successful. Four of these are associated with the Materials Research Laboratory (MRL), and another one is also in the field of materials science. Compared to those at the other universities, the programs at Penn State are relatively inexpensive, with only three or four requiring annual membership contributions of $20,000 or more. Five of the programs permit membership with an annual contribution of $5,000 or less.

The Cooperative Program in Metallurgy was the first of the presently existing programs to be established, and it apparently went through a long period of dormancy. It is not really a liaison program; its purpose is simply to secure funding from the Pennsylvania steel industry to support graduate students in metallurgy.

Soon after its formation in 1962, MRL began its Industrial Coupling Program as a general-purpose or umbrella program that performs much the same functions for the laboratory that Berkeley's College of Engineering ILP does for that college. The MRL and its faculty are both physically and philosophically separated from the rest of the university, although, with the new emphasis Penn State is placing on relations with industry, the university may be moving closer to the MRL position. In contrast with their colleagues in most of the other disciplines at Penn State, the faculty of the MRL have actively sought industrial funding and have been little concerned with the distinction between basic and applied science. The MRL began its first focused program in 1978. In a turnabout of the usual pattern, MRL secured research funding from industrial sponsors and then convinced NSF to provide additional funding to carry out the work of the Center for Dielectric Studies, which is really more a research consortium than a liaison program.

This successful program was followed in 1985 by the Consortium on Chemically Bonded Ceramics and in 1987 by the Diamond Consortium described earlier in this chapter.

Ten of the 14 liaison programs presently existing at Penn State were begun after 1979. At least six attempts to start programs in the early 1980s failed, including programs in the fields of polymer science, ceramics, nondestructive testing, and microelectronics.[36] Penn State provides no centralized control, administration, or oversight of its programs, although the Office of Industrial Research and Innovation under James Lundy serves as a clearinghouse for information on the programs and occasionally also acts as a source of advice and help in starting programs.

The relative success of the MRL programs and the low level of success of similar programs in other fields hint at the underlying reason that liaison programs at Penn State are not particularly important to the university's efforts at relating to industry: lack of industrial interest in university research. A related reason may be that in fields that are particularly strong at Penn State, especially those related to agriculture and the extractive industries, individual faculty members and departments already have strong ties with industry and do not need liaison programs to facilitate this interaction. In other cases the industry (steel and extractive industries, for example) is in such poor financial condition that programs cannot be sustained. The geographical isolation of Penn State's University Park campus may be another contributing factor. It is also possible that as a public institution, Penn State has not had to emphasize development efforts as much as other universities have. Finally, although Penn State is ranked somewhere between fifteenth and twentieth on the list of research universities in amount of research funding, it may not have a national reputation in a broad enough spectrum of science fields to compete effectively for corporate memberships.

It will be interesting to see what happens to Penn State's liaison programs in the wake of the university's new emphasis on the importance of relating to industry. Development and growth of such programs will be one sign of Penn State's success in obtaining support from industry. If liaison programs are not successful at Penn State, some other organizational form will have to fulfill the function of facilitating university-industry relationships.

Why Programs Are Started and How They Are Managed

Despite the considerable variations from university to university in the importance of focused liaison programs to university relations with industry and in their position within the organization of the institution, no clear institution-related pattern emerged on two other dimensions, reason for starting and internal management. This section describes these two dimensions.

WHY PROGRAMS START AND VARIATIONS ON THEIR INCEPTION

Although it is often listed as a main purpose for the establishment of liaison programs, a desire to facilitate technology transfer is rarely the propelling force either for the starting of a program or for its continuation. By far the most frequent reason given for the establishment of a program is the generation of financial support from industry for research, graduate students, or junior faculty. Some programs are used as an active tool for soliciting such support, while others are passive means of organizing a department's development efforts and serve research patrons who would be making contributions whether the programs existed or not.

Quite a number of liaison programs have been organized with seed money from NSF or other federal or state agencies. This money often provides the level of funding needed to attract industrial patrons who seek leverage for their contributions. Of the 31 programs for which information regarding the initiation of the program was obtained, ten appear to have been started with such seed funding.

In some cases a liaison or consortium program was created because of the need to form an interdisciplinary unit outside departmental boundaries to address a particular set of research problems. Twenty-two of the 60 programs examined on this dimension appear to be interdisciplinary in nature. As shown with Berkeley's CAD/CAM Consortium and Stanford's CIS program, other programs are developed to fund the building and/or maintainance of a specialized research facility. An interesting variation on this is M.I.T.'s X Consortium, a true consortium in which the industrial sponsors all participate in the development of M.I.T.'s X software project. The X software promises to provide authors of a wide range of other computer software with "windows" capability. Members of the consortium have the opportunity to influence the direction of the further development of the software, and each has a license to use the software and its future versions in other programs.

MANAGEMENT AND OVERSIGHT

Some mention of the way the programs were administered and controlled was made in the description of liaison programs at each of the four universities. In most cases, the university administration exercises little control or oversight of liaison programs, although at M.I.T. and Stanford the initiation of programs must be approved by a provost or vice president. The faculty are usually in complete control of the programs, typically handling each through a rotating faculty director and a faculty committee. Occasionally the director must report on the program to the department chair, the dean of the department, or the director of the research center. Liaison programs frequently have small staffs to serve members and provide continuity in maintaining relations with them. In many programs, usually in conjunction with the annual review of research, representatives of the members are convened in a board to advise the faculty on the research agenda and on industry needs. Although these boards rarely have more than advisory powers, they often exercise significant influence. The faculty-controlled management of liaison programs raises several issues that are discussed in the next two sections.

Faculty Perceptions of Liaison Programs

In the faculty interview part of this study, faculty members were asked their opinion of the effectiveness of liaison programs in facilitating technology transfer. This question was often interpreted by the faculty as referring to the overall effectiveness of the programs in facilitating relations with industry, not just to effectiveness in technology transfer. Of the 90 faculty members in science and engineering who were interviewed, 73 responded to the question. Of those answering, 48 (65 percent) gave responses that were considered positive or very positive, 17 (23 percent) gave neutral responses, and eight (12 percent) gave negative or very negative responses. There were no discernible patterns in responses either by university or by discipline (biological sciences, electrical engineering and computer science, materials science, other engineering, and other science). However, it should be noted that at Penn State, where there are relatively few liaison programs, there were fewer responses of any kind to the question, and at both Stanford and Berkeley the responses in the biological sciences had a higher concentration of negatives than those in any other category.

Many of both the positive and the negative responses focused on the organizational or bureaucratic nature of the programs. Many faculty members saw the usefulness of the programs in establishing a framework that faculty and departments could use to relate to industry. At most of the universities, especially M.I.T., the programs were seen as key elements in a culture that supported relations with industry and technology transfer, although few saw the programs as a direct means through which technology was transferred. The negative side was also voiced, however: some faculty members saw the programs as a superficial bureaucratic substitute for what should have been a more direct and meaningful interaction.

It was also clear that liaison programs were not effective in some fields. For instance, in agriculture (at Berkeley and Penn State) many mechanisms for industry to relate to the university already existed and often were subsidized by the state government. In other fields, such as civil engineering, the industry was not particularly attuned to research, and internal competitive pressures were so great that the industry did not respond well to liaison programs.

A benefit commonly cited for liaison programs was the opportunity for students to present their research to industry representatives. Faculty considered the practice students got in preparing and giving presentations to be a valuable educational element and a way of honoring the best students. It also provided the students with industry exposure that often resulted in summer employment, work-study opportunities, master's and Ph.D. topics, or job offers.

Finally, many faculty members noted the extensive amount of faculty time required by successful programs. Those with positive feelings toward the programs accepted this as part of the price that had to be paid for industrial support. Most of the negative comments about the programs came from those who resented the amount of time the programs required. Two faculty members resented colleagues' subtle arm twisting to participate in the programs.

Despite a number of reservations, the overwhelming reaction from the faculty interviews about liaison programs was positive. The programs usually were considered to be effective mechanisms for relating to industrial supporters. They were considered less effective in actually carrying out technology transfer, a mission that depends on a degree of active participation and interest by industry that most faculty members felt was lacking.

Issues and Implications

The recent growth of liaison programs and their membership raises many issues that relate to the subject of this book—the role of technology transfer in research universities. The final section of this chapter examines the most important of these issues.

FACILITATED ACCESS

The basis for membership benefits in most liaison programs is facilitated access by member firms to the results of research, faculty, and students of the university. In some fields, particularly those in which the pace of technological advance is rapid and patent protection often is not viable, such as electrical engineering and computer science, this membership benefit can be very valuable indeed, allowing member firms to "get the jump on" a new development or at least not be out-jumped by the competition. In fields where change is not as rapid, facilitated access is less valuable and therefore forms a more tenuous justification for membership in a liaison program. In these fields, what passes for facilitated access is really available to any corporation aggressive enough to pursue the issue.

Whether valuable or not, the main problem with facilitated access is that it seems to run contrary to the popular notion of what a university stands for. Facilitated access for those willing to pay implies denial of access to firms unable to pay the membership fee. Does the development of a liaison program mean that faculty members must not talk to nonmember firms? Does it begin to put a price tag on informal interactions with industry? Fee requirements often exclude exactly those small, start-up, technology-intensive companies that would benefit most from membership. As noted, several efforts were made to solve this problem by providing a sliding-scale fee structure or by developing programs devoted to small companies, such as the M.I.T. Associates Program.

The issue of facilitated access was clearly raised in regard to membership of foreign (particularly Japanese) firms in liaison programs. In some programs, particularly in the electronics and biotechnology fields, there is considerable pressure to exclude foreign firms, and some faculties have considerable sympathy with this view. This study found some evidence that NSF, a leading provider of seed funding and continuing support for industry-university partnerships, informally but actively discourages universities from admitting foreign companies into some liaison programs or research consortia. In several cases the U.S. members themselves made it clear that admitting foreign companies would result in the termination of their membership. The specter of foreign companies capturing American university research prompted a recent General Accounting Office report that showed that foreign companies contribute only 1 percent of all university R&D expenditures and that five universities, including M.I.T., accounted for half of those expenditures.[37]

In mid-June 1989, Representative Ted Weiss, chair of the House Governmental Affairs Subcommittee on Human Services, publicly attacked the M.I.T. Industrial Liaison Program's ties to foreign, especially Japanese, companies (about one-third of ILP's members are foreign, and 17 percent are Japanese). The subcommittee staff analyzed the activities of 337 M.I.T. faculty members between 1984 and 1988 and found that the majority of their contacts were with foreign companies. This evidence seemed to run counter to the claim, made in congressional testimony by the president of M.I.T., that the ILP was the kind of program that could restore American competitiveness. Weiss also attacked the National Institutes of Health and the National Science Foundation for not being more effective in preventing the flow of the results of publicly funded research to foreign companies.[38]

Weiss's attack was countered vigorously by M.I.T. and other universities, which saw it as a threat to the traditional openness of universities and cited the virtual impossibility of keeping research under wraps. Going even further, Carnegie-Mellon University announced in July 1989 that it was abandoning its reluctance to work with Japanese firms and even would begin to seek them out. "I do not feel that American firms are responsive enough, and I don't believe we can get the kind of support we want and need strictly from American firms," complained Richard M. Cyert, president of Carnegie-Mellon.[39]

Many of the faculty interviewed in this study remarked on the thoroughness and intensity with which Japanese firms pursued member benefits. In the words of one faculty member, "Japanese industrial representatives do their homework," coming to meetings well versed in the literature and bringing specific questions and problems, whereas representatives of American firms characteristically take a passive "What have you been doing lately?" attitude. Foreign firms are among the heaviest users of the visiting scientist benefit; American firms, concerned with the cost, less frequently send their own scientists to the university, even when membership fees are quite high and obviously anticipate inclusion of a visiting scientist.

American universities do often recognize the necessity of excluding foreign firms from access to university research. For example, a May 1989 announcement of the formation by M.I.T., IBM, and AT&T of a consortium to advance research on superconductivity contained a statement that foreign companies would not be allowed to join the consortium.[40] The systematic exclusion of these potentially active and supporting foreign firms is, for some, a chilling sign that commercial and political concerns are beginning to reshape the traditional values of the university.

Another problem with facilitated access is that the creation of special relationships with member firms sometimes places the university in embarrassing or contradictory positions. For instance, in one case a written agreement promised program members the right to license results of the research they supported; however, the language of this promise was vague and in fact contravened university patent policy. During the establishment of Stanford's CIS there was considerable discussion about the issue of patents, copyrights, and licensing.[41] In the case described in chapter 8 involving Penn States's Diamond Consortium, the university licensed some diamond technology to a company in which the university owned stock. This action created consternation among members of the consortium, who felt that they had a right to use the technology by virtue of their support of the research. In yet another case, a member of a research consortium sought a separate contract with the university in the same research area but outside the consortium and with a preinvention license clause. The scientist involved was unable to guarantee that the research projects could be separated enough to protect everyone's interests and had to refuse the contract.

These instances and others illustrate a problem that universities face in many aspects of technology transfer. Pressed for funds, the universities naturally seek to trade on the financial value of their services or products (research, graduates). Patents are the most direct and obvious type of commercially valuable product, but facilitated access through liaison program membership is also a product for which companies are willing to pay. The further these practices go toward the direct exchange of value for money, the less like universities and the more like commercial concerns the institutions involved become. Drawing lines becomes important.

NEW ROLES FOR FACULTY MEMBERS

The amount of faculty time required to relate to liaison program members and administer the programs acts to pull faculty into a new role. Relating to industrial sponsors is considerably different from relating to government patrons of research in that industrially supported research generally requires greater interaction between the sponsor and the university researcher. The type of research is also different, tending to be more focused and applied. Thus what sometimes appears to faculty members to be an easy and inexpensive way of garnering research support from industry often begins to influence the research agenda substantially and to require long-term major involvement, at no additional compensation, of a considerable number of faculty members. Very few universities have adjusted the traditional academic reward system to encourage this kind of activity. Therefore, the programs that do exist are evidence of a faculty support for university involvement with industry and, by implication, technology transfer.

INCREASED COMPETITION

As noted earlier in this chapter, the proliferation of liaison programs has led to increased competition among the programs and probably to the decline in membership in some programs. Contributing to the problem is the fact that the programs tend to be clustered in just a few subject areas—electronics, materials science, and biotechnology. This increased competition probably will have several results. First, membership solicitations will become more sophisticated in their marketing appeals. They may also draw on the principle of market segmentation, seeking members in increasingly specialized fields. As this happens, universities' research agendas will tend to become more goal oriented and focused. Most universities prefer to keep such goal-oriented projects as only a small fraction of their total research. More focused research agendas also will tend to make strong programs stronger, but they will not improve weak programs.

Being more market oriented also means being more subject to the vagaries of market forces, including economic cycles and research fads. For instance, Penn State programs in enhanced oil recovery and coal research have declined recently because of a slump in the related industries. This unfortunate decline in interest is likely to be sorely felt when the next energy crisis occurs.

INDEPENDENCE OF SUPPORT STAFFS

Although liaison programs exist primarily because of the support and effort of faculty members, many programs have developed nontenurable staffs to provide services, maintain contact with member firms, and carry out many of the administrative tasks that would otherwise fall to faculty members. In this, liaison programs are similar to some of the other transfer mechanisms treated in this study, most obviously to patent offices and continuing education providers. Such staffs after awhile begin to have a life of their own and to demonstrate independence from other parts of the university. For instance, some liaison programs have their own programs of continuing education that are run separately from regular continuing education units in the university. Program staffs also may show their independence in publications, marketing policies and practices, and even patents and licensing. This trend has implications for the next issue.

LACK OF UNIVERSITY OVERSIGHT

As mentioned earlier, few liaison programs are subjected to review and oversight by university administration. Liaison programs will present an increasing challenge to university administrators as these administrators try to coordinate the activities of the programs, provide information about them and explain their operations to companies and outside constituencies, and exercise sufficient control over them to ensure that major errors are not committed. Each program represents a public relations effort that should be of concern to the university administration. At the same time, administration and control of these programs should not be so heavy-handed that it gets in the way of the one-on-one interactions and close associations between faculty and industry members that represent the fundamental value of these programs.

The growth of focused liaison programs at M.I.T. and the recent organizational rearrangements in the institute's ILP program provide further evidence of the merging of a number of functions that once were organized separately in universities. University efforts to attract donations from corporations cannot and should not be divorced or organizationally separated from other forms of university-corporate interaction, including those related to technology transfer. Universities, and perhaps some of their larger corporate patrons as well, have shifted their efforts from attempts to encourage specific forms of corporate contributions—gifts, research contracts, payments for technology licenses, liaison program fees—to attempts to establish and maintain enduring relations based on the many ways universities and corporations interact. The way in which universities are organizing their relations with industry, exemplified by the recent changes at M.I.T. and by the organization of liaison programs at U.C. Berkeley, is beginning to reflect this trend.

Conclusion

This chapter has shown that liaison programs are a relatively recent development in American research universities, with the most rapid growth in the number and membership of such programs occurring since 1979. These programs serve a number of functions in providing a framework for university, department, and faculty interaction with industry. Although for the most part technology transfer between the university and industry happens only indirectly as a result of liaison programs, these programs, by providing a mechanism whereby industrial scientists are given access to the network formed by academic scientists, facilitate the one-on-one interactions that are most productive of technology transfer. On the other hand, liaison programs have been criticized for not involving enough of the nonscientific decision makers of corporations—the people who will decide which technology to pursue commercially.

The growth of liaison programs is strong evidence of a shift toward a closer identification of universities and the research they produce with the commercial and economic concerns of the nation. This shift will not occur in all fields or at all universities, but where it is occurring, the existence of viable liaison programs is a reliable index of the degree to which the commercialization of research is becoming a part of faculty and departmental roles.

Notes

1. Lois S. Peters and Herbert I. Fusfeld, *University-Industry Research Relationships* (Washington, D.C.: National Science Foundation, National Science Board, 1983), 31.
2. Kay Tamaribuchi, "Effectively Linking Industry with a University Resource" (Cambridge: Massachusetts Institute of Technology, Industrial Liaison Program, 1982), 6.
3. Peters and Fusfeld, *University-Industry Research Relationships,* 22.
4. Nancy L. Hay, "Stanford University Industrial Affiliates Programs, 1987/88" (Stanford, Calif.: Stanford University, Office of Foundation and Corporate Relations, 1989).
5. Kay Tamaribuchi, interview with author, Cambridge, Mass., 27 May 1986.
6. Peters and Fusfeld, *University-Industry Research Relationships,* 92.
7. Ibid.
8. Ibid., 81.
9. *Information Technology and R&D: Critical Trends and Issues* (Washington, D.C.: U.S. Congress, Office of Technology Assessment, 1985), 182–184.
10. David R. Lampe and James M. Utterback, "The Tradition of University-Industry Relations at M.I.T." (Cambridge: Massachusetts Institute of Technology, Industrial Liaison Program, 1983), 1.
11. Ibid., 2.
12. Massachusetts Institute of Technology, *Policies and Procedures, 1940,* M.I.T. Achives, 13.
13. Lampe and Utterback, "The Traditions of University-Industry Relations at M.I.T.," 3.
14. Kent Smith, interview with author, Cambridge, Mass., 19 September 1986.
15. Massachusetts Institute of Technology, "Working with the Industrial Liaison Program: A Guide for Faculty and Research Staff of the Massachusetts Institute of Technology" (Cambridge: Industrial Liaison Program, undated), 13.
16. Massachusetts Institute of Technology, *Policies and Procedures, 1979,* M.I.T. Archives, 208.
17. Ibid.
18. Ibid.
19. Massachusetts Institute of Technology, "Industrial Liaison Program," undated report of activity, 1983–1986.

20. Massachusetts Institute of Technology, "Working with the Industrial Liaison Program," 4.
21. James D. Bruce and Kay Tamaribuchi, "M.I.T.'s Industrial Liaison Program," *SRA Journal*, 12 (1980): 13.
22. James D. Bruce, "University-Industry Interactions, the M.I.T. Experience" (Cambridge: Massachusetts Institute of Technology, Industrial Liaison Program, undated), 4.
23. Massachusetts Institute of Technology, organization chart of the Industrial Liaison Program obtained by author from the M.I.T. Industrial Liaison Program Office in September 1986.
24. Massachusetts Institute of Technology, "M.I.T.-Industry Polymer Processing Program," undated pamphlet.
25. Timothy G. Gutowski, interview with Sharon Graves Floyd, Cambridge, Mass., 22 July 1986.
26. Smith, interview, September, 1986.
27. Tamaribuchi, interview, May 1986, and George Kenny, interview with Sharon Graves Floyd, Cambridge, Mass., 15 July, 1986.
28. Carol E. Van Aken, letter to author, 31 March 1988.
29. Ibid.
30. Tom Moebus, interview with author, Cambridge, Mass., 19 September 1986.
31. Except where noted, all the information presented in this section came from the previously cited report by Nancy L. Hay, "Stanford University Industrial Affiliates Programs, 1987/88."
32. John Linvill, interview with author, 3 October 1988.
33. Frank Guinta, interview with author, 25 January 1988.
34. University of California, Berkeley, "Berkeley Microelectronics Affiliates" (Berkeley: College of Engineering, University of California, 1987).
35. Guinta, interview, January 1988.
36. James Lundy, interview with author, 11 September 1986.
37. United States General Accounting Office, "R&D Funding: Foreign Sponsorship of U.S. University Research" (Washington, D.C.: General Accounting Office, 1988), 1.
38. David L. Wheeler, "Lawmakers Hit Universities for Selling the Results of U.S.-Financed Research to Foreign Companies," *Chronicle of Higher Education*, 21 June 1989, A1, A16.
39. Scott Jaschik, "Frustrated by Tepid Response of U.S. Business, Carnegie-Mellon Says It Will Encourage Japanese Links to Its Federally Financed Research," *Chronicle of Higher Education* 5 July 1989, A14, A22.
40. Andrew Pollack, "U.S. Giants in Research Consortium," *The New York Times*, 24 May 1989, C8.
41. *Information Technology R&D: Critical Trends and Issues*, 182–183.

Bibliography

BRUCE, JAMES D. "University-Industry Interactions, the M.I.T. Experience." Cambridge: Massachusetts Institute of Technology, Industrial Liaison Program, undated.

————, and Kay Tamaribuchi. "M.I.T.'s Industrial Liaison Program." *SRA Journal* 12 (1980): 13–16.

BUDNICK, FRANK S., and RICHARD MOJENA. "University-Industry Cooperative Research Centers: The Rhode Island Experience." *Management of Technological Innovation: Facing the Challenge of the 1980s,* Washington, D.C.: Worchester Polytechnic Institute, 1983.

HAY, NANCY L. "Stanford University Industrial Affiliates Programs, 1987/88." Stanford, Calif.: Stanford University, Office of Foundation and Corporate Relations, March 1989.

Information Technology and R&D: Critical Trends and Issues. Washington, D.C.: U.S. Congress, Office of Technology Assessment, 1985.

JASCHIK, SCOTT. "Frustrated by Tepid Response of U.S. Business, Carnegie-Mellon Says It Will Encourage Japanese Links to Its Federally Financed Research." *Chronicle of Higher Education,* 5 July 1989, A14, A22.

LAMPE, DAVID R., and JAMES M. UTTERBACK. "The Tradition of University-Industry Relations at M.I.T." Cambridge: Massachusetts Institute of Technology, Industrial Liaison Program, 1983.

MASSACHUSETTS INSTITUTE OF TECHNOLOGY. "Industrial Liaison Program." Undated report of activity, 1983–1986.

————. "M.I.T.-Industry Polymer Processing Program." Undated pamphlet.

————. Organization chart of the Industrial Liaison Program obtained by author from the M.I.T. Industrial Liaison Program Office in September 1986.

————. *Policies and Procedures, 1940.* M.I.T. Archives.

————. *Policies and Procedures, 1979.* M.I.T. Archives.

————. "Working with the Industrial Liaison Program: A Guide for Faculty and Research Staff of the Massachusetts Institute of Technology." Cambridge: Massachusetts Institute of Technology, Industrial Liaison Program, undated.

PETERS, LOIS S., and HERBERT I. FUSFELD. *University-Industry Research Relationships.* Washington, D.C.: National Science Foundation, National Science Board, 1983.

POLLACK, ANDREW. "U.S. Giants in Research Consortium." *The New York Times,* 24 May 1989, C8.

TAMARIBUCHI, KAY. "Effectively Linking Industry with a University Resource." Cambridge: Massachusetts Institute of Technology, Industrial Liaison Program, 1982.

UNITED STATES GENERAL ACCOUNTING OFFICE. "R&D Funding: Foreign Sponsorship of U.S. University Research." Washington, D.C.: General Accounting Office, 1988.

UNIVERSITY OF CALIFORNIA. "Berkeley Microelectronics Affiliates." Berkeley: College of Engineering, University of California, 1987.

WHEELER, DAVID L. "Lawmakers Hit Universities for Selling the Results of U.S.-Financed Research to Foreign Companies." *Chronicle of Higher Education,* 21 June 1989, A1, A16.

10

Continuing Education and Technology Transfer in Four Universities

This chapter examines the relationship of continuing education (CE) to technology transfer in the four research universities in this study. In contrast to technology licensing, university equity ownership of start-up companies, and liaison programs, CE usually is not associated with the research mission of the university and therefore, except when the university seeks to employ a very broad and inclusive definition of the term, CE is not identified as part of university technology transfer efforts. Even when CE does deal directly with university research, for instance when universities organize seminars to disseminate the results of university research or when consulting faculty members hold workshops on research findings in companies, the activity is identified more with knowledge transfer or public service than with technology transfer.

This chapter shows that CE has clear relevance to university technology transfer efforts. This revelance is described under two categories which might be named *functional* and *organizational*. Under the functional category, this chapter illustrates that CE, as presently carried out in the four selected universities, while it plays a small direct role in technology transfer, plays a large indirect role in the ongoing relationship between the universities and industry and hence the atmosphere of cooperation and interaction that is so necessary for technology transfer to occur.

The organizational relevance of CE to technology transfer views the way CE is organized in universities and the relationship of faculty members to it as illustrative of the way in which other technology transfer efforts are or might be shaped and the way in which technology transfer might influence existing university organization. Most universities have provided CE for a long time and have well-established organizational structures to support and deliver CE. These relatively mature organizational units, devoted primarily to relating to an external constituency, give some indication of the way universities might organize other forms of external relations, including those connected with technology transfer.

Another aspect of the organizational relevance of CE to technology transfer is the fact that, like many of the activities associated with technology transfer discussed in previous chapters (invention disclosures, patents, and interaction with industry through liaison programs), CE usually is not considered by faculty members as a particularly important part of their roles. Certainly most universities do not recognize CE in faculty reward systems. An examination of the way CE is carried out under these conditions and what faculty members think of CE thus can provide some insight about what is happening and is likely to happen to other technology transfer-related activities that depend upon faculty involvement.

There is one other way in which technology transfer and CE are related under the organizational category. It is clear that concerns about technology transfer, about closer relations with industry, and about the commercialization of research are having an effect on the definition, organization, governance and control, and role of CE units in research universities. Universities are adding functions to and taking functions away from centralized CE units because of such concerns and because faculty members are accepting CE as more central to their roles. This chapter examines present and potential changes closely, drawing conclusions about both the degree of the effect of technology transfer on the process of change in CE and the implications of these changes for CE administrators.

The Functional Role of Continuing Education in Technology Transfer

Although most of the activities of CE units would be more correctly categorized as knowledge transfer and deal with subjects not directly associated with university research or technology, close examination shows that at least some CE activities do either involve technology transfer or directly foster it. The CE offerings of the four universities selected in this study included technology transfer activities of the following types.

CONVENOR OF CONFERENCES

The CE organization often acts as a convenor of conferences for researchers interested in special technological topics. Usually the CE organization provides logistical support rather than initiating the conference. Most universities have some form of support service to aid faculty members who are acting for professional societies in arranging international conferences. An example is U.C. Berkeley Extension's role in the presentation every two years of the International Conference on Solid State and Integrated Circuit Technology.

DISSEMINATOR OF RESEARCH RESULTS

For the most part, keeping scientists and engineers up to date means disseminating the results of the latest research to them. In many universities CE is used to do this and therefore acts as a supplement to the publishing of research. Updating courses are attended largely by scientists and engineers from industry; their companies invest a considerable amount of money to have them attend the courses and expect to achieve some financial return for the investment. Examples of courses that provide updating on specific advances are the M.I.T. summer program called "Advances in Controlled Release Technology: Polymeric Delivery Systems for Pharmaceuticals and Other Bioactive Agents," and Berkeley's "Barrier Metals for Semiconductor Fabrication."

PROVIDER OF BACKGROUND INFORMATION

CE also may become involved when rapid technological or intellectual advances have left significant numbers of industrial people so far behind that they cannot understand the latest research. This has happened recently in the biotechnology area. Examples of programs intended to provide background information are Berkeley's "Genetic Engineering for Chemists and Chemical Engineers" and the seminars in diamond techology conducted by Penn State's Materials Research Laboratory.

PRESENTER OF TECHNICAL SUBJECTS FOR NONTECHNICAL PEOPLE

Technology transfer involves not only scientists and engineers but also a broad range of others who may need to know about aspects of new technologies. These people include patent attorneys, businesspeople, entrepreneurs, and venture capitalists. Examples of courses designed to provide information to such people are Berkeley's "Biochemical Engineering Fundamentals" and "Protein Separations—Principles and Processes" and M.I.T.'s "Fermentation Fundamentals."

CONVENOR OF TECHNOLOGY TRANSFER PLAYERS

Occasionally the CE unit acts as a convenor of all those necessary to transfer a technology successfully from the research laboratory to the marketplace. The same nontechnical people who need to know about the science underlying a particular technology also need to interact with each other and share knowledge about the technology transfer process itself. Examples of programs designed to bring people together for the purpose of technology transfer are Berkeley's "Marketing Biotech Products in an FDA-Regulated Environment" and U.C. San Diego's CONNECT, a program that regularly brings together individuals and businesses involved in or interested in high-technology business developments in the San Diego area.

ADMINISTRATOR OF SPECIAL PROGRAMS

Because CE organizations in universities are flexible and market oriented, they are often assigned responsibilities that fall outside what one would normally call CE. Obviously these may include programs closely associated with technology transfer. Several examples of such programs, including Penn State's PENNTAP program, M.I.T.'s Center for Advanced Engineering Studies, and Berkeley's Pacific Asbestos Information Center, are discussed later in this chapter.

The nature and extent of CE activities associated in some way with technology transfer vary from campus to campus, and some types of activities may not be present in some universities. However, in the four universities of this study and most other universities, there are sufficient activities of one sort or another to qualify CE as an active part of a university's technology transfer effort. Often CE's capabilities as an agent of technology transfer are not recognized either by the faculty or the administration of the university.

History and Organizational Setting of Continuing Education at Four Research Universities

The functional intersection of CE and technology transfer in universities is relatively easy to describe and enumerate. The relationship between the organizational aspects of CE and emerging developments in university technology transfer efforts is more difficult to describe. It is embedded in the history and culture of each university, and is revealed in the changes that CE is undergoing now on every campus. A comparison of the histories of and present settings for CE at the four universities in this study supports the notion that the organization of CE and its place in accepted faculty roles is parallel to and helps us predict the organizational form and context of university technology transfer.

U.C. BERKELEY

U.C. Berkeley has one of the oldest centralized continuing general (as opposed to agricultural) extension organizations in the country.

HISTORY. University extension at Berkeley began in the fall of 1891, when three courses—on Shakespeare, history, and mathematics—were offered by Berkeley faculty in San Francisco. These first U.C. extension courses were the result of the personal enthusiasm of a group of about 12 university professors, an enthusiasm fueled in part by a realization that the university had to meet some of the demands made upon it to become more "popular" and responsive to the citizens of the state. This tension between the elitist standards of the university, which dictated that the faculty not be deeply involved in extension activities, and the pressure that a public university serve popular demands (described in chapter 2) was important to the development of U.C.'s extension at several points in its history. The university's use of the extension unit as a buffer between it and what it saw as dangerous popular demands was perhaps unconscious, but it resulted in a failure to incorporate extension work into faculty and institutional roles and missions. It caused the development of a separate university extension staff and a pattern of employing part-time instructors in extension courses. This pattern contrasts with the development of CE at some other institutions, such as M.I.T. In consciously creating a semi-independent organization, U.C. also created the basis for a tension between the extension unit and the faculty that was periodically manifested in faculty concern over both the quality and the appropriateness of extension activity.

FACULTY RELATIONS. Organizational peripheralization and faculty concern over the quality and appropriateness of the extension unit's course offerings both come into play in a current issue faced by the unit and the university that is reflected on many other campuses today. As the professional schools of the university see the value of CE as a means of relating to industry and professional bodies, they sometimes come into conflict with centralized extension arms, viewing them as barriers to a closer relationship with the external constituency. Although Berkeley Extension has worked out very productive relationships with many units on the campus, notably some departments of engineering, public health, and library science, it also has had a number of jurisdictional disputes with other campus schools or centers. Faculty members who want to be involved in CE, acting upon their tradition of independent scholarship and professional autonomy, oppose the centralized unit that is dedicated to performing this very function. The university administration, which on the one hand wishes to encourage CE and the individual initiative of faculty members and on the other hand wishes to control and organize CE in a way that makes sense both educationally and fiscally, is placed in the middle of this dispute.

ORGANIZATIONAL ISSUES. Other elements of the history and present circumstances of extension at the University of California also are relevant to this study. The extension unit thoughout its history has proven to be an extremely flexible and responsive organization, quickly organizing training efforts during both world wars and responding equally quickly after the wars to demands for increased enrollment that threatened to overwhelm the institution.

The extension arm also has proven to be subject to institutional changes and politics. A case in point is the decentralization of extension activities to the campuses in 1968. With a shrinking state subsidy and a political climate favoring decentralization of authority to the campuses, U.C. delegated responsibility for extension to the chancellors of the individual campuses, and the state of California was divided up into eight service areas, each served by a campus extension arm. A description of the forces and politics behind this decentralization is beyond the scope of this study, but the change has interesting implications because the Patent, Trademark, and Copyright Office (PTCO), one of the last of the university's centralized service functions, currently also may be on the verge of decentralization.

THE CURRENT SITUATION. Today university extension at U.C. Berkeley serves some 55,000 students annually with about 1,800 courses, employs about 200 people (not including teachers), and has an annual self-supporting budget of about $19 million. It has jurisdiction over most of the CE offered by the university. Extension courses cover a wide range of subject areas, including the general areas of engineering, business, science, arts, and literature, as well as more specific areas such as landscape architecture, interior design, and English language instruction. The dean of extension reports to the vice chancellor, the chief academic officer of the institution. The extension staff includes about 30 continuing education specialists who are responsible for creating and presenting the programs.

So far, Berkeley extension has been involved only peripherally in the university's efforts at technology transfer. The unit has been asked to participate in the planning of a conference facility at a proposed real estate development dedicated in part to industrial activity, and it has offered courses directly related to technology transfer in a few fields. Clearly it is not yet considered an active participant in the technology transfer process, however.

PENN STATE

Like the University of California, Penn State is a public, land-grant, state university with a long tradition of CE. Its first programs were "Farmer's Institutes" presented in 1881. It developed an agricultural extension program and, later, separate extension programs in engineering, chemistry and physics, education, liberal arts, and mineral sciences.

RELATIONS WITH DEPARTMENTS. The most interesting contrast between U.C. and Penn State is in the relationship between CE and university departments. At Penn State early CE activities were departmentally sponsored, with engineering sponsoring the most successful program. In 1921 Penn State's faculty successfully opposed centralizing the several extension operations, arguing that a separate entity would result in a decrease in faculty involvement. In 1934, however, President Ralph D. Hetzel instigated the combination of all these extension operations, except for agriculture, into a new Division of Central Extension. (Agriculture was not included because of its unique federal-state sponsorship and because of vigorous opposition from powerful agricultural interests.) J. O. Keller, who had served as head of Engineering Extension, was chosen to head the new division. Keller and a small staff coordinated the activities of the schools and handled routine administrative chores; the school extension operations maintained much of their autonomy.[1]

OFF-CAMPUS CENTERS. At about this time, Penn State formally established off-campus extension centers to provide the first two years of baccalaureate work and to offer courses leading to extension certificates. These "undergraduate centers," as they came to be known, formed the basis of the Commonwealth Education System, which was formalized in 1959 after the Pennsylvania legislature rejected a plan for a separate system of two-year community colleges.

By the end of the 1930s, then, two relationships important in the definition of the role of CE at Penn State had been established. The first was the relationship between the central Division of Central Extension (in 1959 renamed the Division of Continuing Education Services) and the individual schools, in which the division assumed a coordinating and administrative role but the schools maintained control over specific offerings and faculty. The second was the functional and organizational conjunction of extension work and the two-year or community college system.

ORGANIZATIONAL ARRANGEMENTS. The organizational arrangements for CE at Penn State have changed several times over the years. The two-year Commonwealth Campus System was established under a separate administrative unit in 1959. This administrative separation between the Commonwealth System and the Division of Continuing Education Services continued for 20 years, but the campus system remained an important part of extension activities: extension used campus facilities, and extension faculty, although members of the relevant academic departments, taught in and identified themselves with the campus units.

In 1980 the campus system and extension were again administratively unified under a single vice president. The new unit was known as the Commonwealth Education System. Penn State today has a curious system wherein permanent extension faculty are appointed by the academic department (located usually at the main University Park campus), teach at a two-year campus overseen by a campus executive officer, and, when they teach CE courses, have their teaching assignments coordinated (and their salaries paid) by a centralized CE organization also located in University Park. This organizational pattern may be undergoing further change.

THE CURRENT SITUATION. Penn State's Commonwealth Educational System serves about 150,000 CE students per year in 24 defined service areas and over 400 different locations in the state of Pennsylvania. Its staff includes about 70 extension specialists charged with organizing CE programs. The system's CE Division is responsible for a number of undergraduate and graduate degrees, certificates, diplomas, and licensure programs for part-time students as well as conferences, workshops and seminars, and community service programs. Agricultural extension programs, certain programs administered by the College of Business for executives, and a number of other programs fall outside the jurisdiction of this division. The CE Division has an annual budget of about $30 million, part of which comes from state appropriations.

John Leathers, assistant to the vice president and dean of the Commonwealth Education System, describes the organization he directs as a delivery system that exists to serve the colleges and the university. Although the CE staff no doubt exercises considerable creativity and skill in discovering educational needs and developing programs, these aspects are downplayed and are so much under the policy control of the departments that the kinds of conflicts between centralized CE provision and the faculty and departments that occur occasionally at Berkeley are unlikely to occur at Penn State.

The service orientation of the CE Division is likely to benefit it during the reevaluation and reorganization that Penn State and the Commonwealth Educational System are presently undergoing. The administration of President Bryce Jordan, pushed by the state legislature, has taken up the theme of technology transfer and economic development with considerable vigor. Jordan created a new vice presidential position to be placed in charge of areas associated with technology transfer, including the patent office, the contracts and grants office, and the PENNTAP program mentioned earlier (that program previously had reported to the vice president and dean of the Commonwealth Education System). During 1988 the university initiated a study of the Commonwealth System and the CE Division, the first since 1964. As a result of this study, CE on the branch campuses may be decentralized, reporting to the campus chief rather than to the central division. One concern arising from this change is that more of the CE activity of the professional schools may be pulled away from the CE Division. In this case, money rather than academic control of programs is the issue. Some of the professional schools see that in certain areas CE can make money, and they seek to take these areas away from the central division. In turn the CE Division, seeing a part of its base (and therefore its ability to present a balanced and comprehensive program) threatened, seeks to retain program control. Here again the centralization-decentralization issue raises its head.

Another factor is also at work in the Penn State case, however. Interviews with faculty showed mixed reactions to the CE Division. Some faculty members had high regard for the division, viewing it as providing a valuable support function. Others, however, viewed it as bureaucratic, inefficient, and unwilling to risk new ventures because of financial pressures. This appears to be the inevitable fate of a university support organization: some will attack it for being what it is not allowed to be and does not pretend to be, a proactive initiator and financial risk taker.

This brief recitation of the history of and current situation for CE at Penn State illustrates some important points and provides an interesting comparison with developments at the University of California. Although both universities today have centralized CE, there are considerable differences between them, and these differences can be traced to early developments. At U.C., CE, even at its earliest stage, was an organizational entity separate from the departments of the university, run by a series of directors who placed their philosophic stamps on the activity. Regular faculty had little to do with university extension until the philosophy guiding the extension division's activities became too far removed from the academic ideals held by the faculty, at which point they tried to assert control, sometimes successfully.

At Penn State, on the other hand, extension work grew organically out of the concerns of the faculty and departments, becoming an instrumentality of the departments. Although the faculty and the departments have considerable control over CE at Penn State, the centralization of CE has served to remove a part of the activity from faculty concerns. This removal and the accompanying financing agreements negatively affect faculty attitudes toward CE at Penn State. At U.C., most faculty view extension as peripheral to the main mission of the university, but some faculty and some administrators see extension as acting as a buffer between elitist ideals and attempts at popularization. At Penn State, where again faculty see CE as an activity organizationally if not formally separated from faculty activity, extension nevertheless has been used by the university administration to maintain and increase popular support for the institution and has been much more central to its mission.

A comparison of CE at U.C. and Penn State carries at least two important implications for any university trying to establish mechanisms for technology transfer. First, this comparison indicates the importance of establishing an appropriate role for the faculty in the activities. Second, the degree and nature of centralized control exerted by the university clearly is important.

M.I.T

The history and present organization of CE at M.I.T. present a considerable contrast to what was seen at U.C. and Penn State. M.I.T. does not have a centralized CE organization, and the history of its CE development presents a much less coherent picture than did the histories at the other two universities. At first glance it seems odd that an institution with such a reputation for being close to industry and the profession of engineering should not have a more clearly defined organization for CE. On closer examination, however, several explanations emerge.

CHARACTERISTICS OF CE AT M.I.T. From its earliest inception, M.I.T. was dedicated to applying scientific knowledge to practical concerns and thus serving industry. It did not need to develop the mediating mechanism of CE either to foster and encourage such relationships or, as with U.C., to protect itself from them.

The institute's philosophical orientation was reinforced by a geographical proximity to some of the largest and most sophisticated industrial firms in the world. Interaction on educational programs, including the development of early work-study programs, was a natural part of the relationship. The third and probably most important reason that M.I.T. never developed centralized CE is that from the institute's early days, without receiving formal encouragement from the institution or needing centralized logistical support, M.I.T. faculty engaged in many forms of CE. This CE activity arose naturally out of the faculty's close contact with industry. The diffuse nature of much of the CE that goes on at M.I.T. today is encouraged by the complex network of liaison programs (most particularly the ILP) and research consortia, most of which incorporate both formal and informal courses, seminars, lectures, and group discussions, that could be classified as CE.

SUMMER INSTITUTES AND CAES. The two identifiable organized efforts in continuing education at M.I.T. today are the program of summer institutes and the activities of the Center for Advanced Engineering Studies (CAES). Both have relatively short histories. The summer program began in 1949, at the instigation of President James Killian, with the express purpose of increasing interaction with industry. The program was viewed as a complement to the newly established Industrial Liaison Program. CAES began in 1968 with a $5 million gift from Alfred P. Sloan. The desire to establish it grew out of a widely held sentiment that M.I.T. should provide updating education for its graduates.

Today M.I.T.'s summer programs enroll about 2,000 students in 65 to 70 courses each year. Annual income from enrollments in these programs is from $2.5 to $2.75 million; one-third of this goes for the costs of the programs (including the salaries of the small permanent support staff), one-third goes to faculty, and one-third goes to the institute's general fund.[2] The courses are initiated and taught by regular M.I.T. faculty, occasionally supplemented with invited instructors. Although there are some exceptions, the summer courses are typically five to ten days in length and treat highly technical and specialized subjects such as "Transmission Electron Microscopy of Materials" and "Rheologic Behavior of Polymeric Fluids."

CAES has two major divisions—on-campus instruction and off-campus videotape instruction. Both divisions are self-supporting. The videotape instruction division offers about 50 courses and 2,000 lectures. Most of the courses are traditional university ones, though there are a few short courses. The center has its own TV studio, and course instructors are taped there rather than in the classroom in front of students. Although many of the videotape courses are accompanied by manuals, M.I.T. offers no further support for these courses and does not enroll students in them. Instructors are paid a royalty on the sale of the tapes. The videotape division has an annual budget of between $2 and $3 million.[3]

The on-campus instructional program is designed to bring engineers with degrees to M.I.T. to update their skills. Most such students are treated to a customized curriculum lasting up to three months (or even longer) that is fashioned out of existing M.I.T. courses, lectures, and symposia as well as specially arranged courses, meetings, and joint work with M.I.T. faculty. The annual budget for on-campus instruction is between $1 and $1.5 million. Faculty members are compensated for the extra work involved in the program.[4]

THE LOWELL INSTITUTE. A third educational element associated with CE at M.I.T. is the Lowell Institute School. This school was established in 1906 "to provide evening instruction in technical subjects for residents of the Boston area."[5] The school offers about 40 courses per term in the evening on the M.I.T. campus. Courses do not carry credit, but students earn completion certificates for successfully finishing the courses. The school also offers certificate programs, each consisting of eight courses, one in drafting technology and one in electronics technology. Individual courses cover such subjects as elementary calculus, computer literacy, computer language programming, and machine tool fundamentals. The Lowell Institute School is not a direct part of CE at M.I.T., since the M.I.T. faculty are not involved in it. However, the school does apparently fill a need and demonstrates M.I.T.'s responsiveness to the community.

THE CURRENT SITUATION. M.I.T.'s policy of limiting its formal CE primarily to what the faculty can do themselves has drastically limited the scope of its formal CE efforts. Faculty have too many other demands on their time to make CE a high-priority activity. At U.C. and at Penn State, where a regular staff have CE as a permanent assignment and where adjunct faculty can be retained, the scope of CE is considerably larger than it is at M.I.T.

The CE situation at M.I.T. further illustrates certain problems associated with the development of technology transfer mechanisms. Such mechanisms are dependent on faculty involvement and cooperation. However, because faculty have many other concerns, requiring that these mechanisms be administered or controlled primarily by the faculty places limits on the scope and often the effectiveness of the mechanisms. One answer is to hire competent staff to operate supporting structures such as CE organizations or liaison programs, thereby saving faculty time and allowing for greater scope, but this answer can result in an undesirable distance between the faculty and the activities.

Even more important in this context than any of the easily identifiable elements of CE at M.I.T. is the significant amount of CE that goes on outside these organized units. Much of this activity, including organized conferences, meetings, lectures, research updates, and arranged one-on-one meetings, is conducted through the general-purpose Industrial Liaison Program. The focused liaison programs also sponsor a range of activities that can be classified as CE as well. Often faculty consulting contains elements of CE. In these and other ways a considerable amount of CE is done at M.I.T. that cannot be readily seen or quantified. Unlike any other university in this study, M.I.T., by policy and by tradition, makes CE one of the recognized roles of its faculty.

STANFORD

CE at Stanford presents an even less coherent picture than that at M.I.T. A recent inventory of Stanford's CE offerings lists programs sponsored by 15 different departments, ranging from the Stanford Summer Session to the Lively Arts program, which "conducts a lively outreach program in the local public schools, senior centers, hospitals, and community centers where guest artists give lecture demonstrations, workshops and performances."[6]

OFF-CAMPUS DEGREE PROGRAM. The most successful of the various strands of CE at Stanford, and the one most often mentioned by Stanford faculty members interviewed in this study, is the off-campus engineering master's program. This program began as the "honors program" in the 1960s. It was set up in response to requests by a number of companies in Silicon Valley for Stanford to establish a master's program in engineering that company employees could fulfill by studying part time. Students in the program had to meet Stanford's normal admission requirements, could enroll only through their company, and were required to pay fees in addition to regular Stanford tuition in order to fund the additional expense of the specialized curriculum. This program was possible only because of the geographical proximity to Stanford of a large number of high-technology firms and their highly qualified employees. In 1969 the program began to incorporate televised instruction, and by 1974 the Stanford Instructional Television Network was broadcasting over four TV channels to 20 firms within a 40-mile radius of Stanford.[7] Today the program annually graduates about 350 master's students, broadcasts about 60 courses per quarter, and has an increasing student population of engineers who are not seeking degrees but are acquiring professional updating or making career changes.

OTHER PROGRAMS. The university also has at least 13 other identifiable CE programs, which fall roughly into three categories. The first and most important category is programs offered by professional schools for professionals in a field. For instance, the Graduate School of Business offers eight to ten programs per year for executives and high-level managers; several departments in the School of Engineering offer engineering management programs; and the medical school offers programs that carry continuing medical education credits as well as a nursing CE program that includes home-study courses.

Another category of CE at Stanford might be called community-based education. Both the medical school and the Stanford Hospital have community relations departments that sponsor conferences for the public on new medical breakthroughs and on topics related to public and community health issues. The university also sponsors the Lively Arts at Stanford program described previously, a Discovery Travel film series, and a summer session open to the general public.

A third category of programs is offered to alumni and the general public through the Stanford Alumni Association. About 25 conferences in this category are held each year, including travel programs, a series of executive education programs, and the Stanford Publishing Course, a two-week session for magazine and book publishing professionals.[8]

Stanford's newest CE venture is its Continuing Studies Program, a series of liberal arts courses taught by Stanford faculty and open to the general public. The program was initiated in the fall of 1988 with six courses. Stanford hopes to develop the program further and eventually to offer a master's degree through it.

Stanford's history and present provision of CE reflects the organizational patterns characteristic of this university. First, the highly decentralized organization of CE at Stanford is similar to that seen in its liaison programs. CE at Stanford is thus a grab bag of programs directed at a number of audiences and serving a number of purposes. Lacking a centralized provision for CE, Stanford has responded to the need for CE in a loosely organized but organic way, developing programs wherever there was a need combined with faculty or administrative enthusiasm.

Second, Stanford's most successful and sustained program of CE, the televised courses in engineering, rose from the university's relations with industry and its desire to serve industry's needs. The autonomy of Stanford's departments and faculty members and the strong affiliation between Stanford and the neighboring industrial firms are key factors in understanding both CE and technology transfer at Stanford and in making comparisons between Stanford and other universities.

CONCLUSIONS

Issues important in both CE and technology transfer include the appropriate role for faculty in the administration and governance of the activity, the control and oversight exercised by central administration, the consequences of developing (or not developing) a permanent staff for CE, and the role of CE in facilitating relations with industry. In some cases, particularly U.C. and Penn State, CE has been "peripheralized," that is, moved to the edge of the university and conducted primarily by nonfaculty staff. This protects the faculty from having to engage in CE, but it also creates the possibility for conflict between the CE unit and the faculty. To confine CE activity only to that which can be carried on by faculty results in either a broad-based informal CE provision (as at M.I.T.) or a small-scale formal CE program.

Perhaps even more important than the lessons to be learned from these specific issues is the recognition that CE is an important element in the culture of an institution and in its relations with industry and other external constituencies. An examination of the place CE occupies and has occupied in the various universities tells a great deal about how each university and its faculty feel about other kinds of outreach, including technology transfer, and how such activities should be organized.

Faculty Attitudes toward Continuing Education

Faculty attitudes toward CE are important in determining the degree of institutionalization of CE activity. In turn, the way in which CE is organized in the university is an important determinant of faculty attitudes. These relationships also appear in other activities involved in technology transfer.

In the interview phase of this study, faculty members involved in some way with technology transfer were asked the following questions: What role does CE play in the technology transfer process? What role should it play? Of the 90 faculty members interviewed, 77 responded to the questions. It was clear from the responses, however, that most faculty members did not really answer the questions that were asked. Instead of specifically addressing the role of CE in technology transfer, they tended to talk about CE's overall role. Of the 77 responses, 45 (58 percent) expressed "favorable" or "very favorable" attitudes toward CE, 25 (32 percent) responses were classed as neutral toward CE or expressed pros and cons that seemed to balance out, and only seven expressed "negative" or "very negative" views.

Responses were rated by the author on a scale of 1 to 5, with 1 representing a belief that CE had no important role or was not appropriate to the university and 5 representing enthusiastic support for university CE activity.

Differences between universities were not particularly significant. Average response scores ranged from 4.0 at Berkeley to 3.67 at Penn State; the M.I.T. average was 3.85 and the Stanford average was 3.93. The relatively low Penn State score appeared to be associated with negative perceptions of the central provision of CE by those who were not active in CE. Differences by discipline also were not significant, ranging from 3.86 in electrical engineering and materials science to a low of 3.4 in the category called "other science" ("biotechnology related" scored 3.58, and "other engineering" scored 3.73).

A number of themes were repeated frequently in the responses to the questions. First, it was clear that CE was defined quite differently by different faculty members. Some associated CE with external or part-time degree programs; others thought CE included allowing a scientist from industry to work in their laboratories. Three of the respondents interpreted the question to refer to their own CE.

Second, in each of the universities studied, a number of faculty members felt that the university should do more than it was doing in CE, primarily to help keep engineers and scientists up to date in their fields. The most common reason given for not doing more CE was lack of faculty time and the failure of the university to reward faculty for doing CE.

Third, a significant number of those interviewed related ways in which their involvement with CE had helped them in their work or had helped the university. For instance, one faculty member at Stanford told how he had been approached by a large company to prepare and teach a course in a newly emerging technology. He taught the course and then applied for and received a grant from the company to develop the course further so that it could be incorporated in Stanford's graduate curriculum. An M.I.T. professor told how he used his annual summer course to stay in touch with his former graduate students and how these contacts often resulted in consulting work for him and financial support for his university research.

Those who were neutral or negative toward CE often expressed the view that CE took too much time away from more important faculty activity. Some suggested that there were more appropriate providers of CE than major research universities. Again, many of these people had difficulty dissociating the activity from the organizational form it took on their particular campus.

Of all the "transfer mechanisms" discussed in this study, CE ranked second highest, behind liaison and affiliate programs, in average faculty perceived value, scoring 3.73 overall. This support was based primarily on the idea that the university and its faculty owed graduates the opportunity to keep up to date in their fields and the recognition that CE is a natural and accepted way of maintaining relationships with graduates, professions, and businesses.

Implications of Technology Transfer for Continuing Education

The interest in technology transfer, university relations with industry, and the commercialization of research has some significant implications for the provision of CE in research universities. Were continuing educators to understand these implications better, they could be more effective. Consideration of such implications is instructive to those concerned with policies related to technology transfer; it may help them understand how technology transfer can change university organization.

REDEFINITION OF CONTINUING EDUCATION

As technology transfer activities increase and as the number of people in universities whose jobs are devoted to such activities continues to grow, the very definition of CE on university campuses is likely to change. It was clear from the faculty interviews that CE is defined so broadly as to include everything from evening degree programs to discussions between university faculty and industry scientists. This broadening of the definition of CE to include activities beyond classroom instruction indicates that the role of CE is in the process of being reconceptualized in American research universities. This reconceptualization is part of a general pattern of change that includes the new emphasis on technology transfer, relations with industry, and serving university graduates after they have their degrees.

REORGANIZATION

As the definition of CE expands, the way in which the provision of CE is organized within universities also will change. The direction of this change will depend upon local circumstance. In some cases, university extension arms may be asked to assume more diverse functions, including some that are not instructional in nature. For instance, technical assistance programs and business incubator programs may be placed under the administration of the dean or director of CE.

University administrators increasingly are recognizing that their institutions' extension arms can be used to great advantage. First, those arms are generally market-oriented and already possess some of the contacts with businesses and community groups as well as the marketing and promotional skills needed to carry out other technology transfer activities. Second, they are accustomed to serving students (and their employers) as clients in a way that other parts of the institution are not likely to be. Third, they often have useful administrative provisions and dispensations already in place, because of their self-support financing and the recognition that they serve adult students. University extension arms, for instance, often hire part-time instructional staff, spend money on what in other parts of the institution would be considered "entertainment" (such as coffee and doughnuts for coffee breaks or alcohol for cocktail parties), or hire temporary staff without going through extensive bureaucratic procedures. By assigning technology transfer activities to university extension arms, university administrators avoid having to create new administrative structures or make further exceptions to policy.

Another possibility is that in institutions where technology transfer is viewed as an organizing principle, CE may be separated from other units that are more directly related to technology transfer or business development activities. Such separation is apparently occurring at Penn State, where the highly visible PENNTAP program has been removed from the administration of the CE arm and placed in a new category under an assistant vice president. This has been a theme in the history of CE units within universities: programs developed by the leading edge CE arm mature and then are appropriated by schools and departments.

Where centralized provision of CE does not exist, as at M.I.T. and Stanford, there will be increasing pressure to formalize and control CE as the level of technology transfer and industry relations activity grows. This pressure will come both from inside the institution, as organizational untidiness and unnecessary duplication of function become apparent, and from outside, as the market demonstrates unhappiness with administrative confusion in the university.

INTERNAL COMPETITION

Continuing education is such a natural means of relating to external constituencies that just about every program involved in fostering external relations is likely to want to become involved in it. The result can be competition between the official CE arm and other organizations. Liaison programs are clear examples; it is easy to expand the traditional annual review of research into workshops and courses on specialized topics and even further. Alumni associations are another potential competitor for the CE market, as the situation at Stanford clearly illustrates. Of course, internal competition between centralized CE arms and professional schools has been an issue for many years. What is new is the development of organizational entities within and/or associated with professional schools that are staffed by nonfaculty specialists who have as part of their assignment the presentation of conferences and courses.

EXAMINATION AND ASSESSMENT

Increased CE activity, accompanied by increased competition, reorganization, and redefinition inevitably will mean that CE will undergo more intense scrutiny and assessment by the university. New reviews and assessments may be precipitated by a change in leadership, either of the CE unit or of the university administration, or by a financial crisis.

THE AGRICULTURAL EXTENSION MODEL

For the past twenty years there have been initiatives in many states to extend the agricultural extension model to a new "industrial extension" service directed at providing technical assistance to small businesses. Some agricultural extension operations and many university general extension organizations are being asked to cooperate in these efforts. For many reasons, which are described in the next chapter in the section on technical assistance programs, the agricultural extension model is not appropriate for these services, and the CE organization's involvement in these efforts may create a number of difficulties. CE administrators must exercise caution in accepting responsibility for these programs; at the same time, they must remain receptive to the ideal of service represented by them.

Implications of Continuing Education for Technology Transfer

Examination of CE in four research universities suggests a number of implications for technology transfer. For one, the organization of CE in the four universities represents patterns for the organization of other technology transfer efforts, and the history of CE's development provides examples of the possible consequences of various forms that technology transfer mechanisms can take.

INSTITUTIONALIZING PATTERNS

One of many problems facing university administrators as they seek accommodation to internal and external pressures to be involved in technology transfer is the manner in which the activities associated with technology transfer are to be "institutionalized." To appear responsive to current conditions and to create an atmosphere within the university that encourages technology transfer, university administrators are under pressure to produce visible signs that something new is being done. In most cases, it is not enough simply to point out those technology transfer activities that the institution has been engaged in all along, such as publishing the results of research, licensing patents, graduating students, and providing faculty consultations. There must be signs that technology transfer is being institutionalized, or made a permanent part of the university.

In higher education, institutionalization of an activity can take place in one of three ways. First, the activity can be placed in a separate organizational unit, school, or institute. The M.I.T. Industrial Liaison Program is an example of this form of institutionalization. Second, the activity can be incorporated into the faculty role pattern. An example of this is the absorption of sponsored research activity into faculty roles after World War II. Third, departmental activities can be modified to make a place for the activity. The assumption by academic departments of expanded responsibility for institutional governance is an example of this.[9] The histories of CE in the four universities of this study illustrate each of these three methods of institutionalization.

U.C. BERKELEY. U.C. Berkeley Extension represents the separate entity method of institutionalizing an activity. It shows both the advantages and the disadvantages of creating a separate organizational unit to perform an activity. At Berkeley, centralization has resulted in an exciting CE program of broad scope. Through centralization, too, the university administration has been able to exercise control and coordination of CE and to capture some economies of scale. The extension unit is fully accountable both financially and in terms of the educational quality of its offerings. However, centralization has created a distance between CE activity and the faculty, as evidenced by the relatively low participation by U.C. faculty in extension programs. It also has led, on occasion, to faculty concern regarding the appropriateness and quality of the CE offerings. As faculty members, particularly those in professional schools, have come to see CE as increasingly important, they have begun to take more interest in the organization of CE on the Berkeley campus. They seek ways to decrease the distance between university extension and the academic departments. As technology transfer mechanisms are developed, faculty members are becoming more knowledgeable about how university activities designed to serve external constituencies, including CE, should be organized.

PENN STATE. Penn State's provision of CE is a kind of hybrid between Berkeley's separate entity model and the departmentalization method of institutionalization in that it has given academic control of the CE program to the departments while keeping the logistical and service functions of the program in a centralized organizational unit. This arrangement has the advantage of making CE a part of departmental concerns and thus placing it within the powerful departmental structure of the university. It provides faculty members who are interested in doing CE with a convenient and departmentally legitimate vehicle for being involved. At the same time, the centralized service function allows the university to continue to exercise some control over the activity, or at least to know what is going on, and to capture some of the economies of scale in logistical arrangements. These advantages of centralization are particularly important in the area of program promotion and marketing.

There are a number of disadvantages to Penn State's arrangement, however, some of which it shares with Berkeley. Despite the theoretically heavy involvement of the departments, some faculty members still feel dissociated from CE and view the central CE organization as distant and unresponsive. Jurisdictional disputes also have arisen to some extent. Furthermore, as shown even more clearly at M.I.T. and Stanford, when CE is left to departments or faculty, the program tends to be limited to those areas in which faculty or departmental interest is high and to be conventional in its approach to subject matter and course formats. In other words, such a CE program tends to reflect more closely the strengths and weaknesses of the faculty and to be less market oriented than a centralized program might be.

M.I.T. M.I.T. represents the clearest model among these four universities, and perhaps the best example in the country, of the incorporation of CE into the faculty role. CE at M.I.T. is part of a pattern of interaction with industry and the professions and is well integrated with, indeed sometimes indistinguishable from, other activities dedicated to the same end. None of the distance between the faculty and CE activity noted at Berkeley and Penn State is evident at M.I.T.: there are no jurisdictional battles or concerns about quality and appropriateness. CE is simply considered to be a natural part of what faculty members do.

Like all the other situations, however, M.I.T.'s has its disadvantages. CE is only a *part* of what faculty members do, and other parts of their day-to-day activities often are considered more important than CE. As a result, CE remains smaller than it might be if it had a permanent staff that could help faculty members leverage their time and could also reach beyond M.I.T. faculty for instructional staff. M.I.T.'s CE program tends to be narrowly focused and not comprehensive. Because university administration has little control over the activity, it has little ability to quantify and define CE at M.I.T.

STANFORD. In contrast to the other universities, Stanford has not institutionalized CE at all. Except for its most recent venture into CE in the liberal arts, Stanford has undertaken CE as a means to relate to external constituencies, primarily the professions and business, rather than as a form of service to students. Significantly, the largest identifiable programs of CE at Stanford are operated through the Alumni Association and the School of Engineering. CE cannot really be considered a part of the faculty role at this university because the activity is too small, although a number of Stanford faculty members certainly do see CE as an important part of what they do or should do.

Conclusion

Although universities often list CE as an element in their technology transfer efforts, CE is usually a knowledge transfer rather than technology transfer activity. CE is most directly associated with technology transfer when it is involved in presenting conferences designed to disseminate the results of university research or when it treats issues related to research commercialization. While it is expected that this small direct role in technology transfer may grow in importance, the most relevant aspects of CE to university technology transfer will remain its organizational forms, variations, and history, and its relationship to faculty roles.

This description of the role of CE in the technology transfer process and the attempts to categorize the roles played by CE in research universities also show the close relationship among certain activities that are often viewed as separate. CE, technology transfer, relations with industry, and efforts to raise financial support (development efforts) all are part of a general trend of increased awareness on the part of the modern university of the importance of relating to external constituencies.

The relationship between faculty and CE can be viewed as a model for the relationship between faculty and a number of other present or proposed activities related to technology transfer, including the commercialization of research and the aiding of economic development efforts. The four universities of this study illustrate the different ways faculty can relate to an activity that is important but not central to their perceived roles, as well as demonstrating the common problem of adjusting faculty evaluation and reward systems to achieve the desired amount of activity.

Finally, the review of CE has raised some issues related to the way in which all kinds of technology transfer activity might be institutionalized and controlled by university administration. Since most such activity demands the cooperation of the faculty and yet is fraught with potential for conflict of interest and negative public relations, the university must strike a balance between encouraging individual involvement and setting guidelines and monitoring procedures that are detailed enough to prevent abuses or inappropriate behavior.

Notes

1. Michael Bezilla, *Engineering Education at Penn State: A Century in the Land Grant Tradition* (University Park: Pennsylvania State University Press, 1981), 69.
2. F. E. McGarry, interview with author, Cambridge, Mass., April 1988.
3. Shaol Ezekiel and Paul Brown, interview with author, Cambridge, Mass., April, 1988.
4. Ibid.
5. Lowell Institute School, "87–88 Bulletin" (Cambridge, Mass.: Lowell Institute School, 1987).
6. Stanford University, "Continuing Education at Stanford," obtained from Stanford Alumni Association.
7. Howard J. Sanders, "Continuing Education," *Chemical and Engineering News*, 20 May 1974, 29.
8. Stanford University, "Continuing Education at Stanford."
9. William Toombs and Carl A. Lindsay, "Departments and Professions: Institutionalizing Continuing Professional Education" (University Park: Pennsylvania State University, CPE Development Project, 1986), 2.

Bibliography

BEZILLA, MICHAEL. *The College of Agriculture at Penn State: A Tradition of Excellence.* University Park: Pennsylvania State University Press, 1987.

———. *Engineering Education at Penn State: A Century in the Land Grant Tradition.* University Park: Pennsylvania State University Press, 1981.

BRODSKY, NEAL. *University/Industry Cooperation.* New York: New York University, Graduate School of Public Administration, 1980.

BROWN, GORDON S. "Remarks for CAES Dedication." 3 October 1968. M.I.T. Archives, 80-33, Box 46, Center for Advanced Engineering Studies.

CALVERT, STEVEN L. *Alumni Continuing Education.* New York: Macmillan, 1987.

CARLSON, BERNARD M. "The M.I.T.-GE Cooperative Education Course, 1907–1923: A Case of Corporate Ambivalence and Academic Institutional Imperative." University of Pennsylvania, Department of History and Sociology of Science, January, 1980. Unpublished. Available in the M.I.T. Archives.

LOWELL INSTITUTE SCHOOL. "87–88 Bulletin." Cambridge, Mass.: Lowell Institute School, 1987.

LYNTON, ERNEST A., and SANDRA E. ELMAN. *New Priorities for the University.* San Francisco: Jossey-Bass, 1987.

MCCORKLE, C. O. Memorandum to Chancellors. Berkeley: University of California, Office of the President, 28 July 1972.

NOBLE, DAVID F. *America by Design.* Oxford: Oxford University Press, 1977.

PENNSYLVANIA STATE UNIVERSITY. "The Division of Continuing Education." From a document labeled "Source Book, August 6, 1987," obtained from John Leathers.

ROCKHILL, KATHLEEN. *Academic Excellence and Public Service.* New Brunswick, N.J.: Transaction Books, 1983.

SANDERS, HOWARD J. "Continuing Education." *Chemical and Engineering News,* 20 May 1974, 26–38.

STANFORD UNIVERSITY. "Continuing Education at Stanford." Obtained from the Stanford Alumni Association.

TOOMBS, WILLIAM, and CARL A. LINDSAY. "Departments and Professions: Institutionalizing Continuing Professional Education." University Park: Pennsylvania State University, CPE Development Project, 1986.

TRIBUS, MYRON. Report to Alfred A. H. Keil, 30 March 1976. M.I.T. Archives, 80-33, Box 46, Center for Advanced Engineering Studies.

VAN TREES, HARRY L. Letter to President Howard Johnson, 23 November 1970, 5, M.I.T. Archives, 80-33, Box 46, Center for Advanced Engineering Studies.

WICKENDEN, WILLIAM E. "A Comparative Study of the Engineering Education in the U.S. and Europe." *Report of the Investigation of Engineering Education.* Pittsburgh, Pa. Society for the Promotion of Engineering Education, 1930.

11

University Contributions to Economic Development: Technical Assistance Programs, Business Incubators, and Research Parks

The elements of technology transfer examined so far—patent and licensing activities, equity ownership in research-based companies, liaison programs, and continuing education—can be viewed as arising directly from the traditional university missions, teaching and research. They are a consequence of either the university's interest in managing and benefiting from its intellectual property (patents and equity ownership), its attempts to relate more closely to research patrons (liaison programs), or extensions of its teaching function (continuing education). Considered in this chapter are technology transfer activities that are less directly associated with these traditional missions: technical assistance programs, sometimes known as industrial extension services; and business incubators, which provide new businesses devoted to developing technology with inexpensive rental space and technical and business assistance; and real estate developments related to technology transfer, called here research parks. Although these activities may be tangentially associated with traditional university missions and also may be entered into for a university's profit, they are most likely to be viewed primarily as a part of a university's efforts to help develop the local economy.

University Roles in Economic Development

Economic development has been defined as a process of innovation that increases the capacity of individuals and organizations to produce goods and services and thereby create wealth.[1] Economic development efforts by state and local governments have increased over the last two decades as seemingly fundamental shifts in the economy resulted in localized losses of jobs and economic depressions. Such efforts typically concentrate on saving old jobs or generating new ones.

Early local economic development efforts concentrated on luring already-existing businesses from other localities or attracting branches of large corporations. Although this kind of activity continues to be significant, mounting evidence indicates that developing local firms is more effective than "stealing" businesses from other places. Studies in the late 1970s concluded that new jobs were being created faster by new, small companies than by large, well-established ones. Birch found that firms with 20 or fewer employees generated 66 percent of all new jobs; 80 percent of all new jobs were generated by firms no more than four years old; and large firms were more likely to shrink than to grow.[2]

Evidence of this kind was not the only argument for the development of local businesses. Increased competition for branch operations of established companies often required localities to make substantial concessions in the form of tax incentives and taxpayer-financed facilities and services, thus decreasing the benefit given to local economies by new jobs. Localities also learned that branch operations could disappear quickly as a result of decisions from headquarters, "downsizing," and corporate mergers and takeovers.

As state and local governments shifted their attention to nurturing local businesses, *economic development* became increasingly synonymous with *entrepreneurial development*. Economic development efforts often emphasized the attraction or development of high-technology manufacturing companies, both because these companies were thought to be "clean" or nonpolluting and because such companies accounted for about 75 percent of new manufacturing jobs.[3] High-technology companies not engaged in manufacturing accounted for only a small percentage of the new companies created and tended to account for only a small number of professional jobs.

At the same time these shifts in emphasis were occurring, it became clear to universities that small, local enterprises were more effective at developing university research than large companies because they were willing to take a technology at an earlier stage of development and were likely to be more vigorous champions of the technology.[4] This is corroborated by the experience of the Technology Licensing Office of M.I.T., which in 1987–1988 granted more than 60 percent of its technology licenses to small (under 50 employees) firms.[5]

In turn, the examples of Stanford's relation to "Silicon Valley" and M.I.T.'s role in the development of industry along Route 128 encouraged other localities to take an interest in universities near them. Beginning in the early 1960s, this interest resulted in a number of studies of the university's role in the attraction of industry to a locality and in the development of new businesses within that locality.

ROLE IN ATTRACTING INDUSTRY

In the 1960s, studies of the influence of the university on economic development tended to concentrate on universities' influence on the decisions of corporate managers to locate business ventures in a particular area. Most of these studies rated the presence of a major university in an area as being very important in such decisions, but the methods of the studies left some doubt about the validity of their findings. For instance, they did not consider aspects such as the availability of trained manpower, the availability of attractive housing, and the cultural amenities of the area, all of which are related to universities. In 1976, Sirbu et al.[6] found that, although the presence of a university was important in decisions concerning the location of high-technology plants, major universities seldom interacted in a significant and direct way with high-technology businesses. Such businesses were much more likely to have direct ties with local two-year institutions than with universities.

A 1983 study supported the conclusion that the presence of a university was important, but its role in attracting industry remained unclear:

> The presence of a major university, per capita defense spending, and good transportation appear to exert more influence on the location of high-technology manufacturing jobs than do labor force characteristics. . . . Our most important conclusion is that the location and growth of high-tech industry is a very varied and disparate process which will require . . . industry-by-industry analysis.[7]

ROLE IN LOCAL BUSINESS DEVELOPMENTS

The role of universities in fostering new local commercial enterprises is even more difficult to quantify or describe than their role in attracting existing industries. The Stanford-Silicon Valley and M.I.T.-Route 128 phenomena have been the focus of a number of studies that attempt to describe the university's role in the economic development process. At first inspection, the two situations appear similar, and the university appears to have played a significant role in each.[8] In both cases the industry developed in areas in which the nearby university had a strong research base—at M.I.T. in small computers and at Stanford in semiconductors and other scientific fields closely allied with aviation and aerospace. Both developments began just after World War II, and both M.I.T. and Stanford were close to another major university that undoubtedly played a supporting role by adding considerably to the available expertise in a geographical area: Harvard and U.C. Berkeley, respectively.

On closer inspection, however, some significant differences appear between the two developments. First, Stanford's relationship to the development of Silicon Valley appears to be in large measure the result of deliberate policies and actions on the part of the university and its leaders at the time, President J. E. Wallace Sterling and Vice President Frederick Terman. By contrast, the administrations of M.I.T. and Harvard during these years remained aloof from deliberate involvement in local development efforts. Second, while M.I.T. had already developed a strong reputation in electronics before and during the World War II, Stanford's reputation was built after the war, in conjuction with developments in Silicon Valley. Third, the aerospace industry was already firmly entrenched on the San Francisco peninsula just after the war, but no such host industry existed in the Boston area. Fourth, Stanford had excess land that it could devote to attracting industry but M.I.T. did not, although inexpensive, available, and relatively unrestricted land was available along Route 128.

The most important difference between the two developments, however, and also the one most relevant to this discussion lies in the nature of the universities' relationship to start-up companies. A 1970 study of spin-off companies in Silicon Valley noted that only six of 243 such companies founded in the 1960s had one or more full-time founders who came directly from a university.[9] The spin-off activity that was so important in the growth of Silicon Valley came from the companies that originally settled there, not from university sources. Stanford's relationship to the early developments in the area was through its graduates—Hewlett, Packard, and the Varian brothers—and not directly from faculty or researchers who worked at Stanford.

By contrast, many of the people who founded large companies in the Boston area had more direct ties to M.I.T. or Harvard. A study conducted in the mid-1960s identified 200 new technical companies in the Boston area and noted that 167 of them had been started by people who had been employed at M.I.T. or in M.I.T.-adminstered laboratories.[10] For instance, Kenneth Olsen, founder of Digital Equipment Corporation, managed the TX-O project (on which DEC technology was based) at M.I.T.'s Lincoln Laboratory, and the circuit designer for the Lincoln Lab was on the first DEC development team. An Wang, founder of Wang Computer, based his company on work he did as a member of Harvard's Computer Laboratory.[11] Of course, M.I.T. alumni also have been important in starting businesses. A study by the Bank of Boston released in June 1989 found that 636 Massachusetts businesses, having 1988 worldwide sales of $39.7 billion and employing over 192,000 people, had been founded by M.I.T. graduates.[12] Thus, although the administrations of M.I.T. and Harvard did not deliberately set out to aid economic development in their region, their institutions played a more major role in that development than Stanford did in its area.

The point here is neither to minimize Stanford's importance in the development of Silicon Valley nor to exalt M.I.T.'s effect on the economy of Massachusetts, but simply to suggest that the relationship between universities and new enterprises is complex. Because of this complexity, or perhaps because of a failure to recognize and analyze it, attempts to recreate the conditions that spawned Silicon Valley and Route 128 usually have failed; where they have succeeded, they have probably done so partly because of factors not present in either example.

OTHER ROLES AND CONTRIBUTIONS

Most commentators now agree that the presence of a major university is important but not sufficient for the formation of what have come to be called Technology-Oriented Complexes (TOCs), or concentrations of technology-oriented commercial enterprises. Major universities not associated with any identifiable concentration of such enterprises include Columbia, Chicago, Cal Tech—and, for that matter, despite their proximity to such developments, Harvard and Berkeley. On the other hand, universities that are associated with such developments can be found in Texas, Arizona (Arizona State), Utah, and North Carolina.

Universities supply or are strongly identified with a number of elements considered important to the development of a TOC, including government contracts, a concentration of scientists and engineers, cultural and educational amenities (including continuing education opportunities) and trained manpower. However, they either are not associated with or are weakly associated with other important elements, such as a good transportation system, supportive local government, a diverse and cooperative group of organizations employing professionals, and public support of high-technology development.

Universities also contribute spin-off or start-up companies that are partly owned by the institutions and based on university research. However, direct ownership by universities, while an interesting new development, is not yet a significant factor in establishment of TOCs. What is significant, but also difficult to measure, is the kind of spin-off activity just described in relation to M.I.T. University spin-offs tend to be associated with research and development, testing, or consulting rather than with the manufacturing of custom products or the providing of custom services that is characteristic of private industry spin-offs.[13]

University contributions to economic development can be placed in one of three categories: joint research and development, education and training, and entrepreneurial assistance services. A discussion of three activities falling into this last category—technical assistance services, business incubators, and research parks—occupies the rest of this chapter.

Technical Assistance Programs

Many universities have programs designed to assist local or regional businesses by providing technical assistance or business advice. Some of these programs have been inspired by and modeled after the agricultural extension services that have long been a part of university land-grant traditions.

Most university technical assistance programs (TAPs) are designed to serve small to medium-sized businesses in a defined geographical area (sometimes a state) by providing advice either through answering technical questions submitted by business owners or through on-site consultations to solve production or other problems. These consultations can be made by specialists permanently assigned to the TAPs or by consultants, including university faculty, identified by the TAPs as possessing the expertise necessary to solve a particular problem. Most TAPs therefore maintain or are associated with technical reference libraries and have data bases of names and specialties of potential consultants. Many also operate continuing education programs in technical subjects.

Forms of TAPs and the types of services they offer vary widely. The M.I.T. Enterprise Forum is one example of a TAP-like program. In this program a panel of business advisers and other experts critique (*tear apart* is perhaps a more accurate term) a business plan presented to them, thus providing the business's owners with free expert advice. Originally begun in the Boston area, the Enterprise Forum has expanded to other areas through the M.I.T. Alumni Foundation.

Universities are the most common sponsors of TAPs because of their library resources and the expertise of their faculties, although a few TAPs are sponsored by other agencies. A brochure published in 1987 by the National Association of Management and Technical Assistance Centers (NAMTAC) lists 75 members from 39 states, only 8 of which are not associated with a college or university.[14] Included within the classification of TAPs are several hundred small business development centers in community colleges set up by the Small Business Administration, 44 technical and management assistance centers in universities financed by the U.S. Department of Commerce and six more such centers supported by the U.S. International Trade Administration, and 42 centers that offer technical advice to state departments of transportation and are supported by the U.S. Department of Transportation.[15]

Recently there has been a considerable growth in TAPs. From just a few such programs in the early 1960s, the number has grown to several hundred today. The reason for this growth is the emphasis on the support of local industry mentioned in the first section of this chapter. TAPs are sponsored by both major research universities and smaller teaching universities and may serve either broad or limited geographical areas. Only one of the universities selected for study, Penn State, operates a TAP. Several other research universities also operate TAPs, however. In order to gain some perspective on these programs, this discussion describes both the Penn State program and a similar program at the Georgia Institute of Technology (Georgia Tech).

GEORGIA TECH

The Georgia Tech Industrial Extension Service (GTIES)[16] is one of the oldest and most successful of statewide TAPS. Founded 25 years ago by Georgia Tech, GTIES operates through a network of 12 regional offices staffed with resident industrial engineers who have a business background and are familiar with local needs and ready to assist the state's industries in solving technical and management problems. These extension specialists are backed up by regular faculty from the institute; students also work in five of the twelve centers. However, most of the projects handled by GTIES involve on-site consultation by a specialist, are short term, and involve manufacturing processes, facility planning, materials planning, improvement of methods, or cost control. One GTIES specialist reported seeking help from university or outside sources in only 5 percent of the cases.[17] Georgia firms may receive up to five person-days of service free and may negotiate for more assistance if needed.

Although some Georgia Tech faculty view participation in GTIES activities as intellectually unrewarding, and although the GTIES extension agents have little contact with the university, GTIES has proven extremely valuable to the institute. It is popular with both the state legislature and local industry and has benefited the state's economy, enhanced the institute's image, and increased public support for the institution.

PENN STATE

The Pennsylvania Technical Assistance Program (PENNTAP), administered by Penn State, is another example of a statewide TAP. PENNTAP began with an appropriation of $200,000 from the state legislature in 1966. These funds were to be used "to finance research and information dissemination projects to be conducted by Pennsylvania institutions for the benefit of Pennsylvania industry to assure more jobs for Pennsylvanians."[18] This appropriation marked a small shift in the balance of political power in the state away from big business and toward small business. PENNTAP was part of the state's response to an increasingly powerful small business constituency.[19]

Throughout its history PENNTAP has enjoyed a subsidy from the state, which averaged around $150,000 per year but has recently increased to $250,000 per year. To this subsidy Penn State added funding of from two to three times the amount of the state contribution.[20] In addition, PENN-TAP has been able to piece together money from several federal programs. First it received federal money under the State Technical Services Act, and later it was awarded money under the energy extension service (1977), the Department of Commerce university centers program (1981), a special project on coal technology (1981), and the Rural Technical Assistance Program (RTAP) program for local road management (1985).[21] PENNTAP's total annual budget has been between $500,000 and $775,000.

In the view of PENNTAP's director, H. Leroy Marlow, the program has two types of functions, active and responsive. Active functions involve staying in touch with new developments in technology and, where they appear to have benefits for Pennsylvania businesses, disseminating information about these developments through short courses, publications, or other means. For instance, when a technical specialist at PENNTAP learned of some Navy research on the nondestructive testing of airplane tires, he brought Navy and industry representatives together to study the new technology, which was later implemented in tire manufacturing plants in Pennsylvania and other states. Active functions also include the PENNTAP Library Information System, which represents facilitated access to the extensive Penn State library system for technical subjects.[22] In providing these active functions PENNTAP differs from GTIES, which is primarily reactive, the latter simply responding to requests for help rather than actively mounting programs to inform its constituencies of new developments.

Like GTIES, PENNTAP also responds to requests for information. These can be made through any of Penn State's 24 continuing education centers. Since these centers are located around the state, requests usually involve making only a local phone call. The requests are routed to the appropriate technical specialist who responds personally to the client to determine the scope and nature of the problem and then initiates a plan of action to solve it. This plan may involve a visit by the specialist to the site. More often it involves the identification and solicitation of additional experts— faculty members or others not associated with the university.

The key to the success of PENNTAP is the technical specialist, of whom the program has six to eight at any given time. These specialists are permanent employees who devote all of their time to PENNTAP projects and duties. Unlike the specialists in the Georgia program, who are assigned a geographical area in the state and do a great deal of direct consulting to businesses, the PENNTAP specialists focus on specific and different technical fields, such as mineral sciences, electrical and industrial engineering, or library sciences. A specialist handles whatever problems arise in his or her field in any part of the state, often by referring businesses to a university expert or an information source. These specialists hold academic rank (although they are not faculty) and are considered to be members of the relevant academic departments on the University Park campus. This closeness to the academic departments helps the specialists to stay in touch with the latest developments in their fields and also to become familiar with the expertise of the faculty members.

For many years the director of PENNTAP reported through the continuing education department of Penn State to the vice president and dean of the University-Wide System of Continuing Education and the Commonwealth Education System. In a recent reorganization this reporting relationship was changed, and PENNTAP's director now reports to a new associate vice president for Sponsored Programs, Technology Transfer, and Industrial Liaison. PENNTAP also has an advisory council consisting of 15 people from a variety of professional backgrounds. This council recommends policies and priorities for PENNTAP and evaluates the program's efforts.

PENNTAP is widely viewed as one of the most successful programs of its kind in the country. It presently solves from 1,000 to 1,400 technical problems per year, half of which come from small businesses and the other half from public-sector organizations. Director Marlow claims that PENNTAP's annual budget expenditure of $775,000 is returned more than 17 times in economic benefits to the state.[23] The program is consistent with Penn State's new economic development initiatives and has a high national reputation. It is well regarded within the state and by the state legislature, as the continuous funding since 1966 from the state as well as the university attest.

ISSUES RELATED TO TAPS

The preceding examples illustrate some of the issues raised by the existence of TAPs on university campuses. These issues include funding, measurement of effectiveness, promotion, staffing and faculty rewards, and comparison with agricultural extension programs.

FUNDING. Most TAPs depend on public funds for survival. Universities must either fund these programs themselves, as in the case of PENNTAP, or solicit funds from other agencies. The funding must be dependably available over a number of years, which often means that considerable effort must be spent to convince funding agencies of the continuing worth of the programs. Universities view TAPs clearly as public service rather than as part of the institutions' central teaching or research missions. This means that TAPs also have to demonstrate their continuing value to the university.

MEASUREMENT OF EFFECTIVENESS. The effectiveness of TAPs is difficult to measure in terms of dollars or of jobs gained or saved, yet there is considerable pressure for the programs to produce measurable results in order to justify continued funding. This combination of factors causes many programs to resort to questionable methods of measurement. For instance, PENNTAP's measurements of results are based on a form sent to each client requesting information on the number of jobs created and saved and the dollar value of the PENNTAP contribution. Less than 40 percent of the clients respond. Nonetheless, PENNTAP uses the figures it does receive to extrapolate the whole for its reports to the university and to funding agencies.

Job creation statistics are particularly suspect. The fact that a TAP played a part in solving a technical problem does not mean that new jobs associated with the product were created by the TAP. For one thing, alternatives are not considered: What would have happened if nothing had been done? What would have happened if some other kind of help had been obtained?[24] To their discomfort and sometimes embarrassment, universities often are forced to play these spurious numbers games when they sponsor TAPs.

PROMOTION. Successful TAPs spend a significant amount of effort and money in promoting their activity. PENNTAP, for instance, employs an information coordinator to publicize PENNTAP's programs. Promotion is necessary for two reasons. First, potential clients must know about the service in order to use it. Second, engendering broad-based support for the program is necessary to assure continued funding. Such promotional efforts, especially when they trumpet the kind of questionable figures just described, sometimes produce a jarring note in a university. However, as was evident in both examples cited, appropriate public relations for such programs can materially increase the image and standing of the institution in the community and can add to support of other aspects of the university, such as development efforts.

STAFFING AND FACULTY REWARDS. One problem that TAPs have in common with continuing education and other organizational forms associated with technology transfer in universities is the uncertain status of the professional staff who operate the programs and the way in which the institutional reward system is structured for them. Technical specialists normally do not have faculty status, do not teach, and do not perform scientific research, yet they must possess a high level of scientific, engineering, or technical background and other skills as well. They fill jobs with no clear path for advancement and often do not have security of employment. In short, universities rarely have personnel policies and performance incentives appropriate for attracting the best technical specialists.

Although most TAPs depend to some extent on faculty participation, few universities have a reward or compensation plan to encourage such participation. This is partly because of the unsettled question of whether it is appropriate for faculty, especially in research universities, to be involved in solving relatively low-level, applied problems. Burdened with this philosophical question mark and unable to provide a reasonable incentive for faculty participation, TAPs must depend on faculty volunteerism and thus are under severe constraints.

COMPARISON WITH AGRICULTURAL EXTENSION PROGRAMS

In many states, both universities and state legislatures are viewing agricultural extension as an appropriate model for new industrial extension programs that are designed to promote technology transfer from universities and to aid small businesses by providing answers to technical problems. What seems to be needed today is a rediscovery and renewed implementation of the fundamental concept underlying land-grant institutions: a combination of extension with research. After all, the needs of today's knowledge-oriented society are strikingly similar to those of the agricultural society a century ago. Now, as then, there is a need both for new scientific and technical advances and for a way to bring the results of research to the workplace. The latter process became, and still is, the task of agricultural extension.

Nonetheless, there are significant differences between the old agricultural extension and what is possible now in other fields. First, agricultural extension always was heavily subsidized by federal and state funds. No such large-scale subsidies are likely for industrial extension activities. Second, the target audience for agricultural extension—farmers—was easy to identify, even if geographically hard to reach. Potential users of an industrial extension service are much harder to pinpoint.

Third, agricultural extension confined its activities to a well-defined discipline that existed in organizational units within universities, that is, schools of agriculture. A comprehensive industrial extension service, however, must include many disciplines and university departments, from biology to mechanical engineering, and sometimes must arrange for these disciplines to cooperate to solve an industrial problem.

Fourth, the "product" dispensed by agricultural extension often was delivered to the farmer in usable form by an extension agent who was a jack-of-all-trades in the subject area. The advice delivered to industrial users, however, often requires considerable adjustment and implementation before it can become useful, and the problems encountered in industrial extension work span such a range of disciplines that a single agent probably could not be effective in every field. Finally, agricultural interests were much more organized and carried much more political clout than potential users of industrial extension services.

For all these reasons and more, the agricultural extension model is not applicable to the modern situation. Where technical assistance programs have been modeled on agricultural extension services, misunderstandings have occurred. For instance, TAPs often are criticized for spending significantly higher percentages of their budgets on publicity than do their agricultural counterparts. This extra expenditure often occurs because TAPs have to reach a broader intended clientele rather than because they are poorly managed. A General Accounting Office examination of industrial extension operations concluded that "the industrial extension programs we examined are not fullblown replicas of the agricultural extension model. . . . Instead, they use elements of the agricultural model in carrying out their specific objectives."[25]

CONCLUSION

Universities that sponsor TAPs usually do so because of strong external pressure or incentive. These programs are public service activities and rarely have strong alliances with teaching or fundamental research. They require heavy subsidies and therefore must be attentive to the purposes and requirements of funding agencies. When administered properly, however, TAPs can be strong contributors to the image of the university as a valuable resource to a community and an active participant in economic development. They can link small businesses, the segment of industry least able to take advantage of university resources, to the knowledge base of the university. Like many technology transfer mechanisms, however, TAPs exist on the periphery of the university, uncertain of their place and often unsupported by the administration.

Business Incubators

Business incubators fit somewhere between technical assistance programs and university-related real estate developments. Their purpose is to nurture start-up business ventures through their crucial early stages. They provide qualifying businesses with building space at below-market rental rates and offer a variety of services that often include technical or business advice.

Business incubators can be distinguished from more traditional small business assistance centers as the incubators place start-up ventures in physical proximity to each other. This positioning allows the ventures to share facilities, equipment, and services on a *per use* basis. These services may include copying machines, telephone systems, shop equipment, computers, secretarial help, and even skilled technical help. In addition to the services offered through the formally established incubator organization, start-ups in incubators often gain other advantages that association with other start-up businesses can bring. For instance, belonging to a business incubator often brings a good deal of publicity because such incubators are a source of civic pride and the focus of considerable media attention. More important, a concentration of start-up companies often attracts the kinds of special expertise needed by such firms—accountants, legal advisers, venture capitalists, bankers, marketing specialists, financial managers, and so on. These professionals sometimes provide services at below-market rates in an effort to develop clients or in return for stock in the companies.

When incubators are associated with universities, the benefits they provide to new firms are enhanced. University-sponsored incubator firms normally have access to university libraries and other facilities, often including sophisticated scientific equipment, for low fees on a per use basis. They also have access to technical assistance and managerial and business advice. In return for providing MBA students from the university with case study material, businesses in incubators often gain a range of advice and ideas that otherwise would not be available to them.

HISTORY AND GROWTH

The first business incubator appears to have been the University City Science Center, which was begun as a private, nonprofit foundation in 1963 in Philadelphia.[26] With the help of large amounts of funding from public sources and broad-based support from the business and academic communities, the Science Center today has extended well beyond the narrow concept upon which most incubators are based. It encompasses a full-fledged industrial park, manages its own research and development programs (with a budget of $10 million and a staff of 115 scientists in 1986–1987) and joint R&D programs with its partner institutes, universities, and corporations (80 projects involving over 100 companies and 14 universities, with a budget of $27 million in 1986–1987).[27] Its primary purpose, however, remains the generation of new enterprises. Through December 1987 it had served 106 start-up companies, which produced over 6,500 new jobs, and had secured in excess of $20 million in equity in these firms.[28]

Few of the incubators developed since this first one have been so successful. Nonetheless, growth in the number of incubators has been extremely rapid in recent years, although the rate of growth is likely to diminish. The National Business Incubator Association counted 330 incubators in early 1989, up from only 40 in mid-1984, and predicted that by 1991 there would be 660 in the United States.[29] Part of this increase comes from the growing number of incubators sponsored by the for-profit sector, which rose from only 9 percent of the total number of incubators in 1985 to over 25 percent today.[30]

Many incubators were started with little funding or support. Incubators associated with universities typically owed their initiation to the acquisition or sudden availability of an otherwise surplus building near the university. Penn State's first incubator was a former grammar school, for example, and Rensselaer Polytechnic Institute's incubator is housed in what was once a shelter for "wayward girls" run by a religious order.

There are now so many incubators that new ventures are feeling competition. At least four incubators have failed in the last six years, and new entries to the field are having to adopt market niche strategies to attract enough start-ups. Targeting start-ups in a particular industry allows incubators to focus their services on particular needs. For instance, the Spokane Business Incubation Center has established the Kitchen Center, which has a state approved food-processing facility. Texas Woman's University in

Denton is considering an incubation center that would be geared to the university's nursing and allied health departments.[31]

As business incubators have become more numerous, they also have become increasingly sophisticated in their methods of operation. Whereas the first incubators admitted any firm that could pay the rent, most incubators now have well-articulated selection criteria for entry. They also have standards to help determine when a company should leave the incubator in order to assure that the facilities are being used by the companies that can benefit most from them and that the incubator churns out successful companies.

As illustrated by the example of Philadelphia's Science Center, which generated a profit of $100,000 in 1985 from its copying services alone,[32] business incubators, when carefully managed, can be good investments. The rent must be kept low, but charges for services and consulting can be substantial. In addition, there is always the possibility of getting a "piece of the action" of a high-flying start-up.

UNIVERSITY INVOLVEMENT

Because of the unique and comprehensive resources they can offer to beginning companies, universities were important early players in the business incubator field, and they continue to be important today, although at present universities actually own only about 15 percent of all incubators,[33] most of which are part of larger real estate developments.

RENSSELAER POLYTECHNIC INSTITUTE. Rensselaer Polytechnic Institute (RPI) is an example of an early (1980) and successful university sponsor of a business incubator. Located in Troy, New York, RPI was pressed into service to aid the ailing economy of the Albany-Troy area. Tenants in the RPI incubator pay below-market rents and purchase services, including access to equipment or facilities on the RPI campus, on a fee-for-service basis. Most of the tenants find these services a bargain and also credit RPI faculty with helping them in technical matters. Within three years of its initiation, the RPI incubator had spawned three successful companies, one of which, disappointingly, moved out of the area. This relocation raised a question about the ability of the incubator to aid the local economy.[34] The success of the park and its positive effect on the local economy is no longer in doubt, however. By the end of 1988 the park had 45 tenants, who employed a total of 850 people. The value of the land increased from $1.8 million in 1980 to between $15 and $20 million today.[35]

Involvement with the incubator has helped RPI's image in the community, which now views RPI as active and effective in helping the community economy. This is by far the most important advantage of the incubator program to RPI. The glamour of the enterprise also has helped RPI's national image and its attempts to broaden the geographical base of its student body. Its other hope, that incubator tenants would "graduate" to become tenants of RPI's industrial park, has not been fulfilled, however.[36] In several cases RPI has taken stock in start-up companies in lieu of rent, but so far it has not invested its own money in start-ups except in second and third round financing.[37]

PENN STATE. Penn State also has been active in developing incubators, primarily under the Ben Franklin Partnership Program. Building on the success of the University Science Center in Philadelphia, the state of Pennsylvania funded the establishment of more incubators than any other state in the nation (42, as compared to second place Illinois with 26).[38] Funds for these incubators are administered through the Advanced Technology Centers (ATC) of the Ben Franklin Partnership. As administrator of the largest ATC in the state, Penn State has established 15 incubators, which are operated in cooperation with local development districts and private developers.[39]

Not satisfied with all this development, which is far more extensive than anything most other universities would contemplate, Penn State wants to go even further. One element of the economic initiatives proposed by President Bryce Jordan is the establishment of incubators on or near the campus that would serve university people exclusively. October 1989 marked the beginning of the realization of this plan when an on-campus building containing 30,000 square feet was completed. Ten thousand square feet of this building are dedicated to incubator space for faculty enterpreneurs. More incubator space is planned for the university's research park development. It is hoped that by the year 2000 this park will contain over 1.5 million square feet on its 130 acres and that some of this space will be occupied by "graduates" of the present incubator and additional incubators planned for the park.[40]

ADVANTAGES AND DISADVANTAGES. The advantages and disadvantages of university sponsorship of business incubators are similar to those for real estate ventures and technical assistance programs. As at Penn State and RPI, business incubators can be highly visible undertakings that symbolize the university's involvement in the economic development process. Furthermore, when incubated companies are successful, the university sponsoring the incubator stands to gain in several ways. First, it has established a positive relationship with the companies, which may later lead to research contracts, financial support of other kinds including gifts of money or equipment, consulting opportunities for faculty members, and community goodwill. When a university has taken equity in incubated companies, as RPI does, it may gain from an increase in the value of the stock. If the university sponsors an industrial park, incubator tenants go on to become park tenants.

Most of these benefits, however, are indirect and intangible and accrue over a long term. Although incubators *can* be good investments, nothing assures that they will be. Universities must put themselves at risk and expend resources in order to maintain and operate incubators. Like technical assistance programs, business incubators are primarily a service offered by the university to the community. They seldom are closely tied to the university's central missions: teaching and research. Their place in the university therefore often is undefined, subject to misunderstanding by the faculty, and in need of constant explanation and justification. The fact that university-sponsored business incubators nevertheless have proliferated recently is but one more symptom of the underlying trend toward increased involvement in economic development by universities.

Research Parks

The term *research park* is used to describe a real estate development dedicated to serving research-oriented companies. Research parks are distinguished from industrial parks by the kinds of tenants they serve. The Association of University-Related Research Parks defines a university research park as a property-based venture with the following characteristics:

- Has existing or planned land and buildings specifically designed for private and public research and development facilities, high technology and science based companies and support services; and,
- Has a contractual and/or operational relationship with a university or other institution of higher education; and,
- Has a role in promoting research and development by the university in partnership with industry, assisting in the growth of new ventures, and promoting economic development; and,
- Has a role in aiding the transfer of technology and business skills between the university and industry tenants.[41]

The definition does not include freestanding research centers, innovation centers, and incubator buildings that do not have a university connection.

HISTORY

Industrial parks were developed long before university-related research parks. The first industrial park in the country was the Central Manufacturing District, founded in Chicago in 1905. By 1930 the country had eight similar parks, but the idea did not really catch on until after World War II. There were 50 such parks in 1949, and by 1961 there were 1,046 known industrial parks. These parks were an important part of what has been called the "suburbanization of industry."[42]

The first research park in the country was the Bohanson Research Park of Menlo Park, California, founded in 1948.[43] Between 1950 and 1971, 128 research parks were founded, most of them in the early 1960s (79 were founded between 1960 and 1965).[44] Only 81 of these still existed by 1971, however. Some had failed, and others no longer could be classified as research parks because of changes in tenant admission criteria.[45]

The first university-related research park was the Stanford Research Park, founded in 1951. Of the 81 research parks in existence in 1971, 19 were university related.[46] By 1980 the number started had increased to 27, but of these, 16 had failed and five were only marginally successful; only six could be termed unqualified successes.[47] A number of major universities, including Kansas, Iowa State, and Indiana, had abandoned plans for building research parks. The Stanford example was proving hard to duplicate.

The 1980s, however, have seen renewed interest in the development of university lands and university-related research parks. By mid-1989 there were 115 research parks in the country, and university-related research parks housed 2,100 companies and some 173,000 workers.[48]

REASONS FOR UNIVERSITY SPONSORSHIP

Many universities are engaged in some form of real estate development related to research commercialization for the same reasons that they engage in other technology transfer activities described in this book—desires to foster relations with industry, aid in the transfer of technology to industry, and participate in the economic development of the region. In addition, however, universities want to benefit financially from their valuable land holdings. This last reason is a reversal of a university tradition. Although many universities were started with grants of land under the Morrill Act of 1862 (the Land Grant Act) and many, including UCLA and U.C. Irvine in the University of California system, were located in parts of larger commercial real estate developments, for many years people believed that university lands should be reserved solely for university use because the university in effect held its property in trust for future generations of students. This tradition is now breaking down, however, as university lands increase in value and often become crucial elements in localities' land use planning.[49] Further, in recent years real estate investment has become an important part of university endowment portfolio management. In a 1988 endowment survey, 37 percent of the 278 institutions surveyed had real estate investments, typically amounting to 2 to 3 percent of total endowment funds. Fueled by tremendous capital appreciation, almost half of George Washington University's endowment is invested in real estate, and about 16 percent of Stanford's is in real estate.[50]

The financial advantages to universities of research park development remain exceedingly speculative. Nonetheless, the potential advantages of such parks in terms of technology transfer and economic development provide powerful inducements to consider them. Having research-related industries located near the universities that spawned them greatly facilitates a whole range of positive interactions, including the sharing of research facilities and instruments; collaboration on research projects; increased consulting opportunities for faculty members, and the resulting greater availability of professional expertise to industry; fuller utilization and support of university programs such as continuing education and industrial affiliate programs; improved education for students and industry researchers; and—most important perhaps—the earlier recognition of important breakthroughs and the more rapid commercialization of university research.[51] Furthermore, a successful research park can be important in stimulating regional economic development, thus earning political support for the university.

These advantages and more have sparked renewed interest in university-related research parks. Most major universities, including the four in this study, are engaged in or actively considering some form of research park development.

Of the four universities in this study, two—Stanford and M.I.T.—are now or have been active in real estate developments devoted to research-based companies. The other two, U.C. Berkeley and Penn State, currently are making preliminary plans for such developments.

STANFORD

It seems logical to begin with Stanford, the university that started university-related research parks. Stanford's 1885 founding grant from Senator Leland Stanford and his wife specified that the 8,800 acres of the bequest could never be sold. Before 1949 Stanford had not actively used or studied potential uses for its land; the university sought only to generate enough income to offset the property taxes imposed on the portion of the land that was not used intensively for educational purposes.[52] After the war, however, land values increased sharply as the aerospace industry surrounding the NASA-Ames Laboratory began to grow. Stanford's holdings formed an underdeveloped bulge between the foothills and the bay, interrupting the steady advance of housing and industrial development on the San Francisco peninsula. Stanford lands began to be subject to "friendly condemnation" proceedings from state and local governments.

Into this situation stepped the new president of Stanford, J. E. Wallace Sterling. He engaged the board of trustees of the university in a planning effort first by commissioning a preliminary plan by an outside planning firm, and then, in October 1951, by initiating the appointment of a special committee of the board to study land development. Recognizing the importance of faculty participation, Sterling had also appointed in 1950 what was later called the Faculty Advisory Committee on Land and Building Development.

With planning and review well underway, Stanford started its new program of land development in October 1951 by leasing a ten-acre parcel to Varian Associates for 99 years. This lease was followed by leases to Eastman Kodak (1952), Hewlett-Packard (1954), and others. Moving in a different direction, Stanford trustees announced in 1952 that they were considering the development of 60 acres into a shopping center.

As these developments went forward, and particularly when the details of a master plan prepared by Skidmore, Owings, and Merrill came to light, faculty concern increased that the university was moving too fast. The board of trustees responded by increasing the amount of land dedicated to academic use from the 2,300 acres called for in the report to 4,000 acres and assuring the faculty committee that it could continuously review the plan and actions taken under it. Indeed, the faculty committee had much to consider. Land allocations were important, but so, too, were the allocation of income from the leases, the length and terms of the leases, and the criteria for tenancy on university lands. One committee member remembered that many of the faculty committee sessions were stormy and notes that "it took a great many meetings for us all truly to appreciate that the planning of 8,800 acres cannot please everyone, not even the ten members of a committee."[53]

Despite this discord, faculty involvement in planning is one of the primary reasons for the success of what became the Stanford Research Park. Combined with the vision of President Sterling and Vice President and Dean of Engineering Frederick Terman that if the university's plans were to be successful, "a significant number of faculty members must develop and maintain personal acquaintances with key people in local industry,"[54] this early and extensive involvement of the faculty resulted in an unusual degree of integration of the research park with the goals of the university.

Key to this integration is one of the criteria that Stanford uses to select tenants of the park. Prospective tenants must provide a detailed description of the type of business they expect to conduct, and only companies that are clearly engaged in research or the development of research are admitted. This requirement assures some level of community of interest with Stanford and with the other tenants of the park.

By 1983 the Stanford Research Park was fully leased. Its 660 acres housed 80 tenants with a total of over 28,000 employees. Early tenants obtained 99-year leases, but, as demand increased, Stanford began granting leases for no more than 51 years. Leases must be prepaid, at rates that over the years have ranged from $10,000 to $195,000 per acre; these lease payments are added to Stanford's endowment fund.[55] Stanford administers the research park with its own staff.

Stanford Research Park provides a model that many other universities have tried to imitate, usually without success. The failures may arise because Stanford's park grew out of a set of institutional imperatives and a culture supportive of industry-university relationships that do not exist on very many other campuses today.

M.I.T.

M.I.T.'s real estate developments have been much less integrated with institutional goals than Stanford's. They appear to have been primarily business deals rather than a means to build relationships with industry. This is partly because M.I.T. already enjoys strong ties with industry and does not need research parks to further those ties. Also, because of its urban location, the institute has not had large parcels of land available to it for development. Finally, and perhaps most important, private research and industrial parks, primarily along Route 128 and later on Route 495, were developed to serve the many new businesses that were attracted to or which sprang up in the area, so a university-sponsored park would have been redundant. The fact that land along these highways was inexpensive, not bound by restrictive zoning laws, and subject to low real estate taxes was an important factor in the development of these private parks.

Despite this intense private development of research and industrial parks, M.I.T. has been very active in land development near its campus. Like Stanford, the institute maintains an in-house real estate development office, although, with just six people, it is much smaller than Stanford's. In the 1950s M.I.T., in conjunction with the real estate development firm responsible for much of the early development along Route 128, purchased 14.6 acres of land adjacent to its campus in order to expand, and part of this purchase was set aside for the development of a research and development complex called Technology Square. Five buildings were constructed on the site, two of which became the headquarters for Polaroid Corporation and Draper Laboratories. M.I.T. sold its interest in Technology Square in 1971 to finance other real estate ventures.[56]

In 1970 M.I.T. purchased 19 acres adjacent to its campus from the Simplex Wire and Cable Company. During the next eight years the institute added eight acres to the site to make it suitable for moderate-scale, mixed use urban development, and in 1982 it issued a request for proposals to potential developers. These "developer kits" made it clear that M.I.T. was not contemplating a research park, but it did have economic development objectives clearly in view. The objective of the land development was to provide income to M.I.T., tax revenue to Cambridge, and diverse employment opportunities to local workers. "The Institute does not intend to be a prime tenant for any space on the site or to use any land for any of its own academic purposes; nor is there a preference for any specific uses beyond a clear interest in considering the extent to which they will contribute to meeting the objectives stated above."[57] By 1984 a developer had been chosen and the first public meetings on the project were being held.

The Technology Square project was intended to be a research-oriented development and achieved the same result as the Stanford Research Park in drawing close to campus some major research-oriented companies. The Simplex project, now known as University Park, was conceived as a mixed-use development, with research and development being one of several acceptable activities. Indeed, many research and development firms became tenants of University Park. Unlike the Stanford administration, the administration of M.I.T. has made no extraordinary efforts to wed its real estate development policies or activities to the institute's mission or activities. It has treated its real estate holdings strictly as investments, to be handled as treasury assets, and it has not involved faculty in decisions regarding these holdings.

The contrast between M.I.T. and Stanford in their development of real estate reflects the contrast between the two institutions noted in the first section of this chapter. Stanford has proceeded more deliberately, seeking to square its actions with its goals and objectives, including the objective of fostering relations with industry. With strong relations with industry already in place, M.I.T. has been free to set other objectives for its real estate developments, most notably the generation of income and the fostering of positive relations with the surrounding community.

PENN STATE

Penn State's activism in real estate development is recent, coinciding roughly with the appointment of Bryce Jordan as president in July 1983. In January 1987, the university announced plans to build a $7 million graduate center in an already existing industrial park (not owned by the university) in the hope that businesses in the park would consult and hire graduate students studying at the center.[58] Then, in May 1987, President Jordan presented his economic development report to the trustees. This report recommended a program that would "consider the orderly and carefully planned use of University lands for the development of research-office parks, business incubators, a conference center and other potential uses."[59] The extent of the university's role in this was unclear, however: "In general, the university's economic development initiatives will center on the use of passive assets. For example, in the land development initiative, we would anticipate private sector partners where location and land would represent the University contribution."[60]

"PSU Strives to Build 'Silicon Valley' in County," proclaimed the local newspaper on the release of the report, which focused both favorable and unfavorable attention on the issue of the university's use of its lands and its relation to the surrounding community.[61] The controversy intensified in December 1987, when more details of the "orderly" planning process were revealed. Penn State announced that it was seeking funding for the largest development in its history, a complex that would include a 16,000-seat arena, a 300-room hotel and conference center, an alumni center, a visitor's information center, and a 1,200-car garage. The price tag on this complex was estimated to be $70 million. In addition to the complex, a research park was planned for the north side of campus.[62]

Both community proponents and opponents of the university's real estate development had criticisms of the plan. The proponents, primarily businesspeople in State College, expressed the fear that the university was not moving fast enough with its research park idea. The university responded by agreeing to expedite the building of a $1.8 million, 50,000-square-foot business incubator that was to be a part of the research park, without waiting for full development of the rest of the park. Completion of the incubator was scheduled for June 1989. As noted earlier, eventually (October 1989) a 30,000-square-foot building was built, but only 10,000-square-feet were devoted to the incubator.

Critics of Penn State's real estate development found a number of reasons for opposition to university plans. First, they claimed that the university had not consulted enough with the community and other agencies. The university's 45-member economic advisory panel, hardly a secretive body, was accused of "meeting behind closed doors" because it excluded the general public from its deliberations. Second and more important, critics assailed what one editorial in the local newspaper termed the "blank check zoning" privileges that came about because, as a state entity, Penn State is not subject to local land use regulations or standards. Calling the proposed arena-hotel complex the "East Campus Disneyland," the editorial concluded:

> As Penn State behaves more and more like a giant corporation, and less and less like a traditional institution of higher education, it becomes imperative for the community to extend to the campus those controls it exercises over all other developers.[63]

In spite of the opposition, at least a scaled-down version of the original projects is underway, and first $5 million (for planning) and then $28.5 million have been appropriated by the state.[64] The incubator was built with a combination of state and city money and loans against anticipated rental income. However, questions remain both inside and outside Penn State about the university's role in the development. In the words of one Penn State faculty member, "Happy Valley will never become Silicon Valley." Land use issues will continue to be raised by every new plan or development at Penn State.

Under considerable pressure to become more involved in economic development, Penn State has plunged into real estate developments with two major undertakings. This to some extent caught the community and the faculty off guard. Unlike Stanford, Penn State has not had time to go through the difficult process of shifting the academic culture to a more favorable attitude toward real estate development. Located in a small town populated largely by university employees, including faculty members, Penn State's real estate developments cannot avoid having a visible effect on the community. The university cannot isolate its real estate dealings from its other activities in the way possible, for instance, at M.I.T.

U.C. BERKELEY

The failure of U.C. Berkeley to generate significant research-related businesses in the immediate environs of its campus, especially when compared with the spectacular success of Stanford, its neighbor to the south, has been the object of study by both the university and outside scholars. The facts that there is little land available in the immediate vicinity of the Berkeley campus and that the campus is surrounded by a community notoriously opposed to the growth of private industry are not sufficient to explain this failure. In the last few years, at least, some of the larger companies that have been in Berkeley (some of which were formed by U.C. faculty members) have been moving out of town to larger, less expensive quarters.[65] Just a few miles to the north, east, and south of the campus, however, there are communities both favorably disposed to university-related business and possessed of available real estate parcels large enough to accommodate development.

Despite its eminent reputation for research and graduate education, U.C. Berkeley has not been active in economic development. There are several reasons for this lack of activity. First, the university has not possessed a site with market acceptability. Second, university leadership has not emphasized economic development. Third, and perhaps most important, the academic culture of Berkeley, dominated by faculty, is not supportive of economic development initiatives, including real estate development.

This situation may be changing, however. Like most universities, U.C. Berkeley is beginning to feel some pressure to become involved in the economic development of the surrounding communities. While the number of jobs in the nine-county Bay Area (which includes the Silicon Valley counties of San Mateo and Santa Clara) increased by a phenomenal 1.1 million between 1962 and 1982, and while high-technology jobs jumped from 19 percent to more than 48 percent of the total, the communities of Oakland, Berkeley, and Alameda saw relatively little increase despite the advantages of convenient transportation systems (which include the progressive Port of Oakland and full service by the Bay Area Rapid Transit System) and their central location.[66]

One exception to this lack of growth occurred in biotechnology, which was clearly associated with the university. Chiron Corporation, formed jointly by a U.C. Berkeley and a U.C. San Francisco professor, grew from 15 researchers recruited from U.C. laboratories in 1981 to over 300 employees in 1987.[67] It was joined in the East Bay by Cetus Corporation and other biotechnology firms, including Advanced Genetics Sciences, which established its headquarters and laboratories in space rented from the university. These firms were drawn to the area by the reputation of the nearby universities. A 1987 report by Edward Blakely, professor and chair of U.C. Berkeley's Department of City and Regional Planning, saw Oakland and the surrounding communities poised for development of a growing biotechnology industry, with available land, infrastructure, excellent air and ground transportation, and a diverse work force. However, the report cited as one prerequisite of this development the demonstration of a commitment by U.C. Berkeley to the development of biotechnology in the region, for example by establishing an institute dedicated to applied research.[68] The importance of the university to the economy of the region has not been lost on developers, who have approached the university with proposals for joint real estate developments.

All this interest and activity was occurring at the same time that U.C. Berkeley was struggling to reorganize its life sciences departments and to build a large new life sciences building on campus. Even with this new addition, however, the campus was finding itself dangerously short of research space for life sciences as well as for engineering and other research. With this convergence of pressures the university turned its attention to a parcel of land it had owned since the early 1950s, known as the Richmond Field Station. Consisting of 150 acres and located seven miles northwest of the campus on the edge of the Bay in an area undergoing extensive development, including a planned housing development of 1,500 homes and a new freeway extension, the Richmond Field Station had been used by the university to house large-scale research facilities such as the earthquake shaker table, which was the largest such facility in the country.

The university considered three possible alternatives for the development of this site: an extension of the general campus, a research/industrial park, and a research campus. The site was considered too far away to serve effectively as an extension of the main campus, and the research/industrial park idea was considered too risky and the site too valuable for the university to let it go entirely to outside interests.

Development of the site is still in its preliminary stages. Much of the early work has been devoted to refining the concept of the site. In November 1987, a concept paper called the proposed development the Center for Applied Sciences and Technology; a briefing paper issued in November 1988 referred to it as a Complementary Applied Research Campus; and a February 1989 briefing paper called it the Berkeley Research Campus and Knowledge Transfer Center. Some of the changes in the site's name are the result of attempts by planners to engender faculty support for the project. Such support is absolutely crucial to the success of the project, just as it was at Stanford.

As the project is now conceived, the research campus would consist of four main elements: an area devoted to public/private development, a core university research campus, large-scale research facilities, and an area devoted to shared amenities. The area designated for public/private development would be ground leased to outside concerns that share research interests with the university. Lease income from these firms or agencies would help to pay some of the costs of other developments on the site. Also planned are incubator, conference, and teleconferencing facilities.[69]

In contrast to Penn State, U.C. Berkeley is undertaking real estate development in a slow and orderly manner, involving faculty and keeping early ideas about site development flexible. Its internal public relations campaign is not designed to win over the entire faculty, or even a majority of the faculty, to support of the project. Rather, it is intended to identify the requisite number of faculty members who are interested enough in the site and in interaction with potential tenants to make the project successful.

FACULTY PERCEPTIONS

Faculty opinions about the kinds of real estate developments discussed above, gathered from interviews at the four universities of this study and reported in the next chapter, are generally negative. Few faculty members were aware of even the most obvious details of their university's plans for real estate development. Of the 90 faculty members interviewed, only 41 responded to the question about their attitudes toward real estate investment, which was the lowest response rate for any question asked. The average attitude of these 41 subjects was slightly negative; only faculty views on direct investment by universities in research-related firms were more negative. Faculty in "other science" and materials science registered the strongest disapproval, while those in biotechnology, electrical engineering, and "other engineering" registered more neutral views.

Despite these responses, the importance of spatial proximity of the university to related companies was emphasized frequently, particularly at Stanford and M.I.T., where examples of positive interactions with industry, the sharing of facilities and instruments, increased consulting opportunities, and increased joint research were common. In contrast, Penn State faculty members obviously suffer from their relative geographic isolation. They have difficulty scheduling consulting assignments around teaching responsibilities because such assignments usually require from one to two days of travel time, even to cities as close as Chicago. Even at Stanford, however, there was a noticeable difference in opinions of faculty members in different disciplines about the value of geographical proximity to outside firms, with, as one would expect, faculty members in electrical engineering being much more enthusiastic about it than their colleagues in other disciplines.

A few Berkeley faculty members expressed a counterargument, suggesting that spatial proximity actually could be a problem. In reaction to the proposal that a high-technology company might locate itself next to campus, one Berkeley faculty member said: "I'm not sure it's a good idea for a bunch of guys from industry to be hanging around looking for free consulting."

However, the faculty's lack of awareness regarding real estate developments was the most important discovery from the interviews. Not one M.I.T. interviewee knew anything about the institute's development plans for the Simplex project, and Penn State faculty's source of information about the university's proposed plans, after an initial briefing by the president and his representative at a faculty senate meeting, was the local newspaper. Stanford faculty were generally aware of developments there, of course, and faculty at Berkeley, thanks to an extensive campaign by the project planners, were becoming aware of the plans for the research campus development. By and large, however, faculty were unconcerned about university real estate developments.

FACTORS DETERMINING SUCCESS

There is a broad consensus among commentators on the factors that contribute to the success or failure of university-related research parks.[70] These factors can be placed in two categories: factors associated with the site and the surrounding community, which would be present with or without university involvement, and factors associated with the sponsoring university and its administration of the project.

FACTORS ASSOCIATED WITH THE PARK SITE. The key factor in the success of a research park has proved to be its location and size. The first step in planning such a park, then, is to perform a careful evaluation of the proposed site. What are the strengths and weaknesses of the site? How is it situated in relation to transportation? Is there enough land to support the proposed development? What can be done to make the site more attractive?

The second group of site-related factors to consider has to do with the surrounding community. Is the community supportive? Can it supply the services necessary for the project? Are the quality of the community and the quality of life there appropriate for the proposed project? To what extent and in what ways are local governments likely to support the project?

Another indispensable element in planning a research park is market analysis. Can a market for the park be identified and described? Is there sufficient research-based industry in the area? Does the area already have government-funded projects? Does the university or universities have a sufficient research base to attract industry? Is the research base in a growing technological area? What competition will the park have?

The results of all these analyses contribute to answering the most important question: Is the project a sound investment? A preliminary answer to this question can be simple or very complex; it may involve assumptions about financing plans, projected lease terms, and so on. An important factor to keep in mind, especially when feasibility studies proceed in stages, is the amount of up-front investment required to study the project and to acquire the site and/or ready it for development. The greater this initial investment, the greater will be the risk of loss.

FACTORS ASSOCIATED WITH UNIVERSITY SPONSORSHIP. A prime location and ideal environmental conditions are not sufficient to assure the success of a research park. The development and management of the park also must be well conceived and carefully implemented. Again, observers have reached a consensus on the elements of sound university management of research parks and development projects.

First, the development project must be based on realistic expectations. This means that the results of the analysis described in the preceding paragraphs must be carefully studied by planners and decision makers. The analysis should be translated into an achievable plan that is fully communicated to the larger university community. The development also must be based on a sound financial plan that assures financial staying power.

The university should install a simple, effective, and flexible management system to carry out the plan and manage the project. This means careful university oversight, but it also means that the university must delegate authority over day-to-day management to a qualified individual or management group. An element of this delegation is the identification of a strong leader who can remain with the project for an extended period. Every successful university research park or real estate development, including Stanford's and M.I.T.'s, has enjoyed the long-term leadership of a single, dynamic individual whose only job it is to formulate goals and administer the project and who has financial and real estate experience.

The development plan must be flexible, but it also must be based on high standards and on realistic controls and restrictions. Adherence to standards and controls is often difficult, especially when a potential tenant has to be rejected because the tenant does not meet the criteria for admission to the park or because the tenant's development plans do not conform to the university's standards. The park plan also must provide for appropriate "seeding," that is, establishment of a tenant profile that will attract the desired kind of tenants to the park. This often means identifying and making special concessions to a prime tenant, establishing a university presence on the site, or providing low-cost incubation space at or near the site.

Most important, however, from the university's point of view is the requirement that the plan be based on a clear objective, that is, be well integrated with the long-term mission of the university. The integration of Stanford's real estate development with its long-range goal of building a corporate constituency, which was achieved in part through extensive faculty review of the university's plans, has been a significant factor in the Stanford Research Park's success. M.I.T. presents an interesting contrast, however. As noted earlier, the goal of its University Park project is quite plainly to generate income for the university, just like any other endowment investment. In addition, it is clear that M.I.T. sought to protect its borders from a decaying industrial section and to control future developments in the area. In meeting these goals the project serves long-term interests of M.I.T., but these interests are not integrated with the institute's basic functions.

Both U.C. Berkeley and Penn State are in the process of trying to square development plans with their missions and with the ideals of their faculties. Both universities own valuable land that they could view simply as an investment. Instead, they are seeking to develop their property simultaneously for their own uses and for the use of compatible outside businesses and agencies. This combination of internal and external interests creates considerable complexity and will require extensive consultation with faculty before the objectives for the two projects can be stated clearly and communicated to the university community.

Once clear objectives have been established, the university must demonstrate strong support for the project. In addition to providing all the elements described so far, it must offer encouragement of faculty entrepreneurship. This element is crucially important. A university's faculty often are most familiar with the firms that are involved in research in a particular area and are therefore the most likely tenants for a research park. The university, through its faculty and chosen managers, must know its product and be prepared to sell it.

The presence or absence of all these elements should help the university answer the hardest question of all: Does this institution have the commitment and the capacity to be successful with a research park development project?

ROLE IN TECHNOLOGY TRANSFER

Universities' increased interest in real estate development is further evidence of the changes that are taking place in the role of universities in our society. The issues related to these changes have appeared often in previous chapters. As the university steps into real estate development, it is entering unfamiliar territory and therefore must make a number of adaptations. For instance, it must attract and compensate people with the requisite expertise and management capabilities, much as it must do for experts in patents and licensing, for example. For the sake of consistency, it must align its new activities with its traditional mission. As it does so, the mission subtly changes, and so does public perception of the institution's role.

In real estate development more than in any other activity related to technology transfer, the university's self-interest is folded into community needs. Research parks are a highly visible symbol of the university's economic effect on an area. When they succeed, both university and community gain. When they fail, the reputation of the university suffers greatly.

Conclusion

Technical assistance programs, business incubators, and research parks are the three university technology transfer mechanisms most closely identified with regional economic development. As this examination of these mechanisms has shown, most universities move uneasily in this arena and lack the internal expertise necessary to manage such activities effectively. They sometimes are uncomfortable in playing a role so visible in the community and thus subject to community scrutiny and controversy. Despite the risks of financial failure and negative community reaction, however, universities have markedly increased their involvement in these activities because they are often the most visible symbols of the university's potential value to the community and thus the best key to its continued support.

Involvement in these three transfer mechanisms clearly underscores the adaptations that universities are making to increased public expectations about universities' economic value. First, new forms of university organization are taking shape both within existing institutional boundaries and outside, as when university-owned subsidiaries are established to manage economic ventures. Second, these transfer mechanisms add to the growing corps of nonfaculty professionals that universities rely upon to manage and staff such activities.

Third, and most important, these activities, although they remain on the periphery of most universities, exert a subtle but significant force upon university policy and mission. University institutional culture, more policy based and traditional than the culture in most other organizations, requires that all university activities be clearly related to institutional goals and missions. In major research institutions, this is usually accomplished through a process of convincing faculty members that an activity is in their best interests or coincident with the long-term goals of the university, since the long-term success of any university-sponsored economic development activity is related to the degree of commitment of the faculty to it.

In many cases the development of these economic transfer activities precedes formulation of policy concerning them. The university must then go through a process of either adjusting policies or rationalizing the activities to make them consistent with existing policies. In this process policies, and universities, are changed.

Notes

1. SRI International, *The Higher Education-Economic Connection: Emerging Roles for Public Colleges and Universities in a Changing Economy* (Washington, D.C.: American Association of State Colleges and Universities, 1986), X.
2. David L. Birch, *The Job Generation Process: Report of the M.I.T. Program on Neighborhood and Regional Change to the Economic Development Administration, U.S. Department of Commerce* (Cambridge: Massachusetts Institute of Technology, 1979), quoted in Lynn G. Johnson, *The High-Technology Connection: Academic/Industrial Cooperation for Economic Growth* (Washington, D.C.: Association for the Study of Higher Education, 1984), 51.
3. U.S. Congress, Subcommittee on Monetary and Fiscal Policy of the Joint Economic Committee, "Location of High Technology Firms and Regional Economic Development" (Washington, D.C.: Government Printing Office, 1974), 36.
4. Matthew Bullock, "Cohabitation: Small Research-Based Companies and Universities," *Technovation* 3 (1985): 33.
5. John Preston, interview with author, Cambridge, Mass., 7 July 1988.
6. Marvin A. Sirbu et al., *The Formation of Technology-Oriented Complexes: Lessons from North American and European Experience* (Cambridge: Center

for Policy Alternatives, Massachusetts Institute of Technology, 1976), quoted in Johnson, *The High-Technology Connection*, 45.

7. Amy K. Glasmeier, Peter Hall, and Ann R. Markusen, "Recent Evidence on High-Technology Industries' Spacial Tendencies: A Preliminary Investigation" (Berkeley: Institute of Urban and Regional Development, University of California, Berkeley, 1983), 1.

8. Much of the information regarding the comparison between M.I.T. and Stanford in the next paragraphs comes from Nancy S. Dorfman, "Route 128: The Development of a Regional High Technology Economy," *Research Policy* 12 (December 1983): 299–316.

9. Arnold C. Cooper, "The Role of Incubator Organizations in the Founding of Growth-Oriented Firms," *Journal of Business Venturing* 1 (1985): 77.

10. Edward Roberts, "A Basic Study of Innovators; How to Keep and Capitalize on Their Talents," *Research Management* 11 (1968): 254.

11. Dorfman, "Route 128," 314.

12. Bank of Boston, "M.I.T.: Growing Businesses for the Future" (Boston: 1989).

13. Johnson, *The High-Technology Connection*, 51.

14. National Association of Management and Technical Assistance Centers, brochure (Washington, D.C.: NAMTAC, 1987).

15. Judith Axler Turner, "Little Known Extension Services Enable Universities to Help Industry," *Chronicle of Higher Education*, 7 January 1987, 11, 14.

16. The information about the Georgia Tech Industrial Extension Service has been summarized from SRI International, *The Higher Education-Economic Connection*, 19–21.

17. General Accounting Office, "The Federal Role," 37.

18. General Assembly of the Commonwealth of Pennsylvania, HB 301, No. 47-1, An Act, 14 July 1966.

19. H. Leroy Marlow, interview with author, University Park, Pa., 2 March 1987.

20. Robert Scannell, interview with author, University Park, Pa., 16 September 1987.

21. Marlow, interview.

22. H. Leroy Marlow, "Another Perspective on Technology Transfer: The PENN-TAP Experience," in "A Synthesis of Technology Transfer Methodologies, Proceedings of U.S. Department of Energy, Technology Transfer Workshop" (Washington, D.C.: 1984.)

23. Turner, "Little Known Extension Services."

24. Scott Jaschik, "States Trying to Assess the Effectiveness of Highly Touted Economic Programs," *Chronicle of Higher Education*, 3 June 1987, 19, 26.

25. General Accounting Office, "The Federal Role," 37.

26. Karen Karvonen, "Start-Up Support Systems," *US Air*, April 1988, 91.

27. John D'Aprix, "Cooperative Research Corner," *SRA News*, November-December 1987, 8.

28. Karvonen, "Start-Up Support Systems," 92.

29. Buck Brown, "Business Incubators Suffer Growing Pains," *Wall Street Journal*, 16 June 1989, B1.

30. Karvonen, "Start-Up Support Systems," 92.

31. Steven P. Galante, "Business Incubators Adopting Niche Strategies to Stand Out," *Wall Street Journal*, 13 April 1987, 29.

32. Karvonen, "Start-Up Support Systems," 92.

33. "Business Incubators—A Booming Business," *Journal of Accountancy,* June 1988, 18–19.

34. Richard Phalon, "University as Venture Capitalist," *Forbes,* 19 December 1983, 82–93.

35. Arthur D. Little, *University-Related Research Parks in the State of New York: Report to the New York State Urban Development Corporation* (1989).

36. Phalon, "University as Venture Capitalist," 82–93

37. Ibid.

38. Buck Brown, "Business Incubators."

39. Advanced Technology Center of Central and Northern Pennsylvania, Inc., "1985–86 Annual Report" (University Park, Pa.: Advanced Technology Center of Central and Northern Pennsylvania, Inc., 1986), 9.

40. Mary Schultz, telephone interview with author, 9 August 1989.

41. "Association of University Related Research Parks" (Tempe, Az.: n.p., 1987).

42. Charles A. Lee, "Research Park Development from a University Relationships Perspective: A Proposed Framework for Planning," *Planning for Higher Education* 12 (Fall 1983): 33.

43. Ibid.

44. Victor J. Danilov, "The Research Park Shakeout," *Industrial Research,* May 1971, 45.

45. Ibid.

46. Ibid.

47. General Accounting Office, "The Federal Role in Fostering University-Industry Cooperation" (Washington, D.C.: GAO/PAD-83-22, 1983): 10.

48. Nancy Pieretti, "Dartmouth Joins the Joint-Real-Estate-Venture Parade," *New York Times,* 21 May 1989, Y29.

49. Douglas Porter, "Campus Capitalism: Universities as Development Entrepreneurs," *Business Officer,* February 1983, 18.

50. Liz McMillen, "Profits and Perils in Real Estate," *Chronicle of Higher Education,* 22 March 1989, A26.

51. General Accounting Office, "The Federal Role," 12–14.

52. Except where noted, the information about the Stanford Research Park comes from Frank A. Madeiros, "The Sterling Years at Stanford: A Study in the Dynamics of Institutional Change" (Ph.D. diss., Stanford University, 1979), 108–114.

53. Ibid., 113.

54. General Accounting Office, "The Federal Role," 16.

55. Douglas Porter, "The Development of University-Affiliated Research Parks," in *Research Parks and Other Ventures: The University/Real Estate Connection,* ed. Rachelle L. Levitt (Washington, D.C.: ULI-The Urban Land Institute, 1985), 70–73.

56. Perry Chapman, "Developing Unused Campus Land: Risky but Rewarding," *American School and University* 3 (1982): 36e.

57. Massachusetts Institute of Technology, "Simplex Development Area" (Cambridge: Massachusetts Institute of Technology, 1982), 7.

58. Phil McDade, "PSU Strives to Build 'Silicon Valley' in County," *Centre Daily Times,* 6 May 1987, A1.

59. "Board of Trustees Hears Economic Development Report," *Intercom* 16 (21 May 1987): 1.

60. "Senate Discusses Economic Development Initiatives," *Intercom* 17 (17 April 1988): 1, 6, 7.
61. McDade, "PSU Strives to Build 'Silicon Valley.' "
62. Phil McDade, "Trustees Roundup: PSU Expects Development Funding for Arena-Hotel Complex Within One Year," *Centre Daily Times,* 17 January 1988, A12.
63. Bill Welch, "Community Should Withdraw 'Blank Check' to PSU," *Centre Daily Times,* 20 December 1987, B5.
64. McDade, "Trustees Roundup."
65. Laura Evenson, "High-Tech Companies Are Nurtured by UC Berkeley," *San Francisco Business Journal,* 28 January 1985, 3.
66. Virgil Meibert, "High Technology's Eastbay Blind Spot," *Oakland Tribune,* 1 December 1985, B1.
67. Paul Grabowicz and Eric Newton, "Business Park at Center of Debate," *Oakland Tribune,* 12 October 1987, A1, A6.
68. Tom Debley, "Study Sees Oakland, Environs as Hub for Biotechnology," *Berkeleyan,* 18 March 1987, 3.
69. Faculty-Administration Committee on Research Campus Development, "Berkeley Research Campus and Knowledge Transfer Center: A Summary of the Current Feasibility Study Activities for Faculty Review" (Berkeley: University of California, Berkeley, 1989).
70. The material in this section has been taken from the following sources:

    Perry Chapman, "Developing Unused Campus Land: Risky but Rewarding," *American School and University* 3 (March 1982): 36b–36h.

    Gerry Cooke, "Park Feasibility: To Build or Not to Build," *Research Park Forum* 2 (October 1987): 1–3.

    Charles A. Evans, "Park Development: To Be or Not to Be," *Research Park Forum* 2 (October 1987): 4–5.

    Donald Hart, "Critical Issues Facing the Future Development of Research Parks," *SRA News* September/October 1987, 2, 7.

    Charles A. Lee, "University-Related Research Parks: A Michigan Case Study with Selected Comparisons" (Ph.D. dissertation, University of Michigan, 1982), 114, 132.

    James F. Mahar and Dean C. Coddington, "The Scientific Complex—Proceed with Caution," *Harvard Business Review* 43 (January/February 1965): 140–152.

Bibliography

ADVANCED TECHNOLOGY CENTER OF CENTRAL AND NORTHERN PENNSYLVANIA, INC. "1985–86 Annual Report." University Park, Pa.: Advanced Technology Center of Central and Northern Pennsylvania, Inc., 1986.

AMERICAN ASSOCIATION OF STATE COLLEGES AND UNIVERSITIES. *Exploring Common Ground: A Report on Business/Academic Partnerships.* Washington, D.C.: American Association of State Colleges and Universities, 1987.

ASSOCIATION OF UNIVERSITY RELATED RESEARCH PARKS. Brochure. Tempe, Arizona: n.p., 1987.

BANK OF BOSTON. "M.I.T.: Growing Businesses for the Future." Boston: Bank of Boston, 1989.

"Board of Trustees Hears Economic Development Report." *Intercom,* 16 (1987): 1.

BROWN, BUCK. "Business Incubators Suffer Growing Pains." *Wall Street Journal,* 16 June 1989, B1.

BULLOCK, MATTHEW. "Cohabitation: Small Research-Based Companies and Universities." *Technovation* 3 (1985): 27–38.

"Business Incubators—A Booming Business." *Journal of Accountancy,* June 1988, 18–19.

CHAPMAN, PERRY. "Developing Unused Campus Lands: Risky but Rewarding." *American School and University* 3 (1982): 36b–36h.

CHMURA, THOMAS J. "The Higher Education-Economic Development Connection." *Commentary,* Fall, 1987, 11–17.

COOKE, GERRY. "Park Feasibility: To Build or Not to Build." *Research Park Forum* 2 (1987): 1–3.

COOPER, ARNOLD C. "The Role of Incubator Organizations in the Founding of Growth-Oriented Firms." *Journal of Business Venturing* I (1985): 75–86.

DANILOV, VICTOR J. "The Research Park Shakeout." *Industrial Research,* May 1971, 44–46.

D'APRIX, JOHN. "Cooperative Research Corner." *SRA News,* November-December 1987, 8.

DEBLEY, TOM. "Study Sees Oakland, Environs as Hub for Biotechnology." *Berkeleyan,* 18 March 1987, 3.

DORFMAN, NANCY S. "Route 128: The Development of a Regional High Technology Economy." *Research Policy* 12 (1983): 299–316.

EVANS, CHARLES A. "Park Development: To Be or Not to Be." *Research Park Forum* 2 (October 1987): 4–5.

EVENSON, LAURA. "High-Tech Companies Are Nurtured by UC Berkeley." *San Francisco Business Journal,* 28 January 1985, 3.

FACULTY-ADMINISTRATION COMMITTEE ON RESEARCH CAMPUS DEVELOPMENT. "Berkeley Research Campus and Knowledge Transfer Center: A Summary of the Current Feasibility Study Activities for Faculty Review." Berkeley: University of California, Berkeley, 1989.

GALANTE, STEVEN P. "Business Incubators Adopting Niche Strategies to Stand Out." *Wall Street Journal,* 13 April 1987, 29.

GENERAL ACCOUNTING OFFICE. "The Federal Role in Fostering University-Industry Cooperation." Washington, D.C.: GAO/PAD-83-22, 1983.

GENERAL ASSEMBLY OF THE COMMONWEALTH OF PENNSYLVANIA. HB 301, No. 47-1, An Act, 14 July 1966.

GLASMEIER, AMY K., PETER HALL, and ANN R. MARKUSEN. "Recent Evidence on High-Technology Industries' Spacial Tendencies: A Preliminary Investigation." Berkeley: Institute of Urban and Regional Development, University of California, Berkeley, 1983.

GRABOWICZ, PAUL, and ERIC NEWTON. "Business Park at Center of Debate." *Oakland Tribune,* 12 October 1987, A1, A6.

HART, DONALD. "Critical Issues Facing the Future Development of Research Parks." *SRA News,* September/October 1987, 2–3.

JASCHIK, SCOTT. "States Trying to Assess the Effectiveness of Highly Touted Economic Programs." *Chronicle of Higher Education,* 3 June 1987, 19, 26.

JOHNSON, LYNN G. *The High-Technology Connection: Academic/Industrial Cooperation for Economic Growth.* Washington, D.C.: Association for the Study of Higher Education, 1984.

KARVONEN, KAREN. "Start-Up Support Systems." *US Air,* April 1988, 86–94.

LEE, CHARLES A. "Research Park Development from a University Relationships Perspective: A Proposed Framework for Planning." *Planning for Higher Education* 12 (1983): 33–46.

———. "University-Related Research Parks: A Michigan Case Study with Selected Comparisons." Ph.D. diss., University of Michigan, 1982.

LEVITT, RACHELLE L., ed. *Research Parks and Other Ventures: The University/Real Estate Connection.* Washington, D.C.: ULI—The Urban Land Institute, 1985.

LITTLE, ARTHUR D. *University-Related Research Parks in the State of New York. Report to the New York State Urban Development Corporation* 1989.

McDADE, PHIL. "PSU Strives to Build 'Silicon Valley' in County." *Centre Daily Times,* 6 May 1987, A1.

———. "Trustees Roundup: PSU Expects Development Funding for Arena-Hotel Complex Within One Year." *Centre Daily Times,* 17 January 1988, A12.

McMILLEN, LIZ. "Profits and Perils in Real Estate." *Chronicle of Higher Education,* 22 March 1989, A26.

MADEIROS, FRANK A. "The Sterling Years at Stanford: A Study in the Dynamics of Institutional Change." Ph.D. diss., Stanford University, 1979.

MAHAR, JAMES F., and DEAN C. CODDINGTON. "The Scientific Complex—Proceed with Caution." *Harvard Business Review* 43 (1965): 140–152.

MARLOW, H. LEROY. "Another Perspective on Technology Transfer: The PENN-TAP Experience." In "A Synthesis of Technology Transfer Methodologies, Proceedings of U.S. Department of Energy, Technology Transfer Workshop." Washington, D.C.: 1984.

MASSACHUSETTS INSTITUTE OF TECHNOLOGY. "Simplex Development Area." Cambridge: Massachusetts Institute of Technology, 1982.

MEIBERT, VIRGIL. "High Technology's Eastbay Blind Spot." *Oakland Tribune,* 1 December 1985, B1.

NATIONAL ASSOCIATION OF MANAGEMENT AND TECHNICAL ASSISTANCE CENTERS. Brochure. Washington, D.C., National Association of Management and Technical Assistance Centers, 1987.

PHALON, RICHARD. "University as Venture Capitalist." *Forbes,* 19 December 1983, 82–93.

PIERETTI, NANCY. "Dartmouth Joins the Joint-Real-Estate-Venture Parade." *New York Times,* 21 May 1989, Y29.

PORTER, DOUGLAS. "Campus Capitalism: Universities as Development Entrepreneurs." *Business Officer,* February 1983, 18–22.

———. "The Development of University-Affiliated Research Parks." In *Research Parks and Other Ventures: The University/Real Estate Connection,* edited by Rachelle L. Levitt. Washington, D.C.: ULI—the Urban Land Institute, 1985.

ROBERTS, EDWARD. "A Basic Study of Innovators; How to Keep and Capitalize on Their Talents." *Research Management* 11 (1968): 241–266.

"Senate Discusses Economic Development Initiatives." *Intercom* 17 (1988): 1, 6, 7.

SRI INTERNATIONAL. *The Higher Education-Economic Connection: Emerging Roles for Public Colleges and Universities in a Changing Economy.* Washington, D.C.: American Association of State Colleges and Universities, 1986.

TURNER, JUDITH AXLER. "Little Known Extension Services Enable Universities to Help Industry." *Chronicle of Higher Education,* 7 January 1987, 11, 14.

U. S. CONGRESS, SUBCOMMITTEE ON MONETARY AND FISCAL POLICY OF THE JOINT ECONOMIC COMMITTEE. "Location of High Technology Firms and Regional Economic Development." Washington, D.C.: Government Printing Office, 1974.

———. Public Law 89-182. 14 September 1965.

WELCH, BILL. "Community Should Withdraw 'Blank Check' to PSU." *Centre Daily Times,* 20 December 1987, B5.

12

Faculty Attitudes Toward Technology Transfer

This book has examined a number of policies and organizational subunits in four universities and has documented some changes in the volume of activity and organizational structure of these subunits, especially during the last ten years. But are these changes important? Will they have a lasting impact on the universities? The answers to these questions ultimately must come from the institutions' faculty. The degree of impact of technology transfer activities on a university will be measured in large part by the extent to which such activities are incorporated into the regular duties and work routines of its faculty members, and this, in turn, depends on faculty attitude.

No study of technology transfer and universities would be complete, therefore, without some exploration of the attitudes and comments of those faculty members most involved in technology transfer. This chapter summarizes the results of interviews of faculty members that were designed in part to determine their opinions on a number of issues related to technology transfer and the changes they believe technology transfer will make in the nature of research. Among other things, it shows that both faculty members and industrial sponsors are becoming more sophisticated and knowledgeable about university research sponsorship. Contrary to what one might expect from journalistic accounts of conflicts of interest and abuses associated with technology transfer issues, most of the faculty members interviewed did not believe that technology transfer presented a significant threat to the university or the values upon which it is based, although many expressed the view that constant vigilance was necessary to protect the university.

Interviewee Selection and the Interview Process

In this survey 104 faculty members at the four universities of this study (U.C. Berkeley, Stanford, M.I.T., and Penn State) were interviewed. They were selected from two general categories, *involved* and *uninvolved*. Involved faculty were defined as faculty members in fields of science or engineering that were most likely to be involved with commercializable research or with heavy industrial sponsorship or interest. Involved subjects were placed in one of five categories: biotechnology, electrical engineering, materials science, "other" engineering, and "other" science. An attempt was made to find faculty members in each of these categories who had discovered something of commercial value and/or who had received significant amounts of research funding from industry. No attempt was made to select a random sample.

Subjects were interviewed using a structured technique. Exhibit 12-1 shows the protocol that was used for involved faculty members. Interviews were conducted by the author or by others who tape recorded the interviews. The author transcribed and scored the interviews and then summarized the comments in a number of tables. Exhibit 12-2 shows the scoring matrix used to score the interviews. Because the scores assigned by the author to the responses to the questions were not checked by anyone or validated in any way, the values associated with faculty attitudes are not statistically reliable. Rather, they are a numerical expression of the author's subjective judgment.

A much smaller sample of *uninvolved* faculty members at each institution also was interviewed. These faculty members were selected from fields not likely to produce research of direct commercial value or to have significant involvement with industry. Attempts were made to find faculty members who had some familiarity with the issues surrounding technology transfer, however, either through work on university committees or through their own scholarly work or interest in technology transfer. The objective of the interviews was to determine whether these faculty members saw any significant changes in their relationships with faculty members who were involved with research of direct commercial value and whether they saw any significant dangers or opportunities in increased interest in, and activity related to, technology transfer. The results of the interviews with these faculty members are described in a later section of this chapter.

Characteristics of Interviewees

As Table 12-1 indicates, interviews were completed for 90 involved and 14 uninvolved faculty members. Most interviewees had been with their universities for considerable periods; Table 12-2 shows that the average tenure for the group was over 20 years. Most interviewees (61 of 89) were members of some kind of extradepartmental unit.

Table 12-3 shows the average number of inventions reported by those interviewed broken down by university and by field. The overall average was four ideas of commercial value per interviewee, but there was a large variation, ranging from no such discoveries (13 of 90) to one Stanford professor who had been granted 54 separate patents. M.I.T. and Stanford faculty reported significantly higher numbers of inventions than faculty from U.C. Berkeley or Penn State.

Most of these inventors had seen their inventions actively developed. Of the 71 inventors for whom data were collected, 45 had received patents, and of these, 33 had seen their patents licensed to firms and 23 had actually received royalty income.

Of the 90 *involved* faculty members interviewed, 77 reported receiving research support from industrial sponsors. For the most part, such research sponsorship came with few formal strings attached. Twenty-nine faculty members reported receiving sponsorship in the form of outright gifts. Only 19 reported significant formal requirements from industrial sponsors that the research bear specific results or be conducted along narrowly prescribed lines.

As these few statistics indicate, the involved faculty members interviewed in this study were, for the most part, people who had a long history of active participation in university technology transfer and considerable familiarity with the issues surrounding it. Their answers to and comments on the questions in the interview protocol are summarized in the following sections.

Attitudes Toward Patents and Intellectual Property

The 90 involved faculty members reported a total of 365 inventions or discoveries of potential commercial value. To analyze the degree to which these faculty members were willing to become involved in the commercial development of an idea, a scale corresponding to the normal steps in the development process was created and each interviewee was rated on this scale for the invention or discovery that went the furthest toward commercialization. Steps on the scale were "did nothing," "published findings," "gave to sponsor," "submitted invention disclosure," "applied for patent," "obtained patent," "licensed invention," and "received royalty income or started company." Table 12-4 shows the frequencies of these ratings for all of the interviewees and the average rating by university and by field for those who initiated some action.

As one might expect, most of the subjects had at least a vague knowledge of the patent process. Of the 86 involved faculty members who were rated on their knowledge, 29 displayed only a vague knowledge, 21 were

EXHIBIT 12-1   Interview Protocol for Involved Faculty

Record locator: _____

1. University: _____

2. Discipline: _____

3. Faculty status: _____

4. How long at university: _____

5. Affiliated with extradepartmental unit: _____

6. Have you discovered or invented something that you thought or think might have commercial value? _____

6a. How many? _____

6b. What did you do with it (or them)? _____

_____

6c. How far did the invention (or inventions) go toward commercialization? _____

_____

6d. How did you feel about the process? _____

_____

7. Have you received research funding from industry? _____

7a. What form did it take (gift, contract, etc.)? _____

7b. How was this funding different from other kinds of funding? _____

_____

7c. Were there any problems associated with industrial funding? _____

_____

7d. How do you feel about industrial funding? Will you continue to seek it? _____

_____

8. How has the university encouraged technology transfer? _____

_____

8a. How effective are liaison programs in technology transfer? _____

_____

8b. How effective is the patent office? _____

8c. How effective are real estate developments related to technology transfer in promoting technology transfer? _____

EXHIBIT 12-1 *continued*

8d. How do you feel about university direct investment in start-up or spin-off companies developed to commercialize university research? _____

_____

8e. What role does or should continuing education play in the technology transfer process? _____

_____

9. Have you noticed any changes in the nature of or atmosphere surrounding research in the last ten years? _____

_____

9a. Have you noticed any changes in your relationship with your research colleagues? _____

_____

9b. Have you noticed any changes in relationships with your students? _____

_____

9c. Has your attitude toward your university changed in the last ten years? _____

_____

9d. Have you noticed changes in the attitude of funding agencies toward the university in the last ten years? _____

_____

10. Should any university policies concerning technology transfer be changed? _____

_____

11. Do you have any other comments about universities and technology transfer? _____

_____

12. Do you know of anyone who has left the university in order to develop university research into a commercial product? _____

_____

EXHIBIT 12-2   Scoring Matrix for Involved Faculty

6c. How far did invention go?

        0 No invention, did nothing
        1 Published findings, gave to sponsor
        2 Submitted invention disclosure
        3 Applied for patent
        4 Obtained patent
        5 Licensed invention
        6 Received royalty income, started company

Knowledge of Process		6d. Attitude toward Process	
No knowledge	0	No attitude	0
Vague knowledge	1	Very negative	1
Reasonable knowledge	2	Negative	2
Very knowledgeable	3	Neutral	3
Extremely knowledgeable	4	Positive	4
		Very positive	5

7a. What form did funding take?

        0 Did not receive industrial funding.
        1 Received gift funding (no strings attached).
        2 Sponsor participated in determining research.
        3 Sponsor dictated research.
        4 Sponsor got preinvention license.
        5 Sponsor got preinvention ownership.

Knowledge of Process		7d. Attitude toward Process	
No knowledge	0	No attitude	0
Vague knowledge	1	Very negative	1
Reasonable knowledge	2	Negative	2
Very knowledgeable	3	Neutral	3
Extremely knowledgeable	4	Positive	4
		Very positive	5

8. How effective are:

Liaison programs	No opinion 0
Patent office	Completely ineffective 1
Real estate	Slight ineffective 2
Direct investment	Moderately effective 3
Continuing education	Usually effective 4
	Very effective 5

TABLE 12-1   Faculty Interview Subjects: Number by University and Field

	U.C. BERKELEY	M.I.T.	PENN STATE	STANFORD	TOTAL
Involved faculty:					
Biotechnology	4	3	5	4	16
Electrical engineering	5	5	4	7	21
Materials science	3	4	5	5	17
Other engineering	5	6	5	4	20
Other science	3	3	6	4	16
Subtotal	20	21	25	24	90
Uninvolved faculty	4	3	5	2	14
Total	24	24	30	26	104

TABLE 12-2   Faculty Interview Subjects: Average Number of Years Employed by University

	U.C. BERKELEY	M.I.T.	PENN STATE	STANFORD	TOTAL
Total	24.95	20.14	14.94	22.08	20.28
Biotechnology	21.75	16.33	11.10	16.50	16.09
Electrical engineering	24.00	22.20	15.50	27.43	23.10
Materials science	16.00	20.25	10.40	20.00	16.53
Other engineering	27.20	18.00	18.40	22.75	21.35
Other science	36.00	24.67	18.67	20.25	23.44

TABLE 12-3   Faculty Interview Subjects: Average Number of Inventions

	U.C. BERKELEY	M.I.T.	PENN STATE	STANFORD	TOTAL
Total	2.90	4.71	2.96	5.58	4.06
Biotechnology	4.00	6.00	3.80	3.50	4.19
Electrical engineering	4.40	4.60	1.50	3.57	3.62
Materials science	1.33	5.00	3.40	3.20	3.35
Other engineering	2.60	4.33	3.00	15.75	5.65
Other science	1.00	4.00	2.83	4.00	3.00

rated extremely knowledgeable, and the rest (36) were rated between these two extremes. Interviewees also were rated on their attitudes toward the patent process on their campus. Table 12-5 shows the frequencies of ratings for those responding and the average ratings by university and by field.

Most faculty members, even those who had been quite successful at commercializing inventions, viewed the patent and licensing process as inherently difficult and time consuming, a distraction from the more impor-

TABLE 12-4   Faculty Interview Subjects: How Far Inventions Were Taken

VALUE	ACTION	FREQUENCY OF RESPONSE
1	Published, gave to sponsor.	8
2	Submitted invention disclosure.	8
3	Applied for patent.	10
4	Obtained patent.	12
5	Licensed invention.	10
6	Received royalty income.	23
Total		71

Average Value by University	
UNIVERSITY	AVERAGE VALUE
Total	4.08
U.C. Berkeley	3.56
M.I.T.	4.44
Penn State	3.35
Stanford	4.77

Average Value by Field	
FIELD	AVERAGE VALUE
Total	4.08
Biotechnology	4.46
Electrical engineering	4.25
Materials science	4.28
Other engineering	3.87
Other science	3.30

tant duties of research and teaching. Quite a few reported at least one instance in which they had discovered or thought of something that might have had commercial value, but, because they wished to avoid endless consultations with patent attorneys, patent office people, and potential licensees, they either did not report the invention or simply published their findings in the hope that the invention or discovery would be recognized and commercially exploited by someone else. Indeed, publishing an invention or discovery was seen as a convenient way for a faculty member to get rid of a potentially patentable idea and all the headaches it might bring because few in the university would be willing to criticize a publication and, in most cases, publication effectively bars future patenting of an invention. A few faculty members reported giving their inventions to an industrial sponsor of related research out of a feeling of moral obligation.

"Doing nothing" and "giving it away" were much more common responses to inventions at U.C. Berkeley and Penn State, where the burden of going through what was perceived as an ineffective patent office was added

TABLE 12-5   Faculty Interview Subjects: Attitudes toward the Patent Process

Value	Action	Frequency of Response
1	Very negative	15
2	Slightly negative	12
3	Neutral	20
4	Positive	28
5	Very positive	7
Total		82

Average Value by University	
University	Average Value
Total	300
U.C. Berkeley	2.28
M.I.T.	3.15
Penn State	2.48
Stanford	3.91

Average Value by Field	
Field	Average Value
Total	3.00
Biotechnology	3.56
Electrical engineering	2.85
Materials science	3.13
Other engineering	2.88
Other science	2.53

to the inherent burdens of the patent process. Table 12-4 shows that inventions were taken further toward development at Stanford and M.I.T., and Table 12-5 shows that faculty at these two universities were more favorably disposed toward the patent process than were faculty at U.C. Berkeley and Penn State. Even at the latter two universities, however, the majority of faculty members understood and were willing to fulfill their obligation to disclose inventions and to aid in the development process if called upon to do so.

Most of the interviewees who had sophisicated knowledge of patents and the patenting process proved to share certain opinions. First, few expected their ideas to make a lot of money for themselves or their universities. Most understood the complexities of the patenting process and the difficulty of capturing in a single patent or group of related patents a technology that would eventually produce significant revenue.

Even more important, most knowledgeable faculty members recognized the very great distance between ideas from the university, even those that do result in patents, and the marketplace. They understood that most development efforts require large capital investments and that investors who risk money to develop new technologies expect to see high potential returns. A few felt that the university should take a more active role in bringing seminal ideas to the point at which commercial concerns might be interested in taking a chance on them. Some also felt that, at least in some rapidly advancing fields, patents are not effective in protecting intellectual property.

Third, few faculty members viewed patents in themselves as effective in disseminating research or as worthy of consideration in promotion and tenure decisions. Most, therefore, did not object to the fact that faculty reward systems do not consider patents or the generation of commercially valuable intellectual property as contributions to be recognized in such decisions.

Despite these reservations, a significant number of those interviewed could be classed as enthusiastic champions of something they had invented or discovered. At least eight of the 90 interviewees had already decided to leave or were contemplating leaving the university in order to pursue the commercial development of a new technology. For the most part, the incentive for leaving appeared to be not the potential financial rewards but rather the recognition that without the attention of the inventor, a good idea would go undeveloped and unnoticed.

Attitudes Toward Faculty Members Who Leave the University

The last series of questions of the interview protocol asked faculty members if they knew of a colleague who had left the university to pursue the commercialization of research that began on campus and, if so, what they thought about it. The intent of the question was to determine how tolerant faculty members were toward involvement with commercial development and whether there were any differences among universities in such tolerance.

Of the 75 faculty members who responded to the question, 59—including over 90 percent of those interviewed at M.I.T. and Stanford and two-thirds of those interviewed at U.C. Berkeley and Penn State—knew of a colleague who had left the university to aid in the commercial exploitation of university research. In addition, many interviewees had themselves or knew of colleagues who had been involved in the founding of companies while still on the faculty. This was particularly true at Stanford, which, unlike the other three universities, allows faculty members to be operating officers of outside companies.

Despite the apparent differences in policies, traditions, and cultures among the universities, faculty members expressed general unanimity of opinion on this subject. First, most recognized that the most effective form of technology transfer associated with universities is the substantial involvement of the inventor in development work. When a professor gives up a tenured position and places personal assets at risk in order to develop a product, he or she presents the perfect profile of the sort of technology champion that venture capitalists look for. Most of those interviewed felt that the decision to leave the university under such circumstances was a matter of personal choice, and they did not feel betrayed by or jealous of those who had made that decision.

Second, most faculty recognized that those who are involved deeply in commercial development or in the founding of companies cannot perform the full measure of their duties as faculty members. Because the risk of new ventures is so high, academic researchers try to keep faculty positions as long as possible while a new development gets started. In a common pattern, a professor first begins to work for or help develop an outside firm on "consulting time," which is usually set by the institution as one day per week but which often is stretched to include the two weekend days. When the outside activity proves too time consuming and begins to require a greater effort, the faculty member usually requests a leave of absence for up to a year. In all four institutions, this leave usually can be extended by another year with ease.

In all universities, interviewed faculty members expressed disapproval of those who take advantage of the university by remaining on the faculty while being substantially engaged in outside activity. Almost everyone agreed that two years was the maximum leave of absence that the university could tolerate and that even this length of leave could hurt graduate students deprived of advisers and make it difficult for a faculty member to remain current in research.

Faculty also expressed general agreement that once a faculty member formally left the university, coming back would be difficult. Those who leave can come back only if as faculty vacancy occurs and if they successfully compete with other applicants. However, those who do leave apparently are not regarded with any particular animosity and, in fact, often retain some association with the university.

From these interviews it appears that, although the incidence of faculty leaving to found companies or engage deeply in commercial development of research is greater at M.I.T. and Stanford, faculty attitudes toward such activity are not markedly different among the four institutions.

Attitudes Toward Industrial Sponsorship of Research

As previously indicated, 77 of the 90 faculty members interviewed had received some form of industrial funding. Only five of the 90 displayed unfavorable attitudes toward industrial sponsorship. Because industry funding comes in many different forms, ranging from unrestricted gifts to very restrictive research contracts, it is difficult to interpret some of the comments regarding industrial funding. Nevertheless, a definite pattern of responses emerged from the interviews.

The advantages, disadvantages, and problems associated with industrial funding were most often brought out through a comparison with funding from the federal government, although, like industrial funding, federal research support comes in a variety of forms and with conditions and requirements that vary dramatically from agency to agency. A succinct summary of the differences was provided by one of the interviewees: "The government is more interested in reporting; industry is more interested in results."

Most faculty members saw a number of advantages in industrial, as compared to government, sponsorship of research. First, industrial funding usually was easier to apply for, since most of the time it came through contacts the faculty member already had and did not require elaborate proposal writing. Usually all that was required was an understanding between the industrial research office and the faculty member. Once awarded, the money came quickly to the researcher. In general, throughout the contract there was greater flexibility in industrial than in federal funding, although in some industrial contracts the work statement was fairly narrow. Industrial funding in the form of unrestricted gifts was highly prized because it provided much-needed discretionary funding that could fill in the gaps in a research program. With both gifts and contracts, industrial reporting requirements were usually less formal than federal ones.

A number of characteristics of industrial funding could be considered either advantages or disadvantages, depending on the circumstances and the personality of the researcher. Most viewed industrial funding as more focused and more applied than federal funding. Some saw this as a disadvantage because it tended to pull them away from basic research, which they found more interesting, and because it involved work that was inappropriate for graduate students. Others were excited by the prospect of working on a project that could show tangible short-term results and that would be carried further in a corporate research laboratory. Another commonly mentioned difference between industrial and federal support was the closer monitoring of research by industrial sponsors through technical representatives. Some faculty resented this "breathing down my neck all the time," while others found the interaction stimulating and formed collegial relations with their industrial project managers.

There were some clear-cut disadvantages to industrial funding, however. First, getting started with a company often was difficult. Even after having "sold" a company's technical people on a project, faculty members sometimes found it hard to locate the decision maker in the company and get funding released. Research funding also was sometimes held up when companies unaccustomed to dealing with universities engaged in lengthy negotiations over property rights to inventions, or, to a lesser extent, proprietary information and publication review.

Second, industrial funding tends to come in smaller amounts and with shorter performance periods than government funding. This means that it often is not appropriate for graduate students, who usually need support for two or three years. Industrial funding also sometimes takes more administrative time and effort when a number of small grants or contracts is involved.

Third, industrial contract funding tends to be more applied and results oriented and come with more strings attached than federal funding. This means that funding relationships are harder to sustain and that industrial funding is a less reliable source of research support, since failure to produce results, changes in company priorities, and economic downturns all are likely to cause such funding to dry up suddenly. "They think they own you" was a phrase used by several interviewees to describe unhappy relationships with industrial sponsors. The other side of this problem is the danger that one company might form such a close relationship with a researcher that other companies are excluded from access to that person's research.

Finally, dealing with industrial sponsors often is accompanied by a pressure to alter one's research agenda. Several interviewees talked about the balancing act they continually had to perform between delivering what the sponsors wanted and doing research that was fundamental enough to be appropriate to the university. The potential skewing of research agendas toward commercial concerns was particularly dangerous for younger faculty members, who often find it difficult to attract research funding and who need it desperately to attain tenure. Interviewees in this study included two NSF Presidential Young Investigator award winners, who are required under the terms of the award to find matching funding from industrial sponsors. Both resented the time and effort that this search for funding required, and both recounted instances of funding offers conditional on the performance of research outside of their current interests.

Despite these disadvantages, interviewees were overwhelmingly favorably disposed to industrial research sponsorship. The interviews also revealed several other interesting aspects of the current atmosphere surrounding industrial funding of university research.

First, the biotechnology "gold rush" is over, leaving many researchers in this field somewhat disillusioned and wary of industrial sponsorship. Reflecting this disillusionment, interviewees in biotechnology registered the lowest approval of industrial sponsorship of the five fields in this study. The "gold rush" held sway in the late 1970s and early 1980s, when companies were scrambling to become involved in the new technology. Industrial money became plentiful for researchers in biotechnology and many deals were struck between universities and business entities that had been formed to exploit the new research. Many university researchers were dazzled by this interest and to some degree sold out, or saw their colleagues sell out, to the industrial concerns. By about 1984, however, as initially promising developments failed to come to commercialization and as companies began to develop their own capabilities, industrial money in biotechnology began to tighten up once again, leaving university researchers with fewer principles intact and smaller amounts of money. The suddenness of this reversal created a backlash negative reaction to industrial funding, which now is evolving slowly into a more realistic view of industrial funding's role and limitations.

Second, the gift is becoming the form of industrial research funding most preferred by researchers. This is particularly true on campuses such as Stanford, which have high overhead rates that apply to contracts but not to gifts. The contract form of funding is used primarily when the industrial sponsor wants property rights to the research or some other form of deliverable or specific provision. In theory, gifts carry no restrictions beyond the stipulation that the money be used to support the research of a particular individual in a particular field. In practice, however, most of those interviewed reported that gifts were governed by informal understandings between researcher and donor that sometimes prescribed very narrowly the kind of research to be conducted, the time period for the research, and arrangements for reporting results. Most researchers were not troubled by these informal arrangements, viewing them as natural obligations to the donor and as a way around an unfair and burdensome institutional overhead assessment.

Finally, the interviews indicated that both faculty and company research sponsors are becoming more sophisticated and knowedgeable about university research sponsorship. Faculty members are able to articulate university policies regarding property rights, publications, and proprietary information early in discussions with companies, thus avoiding unhappy surprises when funding sponsorship arrangements are subjected to legal review. In turn, companies increasingly understand where the university is unwilling to bend and that, in most cases, informal understandings about property rights are sufficiently strong to protect their interests. It is unlikely that a university would deny a license to a company whose sponsorship resulted in a patent, and most companies who have dealt with universities understand that the university usually is not a particularly aggressive marketer of patents.

It was clear from the interviews that most faculty members felt that industrial sponsorship was a small but important part of the research funding picture and that most of the potential problems associated with such funding could be avoided if proper safeguards were established. Such funding usually was seen as depending on long-term efforts to encourage interaction with industry through a pattern of "transfer mechanisms," examined below.

Perceptions of University Technology Transfer Efforts

Everyone interviewed was asked the question: How has your university encouraged technology transfer? This general question was followed by questions about the subject's attitudes toward specific transfer mechanisms: liaison programs, patent offices, real estate developments, direct investment in start-up companies, and continuing education. The answers to these questions revealed a good deal about how the culture of the four universities related to technology transfer. In the following paragraphs, answers to the first question are summarized for each university. Aggregate data regarding the specific transfer mechanisms then is presented and commented upon.

At U.C. Berkeley most faculty members expressed the opinion that the university was doing nothing to encourage technology transfer. However, most agreed that extraordinary efforts at such encouragement were not necessary or appropriate, except for improvement of the patent office and support for faculty with inventions. Most recognized that the university was doing a great deal of technology transfer through traditional means such as graduating students, faculty consultations, publications, and faculty-formed start-up companies, and they believed that this was enough. A few faculty members perceived a recent shift by the administration in favor of greater involvement in technology transfer activities.

At M.I.T., most faculty members cited the "ambience" of M.I.T. as the element most encouraging to technology transfer. As has been noted in previous chapters, the generally supportive attitude of the administration and the long history of M.I.T.'s involvement with industry allow faculty members to pursue technology transfer activities as part of their normal duties. Many cited the reorganization of the Technology Licensing Office as a recent reinforcement of this supportive culture. Many also listed the Industrial Liaison Program as an important element in technology transfer.

At Penn State, most of those interviewed recognized that the university was involved in deliberate attempts to change its culture to produce increased facilitation of technology transfer activities. Such changes are clearly high on the agenda of the current president, Bryce Jordan, who was appointed with the assignment of bringing Penn State more fully into cooperation with the state in encouraging economic development. The Ben Franklin Partnership has become a symbol of this enforced change, and it was the activity most often cited in responses to this question. The university was virtually forced to become a part of the Ben Franklin Partnership and had to undertake a massive internal informational campaign to bring about faculty involvement in the program. Most faculty are acutely aware that Penn State's culture is still in transition. On the one hand, there are obvious and encouraging signs that the university's administration is behind the cultural shift. On the other, many faculty members feel that the university has failed to deal effectively with patents and patentable intellectual property. This has created a sense of confusion and distress that will take a while to resolve.

The comments of Stanford faculty can be summarized best by quoting one of those interviewed: "The University encourages technology transfer by staying out of our way." As with M.I.T., Stanford's history of interaction with industry and its proximity, geographically and philosophically, to many industrial firms has created an atmosphere conducive to technology transfer.

TABLE 12-6　Faculty Attitudes toward Transfer Mechanisms: Average Scores by University and Field

	By University				
	U.C. BERKELEY	M.I.T.	PENN STATE	STANFORD	TOTAL
Laison programs	3.61	4.05	3.86	3.86	3.85
Patent office	2.86	3.90	2.28	4.00	3.28
Real estate	2.64	2.83	2.33	3.50	2.85
Direct investment	2.81	2.75	2.33	2.47	2.60
Continuing education	4.00	3.85	3.35	3.67	3.37
Total	3.25	3.57	2.87	3.50	3.31

	By Field				
	BIO-TECH	ELECT. ENGINEER	MATERIALS SCIENCE	OTHER ENGINEER	OTHER SCIENCE
Liaison programs	3.20	4.22	3.92	3.80	4.09
Patent office	3.50	3.21	3.21	3.42	3.00
Real estate	3.00	3.00	2.80	3.14	2.37
Direct investment	2.41	2.66	2.36	3.20	2.41
Continuing education	3.58	3.85	3.86	3.73	3.40
Total	3.17	3.46	3.31	3.53	2.88

Scale: 1 = very negative; 2 = negative; 3 = neutral; 4 = positive; 5 = very positive

The faculty in both universities engage in technology transfer activities without the need for formal mechanisms sponsored or supported by the institution. The one exception to this is Stanford's Office of Technology Licensing, which many of those interviewed recognized as being particularly helpful when inventions, patents, or copyrights were involved. Many mentioned liaison programs as being effective in technology transfer primarily because they supported or reinforced individual faculty contacts with industry.

Attitudes Toward Technology Transfer Mechanisms

Faculty attitudes toward particular transfer mechanisms were determined from responses given to the questions that followed the general question about how the university encouraged technology transfer. As noted earlier, the responses to these questions were scored by the author on a scale in which 1 represented a very negative attitude, 2 a negative attitude, 3 a neutral attitude, 4 a positive attitude, and 5 a very positive attitude (see Exhibit 12-2). Table 12-6 shows average attitude scores by university and by field.

Previous chapters on patent offices, liaison programs, direct investment, and continuing education have treated responses to each of these mechanisms in some detail, and the reader should refer to the section in each of these chapters labeled "Faculty Perceptions" for an in-depth analysis. However, the aggregate data shown in Table 12-6 also reveal some interesting patterns.

First, the faculty of all four universities taken together considered liaison programs to be the most effective technology transfer mechanism, followed closely by continuing education. Patent offices came in third, with very high scores at M.I.T. and Stanford and scores in the negative ranges at U.C. Berkeley and Penn State. Real estate as a transfer mechanism also scored in the negative range except, as one would expect, at Stanford, which benefits from proximity to the tenants of its industrial park. Most faculty members viewed university attempts to invest in spin-off or start-up companies with many reservations: direct investment as a transfer mechanism scored the lowest overall attitude rating of any of the elements discussed.

As might be expected from the foregoing chapters, M.I.T. and Stanford faculty showed the most favorable aggregate average views of technology transfer mechanisms, while Penn State showed the lowest, actually dipping into the negative range. There were no surprises in the averages by field, either; the two engineering categories showed the highest average opinion of technology transfer mechanisms, and the "other science" category showed the lowest. The interviewees in the "other science" category tended to be less involved in technology transfer than those in the other categories. The relatively low score for biotechnology can be explained by the previously mentioned backlash reaction to the end of the easy money period of the early 1980s.

In general, faculty members were most favorably disposed toward those mechanisms with which they were most familiar, except for low ratings of the patent offices at U.C. Berkeley and Penn State and high ratings for continuing education at Stanford and M.I.T., where relatively little in the way of formal continuing education presently exists. These aggregate scores are elements in a description of the different environments for technology transfer that exist on the different campuses and in different fields. Assessing and understanding these environments or cultures are important for university administrators as they contemplate changes in policies.

Perceived Changes in the Atmosphere Surrounding Research

Of the 87 involved faculty members responding to the question, 75 (86 percent) saw some change in the atmosphere surrounding research at their university in the last ten years. However, only slightly more than half of those interviewed saw changes in relations with colleagues, relations with students, their attitudes toward their own university, or the attitude toward their university held by outside funding agencies. Descriptions of the nature of perceived changes varied considerably, but several responses occurred with enough frequency to reveal trends.

INCREASE IN APPLIED RESEARCH

Most faculty members saw an increased interest by industry in funding university research and an increased willingness and capacity of universities to accept such funding, which, by its nature, leads to more applied research. Federal policies are also seen to be changing. Federal funding agencies are said to be becoming more results oriented, more specific in defining tasks to be accomplished, and, increasingly, more supportive of projects likely to have economic impact in the relatively near term. Many agencies, including the NSF, have a number of programs that require matching funds from industrial sponsors. Researchers increasingly feel the need to justify research projects on the basis of potential for producing practical applications.

Although these trends were decried by many of those interviewed, a significant minority felt that some applied, focused research was desirable, as long as it remained a relatively small proportion of the total. Projects of this type often were viewed as interesting opportunities to do research with tangible results, to stay in touch with the "real world," and, in the case of industrial funding, to escape the sometimes suffocating realm of peer review.

INCREASE IN "BIG SCIENCE"

Many faculty members reported that it is increasingly difficult for a single scientist to attract research funding. This is apparently the result of two factors. The first, again, has to do with the federal government. Seeking to achieve a greater impact, the government has changed its policies so that instead of supporting individual researchers in a particular field at a number of campuses, it funds large "centers of excellence," concentrating its support in a defined field of science or engineering at just one or a few universities. The second factor is that university research increasingly requires expensive instruments and extensive computer facilities.

The consequences of the move toward "big science" were articulated clearly by many of those interviewed. First, they said, research is becoming more of a cooperative venture. Scientists and engineers, often from several disciplines, have to prepare joint research proposals. Proposals for large amounts of money for the large centers require the support and cooperation not only of research colleagues but also of high-level university administrators and, often, elected government officials as well. Researchers, even sometimes those in different institutions, also have to share expensive research and computer facilities and research instruments. Alternatively, in some fields, universities must borrow expensive instruments from industry in order to carry out research.

Several of those interviewed expressed the opinion that research was becoming "more scientific" because of increased computing power and sophistication of instruments. Others felt that "big science" was leading to increased bureaucracy, which they viewed with mixed feelings. On the one hand, they objected to increased administrative costs and what they saw as bureaucratic obstacles. On the other, they themselves did not want to shoulder administrative tasks.

Finally, several of those interviewed recognized that "big science" introduced long-term instability into university research and increasingly subjected the university to the vagaries of federal research support. The sudden obsolescence of a large research facility or the sudden loss of funding for a center places a university at greater risk than ever before.

INCREASE IN COMPETITION FOR RESEARCH FUNDING

Greater competition for research dollars was a change noted by many faculty members. Although more money is available now than ten years ago, more researchers are also after it. This increased competition has several consequences. First, researchers must spend greater time and effort to obtain funding. Second, particularly in the case of industrial funding, the personality and marketing skills of the researcher have become more important. Third, increased competition poses a greater threat to the independence of the research agenda of the university because researchers are more willing to "go where the money is." Finally, the entire process has become more politicized: political considerations appeared to be more significant than scientific merit in determining some funding award decisions. For example, researchers at U.C. Berkeley, M.I.T., and Stanford all felt that "regionalism" was working against them—that federal funding agencies had established informal quotas for the amount of money that would be given to Massachusetts and California in order to spread research money around and maintain broad-based legislative support.

Comments on these three trends were distributed evenly through the responses of researchers from all campuses and all fields. Some responses did show differences among universities, however. At Berkeley, the move toward "big science" is having particular impact. In some Berkeley departments the culture is not particularly encouraging to cooperative efforts; promotion and tenure decisions place great emphasis on identifiable individual work. Until recently, Berkeley researchers could rely on their individual reputations to attract research support as well. Since 1983, however, Berkeley has failed several times in bids for major research centers. The failure of the Berkeley-led "West Coast consortium" of universities to obtain a major earthquake engineering center has led to a reevaluation of Berkeley's strategy for applying for large programs.

Both younger and older faculty members at Berkeley also noted that there was a decided generation gap between younger faculty members and their more experienced colleagues. Younger faculty members were more interested in applied or practical research and more willing to try to make connections between their research and real-world applications than the patricians of science who, for the most part, felt that university research should concentrate on "fundamental" questions.

At M.I.T., no clear pattern of responses emerged except that the feeling of intensified competition seemed more prevalent. Engineering and some scientific faculty at M.I.T. have only about 65 percent of their nine-month salaries guaranteed by the institution; they must find the remaining 35 percent from outside research funding. This is particularly difficult for younger faculty members and probably accounts for some of the competitive pressures.

At Penn State, a common comment was that faculty members who are interested in applied research or the commercialization of research can now "come out of the closet." The atmosphere at Penn State is currently much more tolerant of such research than it was even five years ago. But several faculty members voiced the view that there was some contradiction in Penn State's dual goals of improving its ranking as a research university and being an important player in the commercialization of research and the economic revival of Pennsylvania.

At Stanford, concern over the high overhead rate applied to research contracts, that is, the percentage of research awards that goes to the university's general fund for overhead costs, (which is in excess of 70 percent), and Stanford's consequent competitive disadvantage in obtaining research sponsorship dominated the responses. This concern over research funding competition was exacerbated by the expectation that, at Stanford as at M.I.T., faculty members are to get a part of their salaries, sometimes as much as 50 percent, from research funding. The overhead rate issue has pitted much of the faculty against the administration, creating considerable resentment.

Concern about instrumentation was also evident at Stanford. Stanford researchers appear to have more opportunities to obtain gifts of equipment from industry or to use industry-owned equipment than researchers at other universities. The disruptive effects of the acquisition of large, expensive research facilities became well known through the controversy that developed in the physics department over the building of the Stanford Linear Accelerator, which eventually resulted in the formation of a separate corporation to operate the accelerator and the separation of the Department of Applied Physics from the Department of Physics.

Differences in responses by field were less distinguishable than differences by university. In biotechnology, however, most faculty members commented on the "shakeout" that occurred around 1984 as the "wild money" from industry began to decrease.

In most of the interviews, the general question about the subject's perception of change was followed by several more specific questions about changes in particular aspects of the atmosphere surrounding research. A brief summary of comments on those specific elements of change follows.

CHANGES IN RELATIONSHIPS WITH COLLEAGUES

Forty-one of 73 respondents (56 percent) noticed some difference in their relations with their research colleagues. Responses on this subject present a rather confusing picture, probably because relationships between colleagues depend on so many factors. However, it is clear that the often-expressed fears that technology transfer issues would decrease collegiality are largely unfounded, at least among those interviewed in this study. It was evident in every university and every field that faculty felt that collaboration between researchers either was increasing or should increase. Those who saw a decrease in interaction with colleagues cited the increased demands of work as the main cause.

Increased secrecy between colleagues was mentioned only eight times—six times in biotechnology (at least once in each university), once in materials science, and once in "other science." As biotechnology is in the greatest turmoil and is undergoing the most rapid changes at present, it is reasonable to expect that collegial relations would suffer most in this field. On the other hand, faculty members in electrical engineering, which has had the longest and most active history of interaction with industry, saw increased collaboration among colleagues more frequently than in any other field.

The responses to this question do not suggest that collegial relations are threatened by increased commercial interest in university research, at least in the long run. The problems in biotechnology are beginning to sort themselves out, and at least two respondents in the biotechnology category felt that trust between colleagues was being reestablished. In fact, the desire to interact with colleagues is still quite strong, even in that field. Where it has been lost, it is mourned.

CHANGES IN RELATIONSHIPS WITH STUDENTS

Only 35 of 75 respondents answered "yes" to the question: Have you noticed any changes in your relations with your students? The most common changes noted mirrored the changes seen in relations with colleagues. Students are now more likely to work in teams and to have more than one adviser. Many faculty members felt that they had less time to work with students than before. Several mentioned that the higher percentage of foreign students had changed the nature of the interaction between student and professor and that students seemed to work harder and be under more pressure than before.

The most frequently cited and significant change, however, was that fewer students are aiming for academic careers or advanced degrees. Students are more likely now than ten years ago to be interested in applied and practical work and to seek employment with industrial firms upon graduation. This tendency sometimes creates conflicts, as professors must be continually vigilant that student theses and dissertation projects do not deteriorate in quality or treat subjects that are not appropriately academic in character.

CHANGES IN ATTITUDES TOWARD THE UNIVERSITY

Fifty-five percent (42 of 76) of those answering responded "yes" to the question: Has your attitude toward your university changed in the last ten years? Penn State showed the highest percentage of "yes" answers (13 of 15, or 86 percent) and M.I.T. the lowest (8 of 21, or 38 percent). At U.C. Berkeley and Stanford, about half of the respondents reported a change in their attitudes. Overall, changes of attitude were about as often positive as negative. Most of the positive changes seemed to be associated with increasing maturity. "I now feel that I am a part of this place" and "I am now more aware of the incredible intellectual diversity within this university" were typical positive expressions. At Penn State, many faculty members saw the administration of Bryce Jordan as more supportive and effective than previous administrations. This perception may account for the relatively high percentage of attitude changes at Penn State.

Virtually all of the negative attitude changes reported were related to frustrations with bureaucracy and the perception that the university administration was not supportive of research or academic work. As noted earlier, such responses were particularly common at Stanford, where the high research overhead rate has come to be seen as a symptom of what one Stanford faculty member called the administration's "MBA mentality."

CHANGES IN ATTITUDES OF EXTERNAL FUNDING AGENCIES

Forty-one of the 66 responding interviewees said that they saw a change in funding agencies' attitude toward the university. External agencies were seen as wanting the university to do more applied and focused research, being more prone to let political considerations enter into funding decisions, and becoming more interested in fostering competition among potential awardees. However, several faculty members in each university also felt that in the past ten years industry has shown an increased willingness to fund university research and that their university had made adjustments to make it easier for industrial concerns to sponsor research.

SUGGESTIONS FOR CHANGES IN POLICY

In addition to asking what changes were taking place in the atmosphere around research, directly or indirectly as a result of technology transfer, interviewers asked faculty members what changes in the present policies of their institutions they would like to see. Only 33 of 77 named any changes, and many of those were minor. For the most part, faculty members appeared to be very comfortable with institutional policies. Some faculty did not distinguish between policies and administrative procedures. For instance, U.C. Berkeley and Penn State faculty suggested improvements in patent administration but no changes in policy. The most common answer to the question was "I don't know what those policies are," followed by an observation that no present policy seemed particularly inhibiting.

At Berkeley, the conflict-of-interest provisions that require researchers to disclose all ties with nongovernmental sponsors of their research raised some comments. Although no interviewee strenuously objected to this policy, some felt that it was unnecessary and overzealously enforced. The most significant criticism of U.C. Berkeley's policies, however, was that the policies were not clear enough to be effective guidelines, particularly with regard to ownership of and association with outside companies.

At the other three universities there was no consistency in comments regarding changes to policies. M.I.T. faculty were almost uniformly satisfied with the institute's policies. Some Penn State faculty felt the university's policies should be liberalized in some way to encourage more technology transfer, but few had concrete suggestions about how this should be accomplished. Stanford faculty again objected to the overhead rate, and a few wanted higher patent royalty shares.

CONCLUSION

Given the differences in institutional setting and academic culture among the four universities, it is surprising that in several areas there was a consensus of opinion about the nature of the changes taking place in the atmosphere surrounding research. Although some of these changes are only tangentially related to technology transfer and the commercialization of research, knowing what faculty members think about important changes in research adds to our understanding of the institutional context of which technology transfer is a part. The most important finding from this series of questions is the consensus that over the past ten years it has become easier for universities and industry to work together in doing research. This finding is consistent with the many other indications of the same trend shown in preceding chapters.

Attitudes of Uninvolved Faculty

As noted at the beginning of this chapter, 14 faculty members not involved in fields likely to produce commercially valuable research were interviewed, primarily to determine whether they were developing negative attitudes toward increased commercialization of research or whether there appeared to be any strains in collegial relations between those who could gain financially from their research and those who could not. These *uninvolved* faculty members were selected from those who might have a general understanding of the issues related to technology transfer, either through service on committees or through their own scholarly work. Interviews with uninvolved faculty were conducted using the protocol shown in Exhibit 12-3.

Although the number of people interviewed in this category is too small to represent fairly the views of all faculty not directly involved with research of potential commercial value, the absence of any serious concern on the part of any of the subjects about the issues surrounding technology transfer is an indication of generally tolerant attitudes. A typical response from a knowledgeable subject was that there was a noticeable trend toward faculty

EXHIBIT 12-3   Interview Protocol for Uninvolved Faculty

1. University: _____

2. Record locator: _____

3. Discipline: _____

4. Length of time at university: _____

5. How have you been involved with or become knowledgeable about the issues surrounding technology transfer and the university? _____
_____

6. Do you feel your university has struck the correct balance between facilitating technology transfer and maintaining its own integrity? Are you uncomfortable with any aspect of technology transfer at your university? _____
_____

7. Do you feel that any of the university policies related to technology transfer should be changed?

   a. Conflict of interest _____

   b. Disclosure of inventions _____

   c. Consulting _____

   d. Financial involvement with companies _____

   e. Ownership of intellectual property _____

   f. Patent and licensing procedures _____

8. What has your university done to encourage technology transfer? What do you know or think about:

   a. Liaison or affiliate programs _____

   b. The patent office _____

   c. Research partnership arrangements _____

   d. Related real estate developments _____

   e. University ownership of equity in companies formed to commercialize university research _____

   f. Continuing education _____

EXHIBIT 12-3 *continued*

g. Foundations and buffer organizations _____

9. What effects do you feel the increased emphasis on the commercial value of university research has or will have on collegiality? _____

_____

10. Have you noticed any changes in the atmosphere surrounding research or in your relationship with your colleagues in the engineering and professional schools in the last ten years? _____

_____

11. What other effects of technology transfer do you see or predict?

_____

involvement in commercial activities but that, except in a few instances, things were not seriously out of line. Nonetheless, the respondents felt that the university would have to be increasingly vigilant in enforcing existing policies and perhaps would need to institute new policies if they became necessary to protect the university.

Several of those interviewed were concerned about consulting policies and practices, an issue that by contrast came up infrequently in interviews of involved faculty. At all of the universities the policy of allowing, even encouraging, consulting one day per week was so vaguely stated and so rarely enforced that the concerned faculty members feared that abuses were commonplace. Consulting policies are abused not only when consulting takes up too much faculty time or interferes with university duties but also when consulting is used to appropriate patentable university intellectual property for the personal gain of the faculty member.

Uninvolved U.C. Berkeley faculty generally recognized that the university was being pushed toward a more active stance with regard to technology transfer. By and large, they did not decry this trend. Several reflected the feelings of the involved faculty members that clearer guidelines and monitoring mechanisms should be established and that there should be a controlled movement to encourage technology transfer and to publicize what the university was doing. Everyone recognized that the College of Engineering was the leader in this movement. One history professor suggested that the humanities were beneficiaries of a $5 million university grant primarily because the administration sought to improve the balance between the rich engineering and hard sciences and the relatively poor humanities.

This same sentiment was reflected in a more general way by uninvolved M.I.T. faculty, who recognized that the success of engineering and science faculty in attracting research funding benefited everyone in the institute. Uninvolved faculty members at M.I.T. exhibited the highest level of comfort with new trends associated with technology transfer. They were well aware of the traditions and culture of M.I.T. and accepted many kinds of ties with industry as part of what the institute stands for.

In several public meetings the humanities faculty at Penn State voiced serious concern over recent efforts by the university to be more involved with industry and with economic development efforts in the state. This concern was reflected in some of the interviews conducted in this study. Penn State's condition of transition was very evident, and it was clear that those interviewed had an understanding of some of the philosophical issues involved. In general, they realized that there were problems and that the university had to maintain a delicate balance between what the state legislature and the governor wanted and the traditional ideals of the faculty. One interviewee put it this way: "We should strive for a broadened scope of activity, but we don't want to be redefined."

Only two people in the uninvolved category were interviewed at Stanford. One summed up the problems as he saw them: "The humanists think the scientists are rich and have it easy, and the scientists think the overhead rate is too high." Again, there seemed not to be any extraordinary present or developing tensions between the two groups.

The concern combined with lack of strongly held negative feelings toward technology transfer in universities exhibited by these uninvolved faculty members is some indication that research universities are prepared to expand their roles in technology transfer and will not meet strenuous resistance even from those unlikely to benefit directly from such an expanded role. However, universities will have to proceed carefully to avoid alienating such faculty members with egregious examples of abuses, conflicts of interest, or negative publicity.

Conclusion

A number of conclusions can be drawn from detailed analysis of these responses and from consideration of them in light of the earlier examinations of institutional history and changes in transfer mechanisms. These conclusions form an important part of the background needed for the development of university policies and practices related to technology transfer.

GREED AND JEALOUSY NOT ASCENDANT

Although most of those interviewed were in fields in which university research has recognized potential for commercial value and the subjects were chosen in part from those who had experience with commercializing their own research, there is little evidence that the prospect of large financial gains poses a serious threat to institutional values or collegial relations.

There were a few exceptions to this in the fields related to biotechnology, but as the excitement of the late 1970s and early 1980s over advances in this field has begun to subside, these instances are viewed as exceptions rather than part of a permanent pattern.

Strong ancedotal evidence also suggests that traditional university values continue to be supported. The interviews disclosed a surprising number of instances in which inventors voluntarily surrendered all patent rights and financial returns from inventions, usually turning them over to their departments to support further research. The general reluctance of faculty members to become involved in the time-consuming and frustrating patent process, intensified on campuses that have inefficient patent offices, is further evidence that faculty are not involved in a gold rush toward commercialization. In the words of one Stanford faculty member: "If I wanted to get rich I wouldn't have come to work for a university."

Although there was widespread resentment toward faculty members who abused the system by devoting substantial time and energy to outside commercial ventures while remaining on the faculty, such resentment did not extend to those who left the university entirely, even when they became extremely wealthy. In fact, in most cases these individuals were viewed as success stories of which the university could be proud, and often some ties between the individual and the university were maintained.

DIFFERENT CULTURES, SAME VALUES

These interviews reinforced the notion that the four universities in the study possessed distinctly different cultures and that these differences could be discovered in part by examining the different faculty approaches to and attitudes toward technology transfer. U.C. Berkeley, insulated by state support and by a history of excellence in fundamental or basic research, to a certain extent, is struggling with the new realities of "big science" and increased industrial interest in research. It is finding some of its policies and transfer mechanisms inadequate to deal with these trends. M.I.T. has a long tradition of interaction with industry and has built an impressive organizational structure to facilitate such relationships. Penn State is in transition, trying to be involved in basic research and at the same time to satisfy its public constituency by being involved in economic development. Its faculty, like its administrative structures, evidence some turmoil and confusion over the issues and over the university's role. Stanford, like M.I.T., has a history of interaction with industry and is comfortable with that interaction. At Stanford, however, industry interaction is not as centralized or centrally supported as it is at M.I.T. Rather, it is largely the responsibility of departments and individual faculty members and is confined narrowly to a few fields of science and engineering.

Despite these different cultures, the framing of the problems and the standards of conduct recognized are quite similar among faculty members of the four institutions. The evidence gathered in this study supports the notion that "segmentation and fragmentation are characteristic of the academic profession, but an integrating effect of overarching basic values also exists."[1] The rules and policies of all four universities are based on the premise that full-time faculty owe their primary allegiance to the university and should devote most of their time and creative energy to fullfilling their university duties. Given this common basic philosophy, it is not surprising that these universities have developed similar rules regulating outside activities of faculty and that their faculties have similar views of what constitutes unacceptable behavior.

CHANGES IN RESEARCH

The most common change in the atmosphere surrounding research cited by the interviewees was the trend toward more focused, applied research. This is a result of both a change in federal policy and an increase in industry support of university research. It also was clear that connections between the university were becoming easier to make because of increased industry desire to fund university research, increased knowledge and sophistication on the part of industry in negotiating with universities, and accommodations made by universities to facilitate industry sponsorship.

However, this change was overshadowed by the trend, again driven both by federal policy and by the realities of modern research, toward "big science." Many of the other trends in university research—the increased politicization, the need for greater collaboration among researchers and institutions and between universities and industry, the increased competition for funds—can be seen as derivatives of the big science movement.

TRANSFER MECHANISMS

In general, the greater the interviewee's familiarity with a particular transfer mechanism, the more likely the he or she was to give a favorable assessment of that mechanism's role and importance. There were two exceptions to this: the patent offices at U.C. and Penn State were viewed less favorably by those who had dealt with them than by those who had not, and continuing education was rated high in importance even by those who had little experience with it.

FACULTY ROLES

Perhaps the most important conclusion to be reached from these interviews has to do with the way technology transfer may be affecting faculty roles. Most of those interviewed realized that technology transfer activities and relations with industry required substantial amounts of faculty time but were important enough to justify it. This reallocation of faculty time represents a subtle shift in faculty roles. Part of this time, as well as the increased time required to obtain federal funds apparently is being taken from time that otherwise would be spent with colleagues and students. Partly to mitigate the effects of this trend, faculty members are supporting the establishment of administrative staffs and structures, such as liaison program administrators and technology transfer offices, that will provide a "new bureaucracy" within the academy to buffer, support, coordinate, and control university and faculty efforts related to technology transfer, relations with industry, and commercialization of research. Support for this kind of bureaucracy is occurring simultaneously with an increase in faculty frustration over other kinds of bureaucracy that appear more removed from direct faculty concerns.

From these interviews it is clear that faculty members involved in technology transfer are helping the university define its role in light of the new interest in industry support and the commercialization of university research. The four campuses in this study are at different stages in the development of a coherent response to the internal and external pressures propelling these new interests, but none of their faculties doubts that some response more favorable toward and more facilitative of technology transfer will result. A new wind is blowing, and to these faculty members its direction is clear.

Note

1. George D. Kuh and Elizabeth J. Whitt, *The Invisible Tapestry: Culture in American Colleges and Universities* (Washington, D.C.: Association for the Study of Higher Education, 1988), 75.

Bibliography

KUH, GEORGE D., and ELIZABETH J. WHITT. *The Invisible Tapestry: Culture in American Colleges and Universities.* Washington, D.C.: Association for the Study of Higher Education, 1988.

13

Conclusions and Recommendations

The American research university is changing in response to internal and external pressures that it become more active in promoting national and regional economic growth. These changes are both exciting and threatening: exciting because they present new opportunities for universities to serve society, and threatening because, as universities move closer to the commercial world, they begin to relinquish their special place in society.

The occurrence of significant changes could be recognized in the late 1970s, when university traditions and long-standing practices began to be overturned. At first the incidents of change appeared isolated, confined to a particular institution or a narrowly defined issue within institutions (for example, the issue of inventor shares of royalty income on inventions). Gradually, however, the interconnectedness of seemingly disparate events has penetrated the collective awareness of enough people that a coherent institutional response can be made. In other words, the changes are becoming institutionalized.

Technology transfer in universities has many aspects. One part is the process of enlarging the constituencies of the university or deepening ties and interactions with constituencies that already exist. Relevant particular constituencies include commercial concerns, which are both consumers and patrons of university research, and the immediately surrounding community, which both demands and supports university efforts to aid local economic development (such as research parks and business incubators). Another aspect of university technology transfer is the increase in efforts by universities to profit from the intellectual property they produce. This is particularly important because such attempts at profit appear to contradict some of the traditional values of the university, from the point of view of both those within and those outside the university.

The meaning of technology transfer in universities is coming into sharper focus as events unfold and as studies such as this one provide increasingly detailed maps of the phenomena involved. The purpose of this book has been to help make explicit the interrelatedness of elements of technology transfer and to illustrate a breadth of change that extends across institutional boundaries and involves a wide range of current issues.

Conclusions from Dimensional Analysis

Preceding chapters examined technology transfer activities and mechanisms along several dimensions, including those of history, institutional setting, organization, faculty attitudes, resources, policy, and academic culture. All these overlap and affect one another. Individually and collectively, they form the basis for certain conclusions that can be drawn from this study. They also form a framework for the recommendations made at the end of this chapter. In the following sections each of these dimensions is examined in turn.

HISTORY

The late 1970s was a watershed for the history of technology transfer in universities. Before that, technology transfer from universities to industry occurred largely through traditional processes: students who graduated and went to work for industry, publication of the results of university research, and faculty consulting. "Knowledge transfer" is perhaps a more accurate description of the result of these processes. Beginning about 1977, however, there was a sharp increase in the number and activity of new transfer mechanisms. University patent offices were organized and reorganized, and they experienced a sharp increase in activity. Many universities began to experiment with new ways of obtaining value from their intellectual property, such as equity ownership. Liaison programs were formed, and membership in them increased. Industrial sponsorship of university research grew, both in absolute volume and as a percentage of the total of research support funding. Universities increasingly were drawn into regional economic development plans through such activities as real estate projects and technical assistance programs. All this activity signaled dramatic change. To the relatively passive knowledge transfer activities of the past, universities added a new active engagement in the commercialization of research and the processes of economic development. This new activity was called by a new name, *technology transfer*—a term that quickly gained currency, even though it was not well defined.

As this new activity increased, so did some differences between departments within the university. Academic disciplines have been affected in different ways by these changes and have begun to develop separate histories with regard to technology transfer, thus increasing fragmentation within the university and decreasing its ability to respond coherently to the changes taking place. Although, as one would expect, there were increased differences between the traditional two cultures of humanities and sciences, a more significant change was the increase in differing attitudes developed between proponents of traditional paradigms of scientific investigation and conduct and the new scientific activists. The activists were more willing to perform applied research, to engage in and profit from research development activities, and to interact with industry.

In isolated cases, as in some departments in the fields associated with biotechnology, the differences between traditional and activist colleagues created internal dissension and a deterioration of collegial relations. In fields with a longer history of dealing with research of relatively immediate commercial value, particularly electrical engineering and computer sciences, the changes brought more resources and greater internal cohesion.

An examination of recent history, then, reveals the complexity of the effects of technology transfer. When this recent history is viewed against the backdrop of the history of technology transfer prior to 1977, however, it becomes clear that many of the new technology transfer mechanisms and activities had precedents that provide examples for current practice. For instance, M.I.T.'s experience with William Walker's Research Laboratory for Applied Chemistry in the 1920s showed what can happen to a research university when applied research becomes too high a proportion of the total research done in the university. The history of the University of California's use of its extension division as a buffer between its elitist ideals and the demands to popularize the university showed how a university unit can be used to mediate between internal and external pressures.

The historical dimension of this study, then, not only highlights the significant changes that have taken place in the last ten to twelve years but also, by providing examples of successes and failures from the more distant past, teaches some lessons that can inform us as we adapt to current changes. The study of the historical dimension is also necessary to an understanding of the other dimensions used here to analyze technology transfer, especially those of institutional setting and academic culture.

INSTITUTIONAL SETTING

An analysis of the circumstances and conditions under which technology transfer presently is taking place in each of the four universities studied in this book has highlighted important differences and similarities. For example, the quality and orientation of its current leadership is clearly an important aspect of the institutional setting at each of the four universities. President Bryce Jordan's efforts to encourage technology transfer and economic development efforts at Penn State provide one instance of the difference that leadership can make when that leadership has defined the issues clearly.

Physical location is another important aspect of institutional setting. Penn State's geographical isolation, especially in comparison to the other universities in this study, has proved to be a significant influence on the nature of the technology transfer activities it conducts. For instance, access by industrial firms to faculty through consulting arrangements or on-site review of research is more difficult at Penn State. On the other hand, Stanford's and M.I.T.'s proximity to industrial firms has been extremely productive of technology transfer activities.

Perhaps the most important aspect of the institutional setting, however, is the set of pressures and demands, both internal and external, that push on an institution and its leadership. At Penn State, the Pennsylvania legislature's demand for university involvement in an economic development effort provides a clear example of external pressures to develop a capability for technology transfer. On the other hand, Stanford's utilization of its ties to industry to achieve its goal of becoming a top-ranked research university is an example of internally generated pressures. Listing and gauging the strength of these pressures is a key preliminary to development of a coherent institutional approach to technology transfer.

The Penn State example also points up another aspect of both history and present institutional setting that is important to this story: the influence of a catalytic event in stimulating a chain reaction of other events and changes that sweeps an institution to a new level of awareness and coordination of technology transfer activities. Rarely is that event as galvanizing and dramatic as the one at Penn State; however, in many cases where multiple changes have occurred, they can be traced back to a single event that becomes a symbol for the whole series.

ORGANIZATIONAL IMPLICATIONS

Previous chapters described many ways of organizing university technology transfer activities, both inside and outside the boundaries of the university. Although great variety exists in the forms this organization may take, the examination did reveal some common features.

First, technology transfer activities that are less consistent with the traditional missions of the university, less in tune with the prevailing academic culture, and less consonant with the public image of the functions appropriate for a university are more likely to be placed in organizations outside the university—that is, in what we have termed "buffer" organizations. Examples include the venture capital partnerships formed by Harvard, Johns Hopkins, and Penn State to exploit commercially the results of university research through equity ownership in start-up companies. Another example is the use by Penn State and other universities of professional outside patent and technology licensing organizations, such as Research Corporation

Technologies, Inc., to handle some aspects of patentable intellectual property. Universities also have set up separate corporations to build and manage research parks and other real estate ventures.

Buffer organizations have the advantage not only of shielding the university from legal and public relations problems but also of helping to isolate certain technology transfer activities from the rest of the university and allowing them to be placed under the relatively independent control of experienced professional managers. This isolation can protect the activities from inexperienced meddling by university administrators and faculty and also can preserve the appropriate relationship between the faculty and the university. For instance, a venture capital partnership managed by experienced businesspeople can deal with faculty inventors in a more dispassionate and objective manner than could a university administrator who must deal with the same faculty members in other capacities. The use of buffer organizations also helps to avoid, though it does not entirely eliminate, charges of institutional favoritism toward certain members of the faculty or certain departments. Finally, such organizations are useful in avoiding university rules, regulations, and established practices that could cripple some technology transfer activities. For instance, university salary scales and compensation practices often are not adequate to attract the level of professional expertise that is necessary for successful management of specialized activities.

Of course, the use of buffer organizations carries some risks as well. These organizations may become so independent that they begin to have a life of their own and make decisions that go against the best interests of the university. A minor but symbolic example of this occurred when the Wisconsin Alumni Research Foundation (WARF), which exists in part to market the results of research from the University of Wisconsin, named one of its most successful products, a rat poison, "Warfarin," thereby linking the university with a product that might have a negative effect on its image, particularly were the product to create negative publicity.

When technology transfer activities are organized within the university itself, a situation sometimes occurs analogous to that arising with buffer organizations. If an activity is considered inappropriate for faculty involvement, either by the faculty themselves or as a result of institutional policy, it may be placed in an organization on the periphery of the university. This serves to identify the activity with the university and at the same time to insulate the faculty from significant involvement with it. The organization of continuing education in some (larger research) universities provides a mature example of this process. At U.C. Berkeley, for instance, with a strong land-grant tradition, continuing education is carried on, by policy, within a centralized service unit; nevertheless, that unit is viewed by a research-oriented faculty and some administrators as a buffer protecting faculty from extensive involvement in continuing education (allowing them to do priority research) while at the same time fulfilling public service obligations of the university. More recent service activities, such as technology licensing, technical assistance centers, and real estate developments, may follow the same pattern.

Peripheralization carries with it some common consequences. Peripheralized activities and organizations, like buffer organizations, tend to assume a life of their own. Control, oversight, and coordination of such activities may become problematic. When activities such as continuing education, technology licensing, and real estate development are staffed by professionals, salary administration and the relation of these new members to the rest of the university and to the faculty become issues. For instance, at the University of California it has been relatively hard to recruit technology licensing personnel to the patent office partly because the salary offered is too low.

An alternative to this peripheralization is faculty involvement in the activities. The extent of faculty involvement in technology transfer activities is one measure of the degree of centrality of those activities to a university's mission, as involvement in technology transfer activities tends to change faculty roles significantly.

In a few universities, including two in this study, the importance of and the need for coordination of technology transfer activities has made technology transfer itself an organizing principle. Penn State has gathered together many university organizations involving technology transfer or relations with industry, including the patent office and the research contract office, under a new vice president. M.I.T., through its Industrial Liaison Program (ILP) and earlier through its Division of Industrial Cooperation, also coordinates and institutionalizes its relations with industry. These efforts stand in contrast to the situation at Stanford, where departmental autonomy is part of the reason that no attempts have been made to centralize or coordinate any aspects of technology transfer.

Each approach—centralization or decentralization—has its strong and weak points. A centralized approach may be organizationally tidy, prevent duplication of effort, present a more uniform institutional face to an external constituency, and render activities more accountable to the institution. In Stanford's case, where such centralization is lacking, 34 liaison programs and several "mega-programs" often approach the same firms to solicit membership. However, a centralized organization tends to put technology transfer activities at a remove from the faculty. Because the function really originates with the faculty, this removal can impair seriously the effectiveness of technology transfer. An example is the development of the "mini-ILP" programs at M.I.T., which were set up because the general purpose ILP was not serving the faculty and departments that had established close ties to industrial firms.

In the four universities in this study the organization of technology transfer activities is moving in both directions at the same time. The goal is to centralize activities in which economies of scale can be achieved and coordination is important and to decentralize activities that depend for their success on individual efforts and external contacts. Staying with M.I.T.'s liaison programs as an example, the general purpose ILP now has a working relationship with the mini-ILP programs. At U.C. Berkeley, the College of Engineering's general purpose (centralized) ILP program was established to foster and serve individual liaison programs formed by the college's departments.

Examining university technology transfer from an organizational perspective reveals not only the characteristics just described but also the considerable organizational change associated with technology transfer that is taking place in research universities today. This is another indication of the importance of technology transfer. When the organization of technology transfer-related activities becomes stable, such activities will have matured and become a permanent part of universities.

FACULTY PARTICIPATION

Because the faculty is the heart of a university, the extent of faculty members' involvement in and concern about technology transfer is an important measure of the significance of technology transfer in universities today. Chapter 12 reported the views of 104 faculty members who were interviewed in the course of this study. In each of the universities studied it was clear that faculty involvement in technology transfer activities is increasing. However, the interviews also indicated that the extent of faculty involvement in and attitudes toward technology transfer activities varied from discipline to discipline. These variations lead to the conclusion that the changes introduced by technology transfer affect disciplines differently and, in some cases, increase the distance between the disciplines.

Faculty members are most likely to spend time on technology transfer activities related to their own interests or their traditional tasks of research and teaching. For instance, they spend more time with liaison programs than with technical assistance programs because they see liaison programs as productive of research funding from industrial sponsors and technical assistance programs as not related to research or teaching. They might also spend a great deal of time helping to develop a patent application because of the prospect of financial return or the satisfaction of seeing an invention come to fruition. As a consequence, activities such as liaison programs are logically placed under faculty control, while activities that are less likely to involve faculty, such as technical assistance programs, are more logically handled by peripheral organizations with separate staffs.

Unfortunately, because faculty members are busy people with many demands on their time, even technology transfer activities in which they are interested are unlikely to get the kind of sustained attention they deserve. The problem is exacerbated by the failure of many faculty reward systems to recognize technology transfer activities as important. These factors have tended to push technology transfer activities away from the faculty and thus represent a barrier to the fuller institutionalization of them. That so many such activities nonetheless enjoy extensive faculty involvement attests to the high value that many faculty members place on them.

The faculty interviews also disclosed that technology transfer was only one of a number of factors causing important changes in the nature of university research today. Because the federal government remains by far the most important patron of university research, changes that it is forcing on universities are much more significant than the technology transfer-related changes discussed. However, because the federal government has instituted programs encouraging technology transfer, the two factors are interrelated. Almost all of those interviewed reported that competition for federal funding is more intense now than ten years ago, and many attributed this to the government's emphasis on "big science," which concentrates some of the funding in "centers." In addition to requiring matching funds from industry, these center programs encourage cooperative, interdisciplinary approaches to research, thus bringing university research closer to the methods, if not the subject matter, of industrial research. Thus federal programs have operated in both direct and subtle ways to increase the involvement of university faculty in technology transfer.

RESOURCE ALLOCATIONS

Another measure of the importance of technology transfer and the changes it is causing in universities is the level of resources that technology transfer activities bring to the university or command from it. Certainly a major justification for engaging in technology transfer activities is their capability for generating income—patent royalty income, research sponsorship, state appropriations, gains from the equity appreciation of start-up companies or real estate developments, and so on. Technology transfer, however, also uses resources, and the amount of scarce financial and other resources allotted to various technology transfer activities is an additional measure of their importance in an institution.

One of the largest resource-related problems has proven to be the generation of start-up support. This problem appeared most strongly in the discussion of funding for patent administration. As with many other technology transfer activities, the payoff for sound patent administration occurs over a long period. Because of scarcity, accounting systems that are based on a year-by-year budget and fund accounting, and uncertain prospects for future return, many universities have difficulty in allocating present resources to fund such activities. Universities face a similar problem when they establish or increase the size of development offices.

In addition to their being uncertain, the returns for some technology transfer activities, such as community goodwill or close association between university and industry scientists, can never be measured in dollars. Without a measurable future return to justify current expenditures, activities have great difficulty in competing against other potential uses for scarce resources, many of which are more closely associated with the traditional university missions of teaching and research. Technology transfer activities need extraordinary institutional commitment to survive such competition.

Allocation of income from technology transfer activities is another issue in the resource dimension. Problems often arise because of unrealistic expectations of an activity's capacity to produce income. Patent royalties provide an example of this. The examination in chapters 4 through 7 indicated that significant returns from university-owned patents are rare and, when achieved, usually depend on a small number of very lucrative inventions. Continuing education is another example of a technology transfer activity from which financial returns not only are expected but also are an important justification for university involvement.

Both patent administration and continuing education suffer from institutional and faculty underestimation of the costs of carrying them out and the risks involved. The activities themselves must contend for the resources they generate, and they often find themselves left with too small a budget to operate at peak effectiveness. Always at issue is the amount or proportion of the resources generated that should go to the university people upon whom the activity depends—in our examples the faculty inventor or the faculty continuing education instructor. Too large a share cripples the activity itself by removing funding that could go to financial growth or improvement, and it may involve questions of unjust enrichment or conflict of interest. Too small a share discourages faculty involvement.

In general, when the university sees a technology transfer activity's primary purpose as being revenue generation, the institution is poorly served and the activity eventually suffers. Few technology transfer activities are real gold mines, and most cannot survive without a degree of integration into the programmatic elements of the university.

One final aspect of the resource dimension bears mentioning here. The importance of the change in the university's view of the results of university research, from a product intended to be disseminated freely to the scientific and lay community to an asset intended to be exploited and profited from, cannot be overemphasized. This transformation of universities' views of their intellectual property has resulted in the expansion of patent and licensing activity and is the impulse behind the current increase in university venture capital investment in research-related start-up companies, to name just a few effects. In moving the university closer to the marketplace and causing it to behave increasingly like a business concern, this transformation of the idea of intellectual property threatens to erode the philosophical basis of the university from both inside and outside the academy.

POLICY

An examination of university policies related to technology transfer has been important to this study because policies both guide current action and reflect the prevailing academic culture of an institution. Occasionally, university policies antithetical to the academic culture are imposed in an attempt to change that culture. A change in policy or the adoption of new policies often signifies that something important has occurred. Many policies related to technology transfer recently have been changed or are in the process of being replaced by new ones, giving further evidence that technology transfer is a significant issue in higher education today.

The most sweeping and dramatic change in policies discussed in this book occurred at Penn State in 1987 when President Bryce Jordan and the board of trustees adopted the university's economic development initiatives, which encouraged technology transfer. These policies contrast in many ways with those on other campuses. First, they are comprehensive; their designers addressed a wide range of technology transfer issues at the same time. By contrast, most universities are changing individual policies on an ad hoc basis as problems arise, acting without the comprehensive overview of the issues that is necessary to create a sound policy base. Second, the Penn State policies also concentrate on what is permissible rather than what is prohibited. Finally, they provide an example of policies that were imposed on the university community by the administration in an attempt to change the academic culture. Little effort was made to obtain faculty support or advice before the initiatives were adopted. The adoption of these initiatives symbolized a new direction and orientation for the institution. On most campuses, by contrast, new policies are adopted in response to changes that have already occurred, and their adoption involves extensive consultation with faculty.

The Penn State example illustrates a number of trends in the development of university policies related to technology transfer. First, it shows growing awareness of the interconnectedness of policies involving technology transfer and a recognition that changes are not as effective when they are made piecemeal in response to specific opportunities or problems. For instance, a policy on conflict of interest is not complete unless policies relating to consulting also are clearly stated. Second, policy changes are occurring within a framework that recognizes their symbolic impact. Penn State's administration changed policies as part of a deliberate attempt to change the academic culture.

When policies need to be changed to bring them into line with a prevailing situation, or when the effects of a change are unclear, extensive consulting and review by a spectrum of the university community are desirable. An example is Harvard's initial rejection and later acceptance of equity involvement in research-related start-up companies. Third, university policies are beginning to address what is permissible as well as what is prohibited. For instance, at M.I.T. it is now permissible, under certain circumstances, for faculty members to receive research support from companies in which they have an equity interest.

Several other trends should be noted. One is a trend toward policies that establish a process by which judgments can be made. These process policies take the place of policy statements which attempt to cover a broad range of circumstances and which, even when long and detailed, fail as guides when applied to specific cases. For instance, take the case of a faculty member who requests a leave of absence in order to work for a company in which he owns stock and which has been formed to exploit commercially an invention made by the faculty member and licensed from the university. This is not an unusual situation, yet so many issues are involved in it that policies usually are inadequate to deal with it. Rather than developing extensive policies in an attempt to address this complex situation and the many others that can arise, M.I.T. has formed a faculty review board to make judgments in such cases. The board can consider a wide range of factors that could not possibly be spelled out completely or quantified in a single policy—the value of the faculty member to the department, the closeness of the outside work to the faculty member's research, the effects on graduate students of the faculty member's absence, and so on. In a similar fashion, Stanford, rather than formulating extensive policy statements about preinvention rights of sponsors of research, has invested the director of the Office of Technology Licensing with authority to make judgments in such cases.

Allied with this trend is the increasing use of the case law-type approach, in which a decision in one case serves as a guide for decisions in other cases. This implies a centralization of decision-making authority, either in a board with continuity of membership or in a designated official.

In some areas it is clear that universities need to develop new policies. This is particularly evident in the case of outside activities of faculty members. Consulting policies are notoriously loose and unenforceable, and to try to extend them to faculty activities involved in forming outside businesses or exploiting intellectual property is straining them beyond their limit. Universities exhibit ambivalent attitudes toward faculty entrepreneurs. On the one hand they are willing to take credit for the jobs and economic benefits that such faculty bring to a community but, on the other, they are fearful that the faculty members' loyalties are divided and that university duties are being neglected. Clarification through policies or through case law-type decisions by boards or officials clearly is desirable.

Because of the diversity of situations related to technology transfer, particularly in business arrangements and in relations with faculty members, policies need to be flexible. The process for changing policies should be clearly established. The ability of a university to relate actions to policies in this area is a major indication of the maturity and degree of institutionalization of technology transfer.

ACADEMIC CULTURE

The last dimension of this analysis, academic culture, is a catchall concept that incorporates all the other dimensions listed here and, furthermore, their interaction. It has been defined as "a shared way of thinking and a collective way of behaving."[1] The idea of academic culture, like the more widely discussed corporate culture, is difficult to define and, in any specific instance, difficult to describe. Throughout this investigation the idea of academic culture nonetheless has been invoked to express that complex of factors that define a particular university and make it unique. The success or failure of a technology transfer activity in a university is partly determined by the degree to which it is consistent with that university's prevailing academic culture. A good example of this is the changes that took place in patent administration at M.I.T. in 1985. That administration was changed because it had been judged to be too legalistic and not focused in its marketing of M.I.T. technology and thus was seen as inconsistent with the highly entrepreneurial nature of the institute. By contrast, the same complaints were lodged against patent administration at the University of California, but to little effect for many years.

Of course, the idea of academic culture is so capacious as an analytical tool as to admit wide interpretations. It has been used in this book to describe differences among disciplines as well as among universities. In fact, the culture of a single university is a composite of a number of cultures. The analytical task of sorting out the effects and interactions of all these cultures mirrors the task of the university administrator who must not only understand these different cultures but also work within them to effect change.

Recommendations for Adapting to Technology Transfer

The concept of academic culture provides a useful beginning to the final section of this book. The recommendations contained herein are directed to a hypothetical top university administrator who is charged with the task of framing and organizing a coherent institutional response to technology transfer issues. In most cases this means changing the university. True and permanent changes in convention occur, to quote the late English social thinker Raymond Williams, "only when there are radical changes in the general structure of feeling."[2] To change the university one must change the feelings of the people who compose it. This rarely can be done at a single stroke or by design. Rather, such change occurs over a long period through a steady accumulation of small changes that are consistent with the direction of an increasingly accepted larger change.

University administrators are likely to find themselves facing the need to be responsive to change in the academic culture and to make policy, organizational, and resource allocation changes to respond to technology transfer. How should such changes be made?

Descriptions of the steps of a planning process are readily available from many sources and need not be repeated here. However, these particular plans must be based on a comprehensive understanding of what technology transfer is and some vision of what the university should be doing. This book has attempted to provide an understanding of university technology transfer. A vision of what technology transfer should be in a particular university, however, is dependent upon so many institution-specific elements that it cannot be provided here. Once university leadership understands what technology transfer is and what it should be on a particular campus, specific plans can be made. These plans might include one or several of the following responses.

"Ignore it, let it happen naturally" is probably the most common response, mostly because it requires little or no effort. It also may be the best response in some cases for that very reason, as well as because it threatens established structures and cultural norms the least. However, this strategy usually is possible only where the pressure to change is not yet very great.

The "public relations" response emphasizes the need for the university to communicate in a positive way its actions with regard to technology transfer. In this response, technology transfer activities are inventoried through reporting and monitoring systems, and accomplishments are dramatized for the best public relations effect. This response is appropriate when external political factors are important. It also can be used to protect the university from unwanted intrusion from external forces by heading off negative publicity.

The "enabling" response attempts to remove barriers to technology transfer where they exist, allowing individual faculty members and internal academic cultures the freedom to do technology transfer activities as they will or as the need arises. This response does not attempt to require or impel participation in technology transfer or to centralize it.

The "service" response goes a step beyond the enabling response in that it establishes an organizational base of subunits, usually on the periphery of the institution, that are designed to serve faculty members who want to engage in technology transfer and also to insulate faculty members from intrusion of technology transfer activities on traditional university functions. These peripheral subunits might include patent offices, general purpose liaison programs, industry relations offices, consulting clearinghouse services, industrial extension services, and so on. All are designed to perform technology transfer activities with a minimum of faculty involvement.

The most difficult but also ultimately the most effective response of all is one that attempts to integrate technology transfer activities with faculty roles. This strategy is most appropriate where a long-term commitment to technology transfer has been made. It requires a sustained and conscious effort on a broad front in order to be successful. It involves the difficult task of changing people's feelings about technology transfer. It also requires adjustment in other traditional faculty duties, since faculty time is a "zero sum" quantity—activities cannot be added to it without others being taken away. M.I.T., where faculty move freely from the academic realm to the world of business, is perhaps the best model of this integrated approach.

In practice, a university may adopt aspects of several or all of these responses—or others—in fashioning an approach to technology transfer. The point is that some attempt should be made to articulate a coherent institutional stance toward technology transfer.

Predictions

This study will ultimately be judged not on the conclusions drawn from dimensional analysisor upon the recommendations made in the preceeding section. Rather, it will be judged on how accurately it predicts what will eventually happen. This section sets forth those predictions with the expectation that they will come true in five to ten years from the date of the publication of this book.

1. Every major American research university will eventually articulate in formal policy and mission statements its commitment to technology transfer efforts and will reflect this commitment expressly in its organizational structure and its resource allocations. The phenomenon called *technology transfer* in this book is too broad and sweeping for any university to ignore.

2. At the same time universities that will be expected to increase their efforts at technology transfer, including the commercialization of university research and increased efforts at economic development efforts only indirectly associated with teaching and research, they will also come under increasing attacks. They will be told by both faculty and others that they are being too commercial, not sufficiently protective of their reputation for objectivity and that they are violating the traditional tacit agreement with the rest of society that they not be commercially oriented.

3. Institutional policy and practice and academic culture will increasingly allow those university faculty members who wish to become involved in activities associated with the commercialization of their research to do so. This will be accomplished without serious damage to the collegial atmosphere or with the notion that faculty owe their primary allegiance to the university. As universities recognize their obligation to serve society through technology transfer activities, institutional purpose and individual interest in this request will converge.

4. Organized units within universities staffed by professionals and dedicated to specific tasks related to technology transfer will continue to be formed and will increase their activity. Technology transfer (or some other expression for the same concept) will become the structuring principle under which these activities will be coordinated and overseen. One major thrust of this new organization will be the coordination of relations with industry. The seemingly disparate actvities

of corporate donor relations, corporate research partnerships, corporate-university economic development initiatives, student employment opportunities, continuing education, and technology licensing will come to be viewed as parts of a pattern of important and unitary interactions with corporations which need to be fostered and maintained over a long term.

5. Every major American research university will eventually become a financial partner in start-up companies begun to exploit that university's intellectual property. This financial involvement will extend beyond the passive ownership of equity in these new companies to some form of active participation in the generation of venture capital. In most cases this involvement will be formally separated from the university through the use of buffer organizations.

6. The total and relative (to the federal and state government) financial contributions of industry to the university will steadily increase. These contributions include gifts, research funding, payments of licensing fees and other direct payments for the use of university property, and membership dues and other special payments for access to the university. In addition, the federal and state governments will increasingly recognize and reward universities for their efforts to interact with industry.

7. Continuing education, both as an activity instrumental in the technology transfer process and as an organizational form within the university capable of facilitating technology transfer efforts or useful as a model for other efforts, will become more important and visible. Depending upon a given university's academic culture, continuing education units will be empowered to undertake significant additional functions, or not. In some situations the development of technology transfer may even contribute to the decentralization of continuing education.

8. Policies governing university and faculty interactions with commercial concerns will become more process-oriented and less proscriptive. This will result in the formulation of special review committees designed both to protect university values and to foster appropriate university commercial involvements.

Conclusion

The river of events related to university technology transfer is flowing so quickly that no book can hope to present a description of it that will remain current for long. Yet the history of these events is now extensive enough that some general patterns can be discerned and some useful conclusions can be drawn.

By examining the history of technology transfer in general and also of particular technology transfer mechanisms, and by documenting recent increases in the volume and importance to universities of technology transfer activity, this book demonstrates that technology transfer is not a passing fad. Rather, it is a broad-based trend that is bringing significant long-term changes to research universities. It is associated with a significant shift in the way the university views itself and in the way its publics view it.

This book attempts to provide the understanding of the complex nature and details of technology transfer that is the first step toward sound decision making in determining university responses. It shows that the technology transfer activities of a college or university can be fully effective only if they are in accord with the institution's history, organizational structure, policies, mission, and academic culture. Only by drawing on a detailed knowledge of technology transfer and of the particular institution can administrators wisely guide the adaptations of the institution to this new phenomenon.

Notes

1. T. Becher, "The Cultural View," in *Perspectives in Higher Education*," ed. B. Clark (Berkeley: University of California Press, 1984), quoted in George D. Kuh and Elizabeth J. Whitt, *The Invisible Tapestry: Culture in American Colleges and Universities* (Washington, D.C.: Association for the Society of Higher Education, 1988), 9.
2. Raymond Williams, *Culture and Society, 1780–1950,* (New York: Harper & Row, 1966), 39.

Bibliography

KUH, GEORGE D., and ELIZABETH J. WHITT. *The Invisible Tapestry: Culture in American Colleges and Universities.* Washington, D.C.: Association for the Study of Higher Education, 1988.

WILLIAMS, RAYMOND. *Culture and Society, 1780–1950.* New York: Harper & Row, 1966.

Index